Recent Progress in
Industrial Crystallization

Recent Progress in Industrial Crystallization

Editors

Heike Lorenz
Alison Emslie Lewis
Erik Temmel

Basel • Beijing • Wuhan • Barcelona • Belgrade • Novi Sad • Cluj • Manchester

Editors
Heike Lorenz
Max Planck Institute for Dynamics
of Complex Technical Systems
Germany

Alison Emslie Lewis
University of Cape Town
South Africa

Erik Temmel
Sulzer Chemtech Ltd.
Switzerland

Editorial Office
MDPI
St. Alban-Anlage 66
4052 Basel, Switzerland

This is a reprint of articles from the Special Issue published online in the open access journal *Crystals* (ISSN 2073-4352) (available at: https://www.mdpi.com/journal/crystals/special_issues/R_P_Industrial_Crystallization).

For citation purposes, cite each article independently as indicated on the article page online and as indicated below:

Lastname, A.A.; Lastname, B.B. Article Title. *Journal Name* **Year**, *Volume Number*, Page Range.

ISBN 978-3-0365-7714-2 (Hbk)
ISBN 978-3-0365-7715-9 (PDF)
doi.org/10.3390/books978-3-0365-7715-9

Cover image courtesy of Somayyeh Ghaffari/Max Planck Institute Magdeburg

© 2023 by the authors. Articles in this book are Open Access and distributed under the Creative Commons Attribution (CC BY) license. The book as a whole is distributed by MDPI under the terms and conditions of the Creative Commons Attribution-NonCommercial-NoDerivs (CC BY-NC-ND) license.

Contents

About the Editors vii

Preface ix

Slobodan Jančić
In Memoriam—Gerda van Rosmalen
Reprinted from: *Crystals* **2022**, *12*, 177, doi:10.3390/cryst12020177 1

Iben Ostergaard and Haiyan Qu
Solubility and Crystallization of Piroxicam from Different Solvents in Evaporative and Cooling Crystallization
Reprinted from: *Crystals* **2021**, *11*, 1552, doi:10.3390/cryst11121552 3

Corin Mack, Johannes Hoffmann, Jan Sefcik and Joop H. ter Horst
Phase Diagram Determination and Process Development for Continuous Antisolvent Crystallizations
Reprinted from: *Crystals* **2022**, *12*, 1102, doi:10.3390/cryst12081102 15

Ina Beate Jenssen, Oluf Bøckman, Jens-Petter Andreassen and Seniz Ucar
The Effect of Reaction Conditions and Presence of Magnesium on the Crystallization of Nickel Sulfate
Reprinted from: *Crystals* **2021**, *11*, 1485, doi:10.3390/cryst11121485 31

Roman Sadovnichii, Elena Kotelnikova and Heike Lorenz
Thermal Deformations of Crystal Structures in the L-Aspartic Acid/L-Glutamic Acid System and DL-Aspartic Acid
Reprinted from: *Crystals* **2021**, *11*, 1102, doi:10.3390/cryst11091102 47

Lie-Ding Shiau
Comparison of the Nucleation Parameters of Aqueous L-glycine Solutions in the Presence of L-arginine from Induction Time and Metastable-Zone-Width Data
Reprinted from: *Crystals* **2021**, *11*, 1226, doi:10.3390/cryst11101226 61

Gina Kaysan, Alexander Rica, Gisela Guthausen and Matthias Kind
Contact-Mediated Nucleation of Subcooled Droplets in Melt Emulsions: A Microfluidic Approach
Reprinted from: *Crystals* **2021**, *11*, 1471, doi:10.3390/cryst11121471 73

Abad Albis, Yecid P. Jiménez, Teófilo A. Graber and Heike Lorenz
Reactive Crystallization Kinetics of K_2SO_4 from Picromerite-Based $MgSO_4$ and KCl
Reprinted from: *Crystals* **2021**, *11*, 1558, doi:10.3390/cryst11121558 95

Izabela Betlej, Katarzyna Rybak, Małgorzata Nowacka, Andrzej Antczak, Sławomir Borysiak, Barbara Krochmal-Marczak, et al.
Structural Properties of Bacterial Cellulose Film Obtained on a Substrate Containing Sweet Potato Waste
Reprinted from: *Crystals* **2022**, *12*, 1191, doi:10.3390/cryst12091191 119

Taona Malvin Chagwedera, Jemitias Chivavava and Alison Emslie Lewis
Gypsum Seeding to Prevent Scaling
Reprinted from: *Crystals* **2022**, *12*, 342, doi:10.3390/cryst12030342 129

Claas Steenweg, Jonas Habicht and Kerstin Wohlgemuth
Continuous Isolation of Particles with Varying Aspect Ratios up to Thin Needles Achieving Free-Flowing Products
Reprinted from: *Crystals* **2022**, *12*, 137, doi:10.3390/cryst12020137 . . . . **145**

Jana Sonnenschein, Pascal Friedrich, Moloud Aghayarzadeh, Otto Mierka, Stefan Turek and Kerstin Wohlgemuth
Flow Map for Hydrodynamics and Suspension Behavior in a Continuous Archimedes Tube Crystallizer
Reprinted from: *Crystals* **2021**, *11*, 1466, doi:10.3390/cryst11121466 . . . . **165**

Marcelo Martins Seckler
Crystallization in Fluidized Bed Reactors: From Fundamental Knowledge to Full-Scale Applications
Reprinted from: *Crystals* **2022**, *12*, 1541, doi:10.3390/cryst12111541 . . . . **187**

About the Editors

Heike Lorenz

Heike Lorenz studied Chemistry at Freiberg University of Mining and Technology and received her Ph.D. from the Otto von Guericke University Magdeburg. After research periods there and at the Max Planck Institute for Dynamics of Complex Technical Systems in Magdeburg, she obtained her habilitation with a thesis on heterogeneous processes based on the example of the combustion of solids and crystallization from solution. Since 1998, she has been Senior Researcher at the mentioned Max Planck Institute and is heading the crystallization research team. In 2010, she was appointed as apl. Professor at Otto von Guericke University. Heike Lorenz has authored or co-authored more than 180 papers, 11 book contributions, and has edited 2 international conference books. She is a member of the European Federation of Chemical Engineering's (EFCE) Working Party on Crystallization and the Crystallization Board of DECHEMA/VDI-GVC's ProcessNet in Germany. Her research interests include crystallization for separation process and product design, process monitoring, and purification of fine chemicals, large-scale industrial and natural products. Since 2019, she is a member of the Editorial Board of *Crystals* and Editor-in-Chief of its *Industrial Crystallization* section.

Alison Emslie Lewis

Alison Emslie Lewis is Dean of the Faculty of Engineering and the Built Environment at the University of Cape Town. She has undergraduate, Master's and PhD degrees in Chemical Engineering from the University of Cape Town and more than 25 years of teaching and research experience. She founded the Crystallisation and Precipitation Research Unit at the University of Cape Town, a University-accredited Research Unit that focusses on Industrial precipitation and crystallization, particularly water treatment and recovery of value from effluent streams and Eutectic Freeze Crystallization.

There is a strong focus on precipitation in hydrometallurgy, particularly Rare-Earth Element recovery, metals and metal salts production and recovery and design and optimisation of precipitation processes.

She has supervised postgraduate degrees on a range of academic and industrially applied research projects and is the author of 81 peer-reviewed journal publications, a patent and 12 book contributions, including being the lead author on Lewis et al (2015)*.

Professor Lewis has been a Visiting Professor at the Swiss Federal Institute of Technology (EPFL), Sion, Valais, Switzerland; the University of Sheffield and the University of Mauritius. She is a Professional Engineer and a Fellow of the Institute of Chemical Engineers (FIChemE). *Lewis, A E, M M Seckler, H Kramer, and G M van Rosmalen. 2015. *Industrial Crystallization: Fundamentals and Applications* (Cambridge University Press).

Erik Temmel

Erik Temmel studied chemical process engineering at the Otto von Guericke University and received his Ph.D. for research on continuous crystallization at the Max Planck Institute in Magdeburg. After several post-doctoral fellowships, e.g., at DSM (Switzerland), he is now heading the R&D team for crystallization at Sulzer Chemtech Ltd (Switzerland) and lectures on melt crystallization at the Technical University Dortmund.

Preface

Crystallization is an important industrial process, a purification technique, a separation process and a branch of particle technology. It also encompasses several key areas of chemical and process engineering.

Products produced using industrial crystallization techniques range from highly engineered nanoparticles and crystals for the rapidly expanding field of battery production to pharmaceuticals, amino acids and proteins, and inorganic salts. Crystallization is also an important technique in water and effluent treatment, reflecting the waste-to-resource movement that is becoming increasingly important and relevant in the field. The overall theme of this Special Issue is the link between industrial crystallization and the underlying theoretical concepts, and how practical understanding in the field is enhanced through applied research.

The articles collected here reflect the breadth of the field, with topics ranging from fundamental aspects, such as the development of phase diagrams, the study of reaction conditions on solid-state behavior, the measurement of kinetics and a study on contact nucleation; to more applied topics such as seeding to prevent scaling, improved downstream processing and a review on the state-of-the-art of crystallization in fluidized bed reactors.

The Special Issue is dedicated to Professor Gerda van Rosmalen, who was a pioneer in this field. She developed the field of industrial crystallization research through her original approach and was regarded globally as a pre-eminent figure in the field.

We thank all authors whose articles are included in this Special Issue for their innovative contributions to the theme, and for their role in advancing the frontiers of the field. Below, the individual articles are briefly introduced in the context of the topics mentioned above.

1. Obituary: **In Memoriam–Gerda van Rosmalen**.
2. **Solubility and Crystallization of Piroxicam from Different Solvents in Evaporative and Cooling Crystallization** highlights the polymorphic behavior of an anti-inflammatory API and links operating conditions during solidification in various solvents with the mechanisms and probability of occurrence of the two relevant forms.
3. **Phase Diagram Determination and Process Development for Continuous Antisolvent Crystallizations** demonstrates an efficient systematic to assess phase diagrams for the purpose of process design and optimization with examples of sodium bromate, DL-asparagine monohydrate, Mefenamic acid and Lovastatin.
4. **The Effect of Reaction Conditions and Presence of Magnesium on the Crystallization of Nickel Sulfate** experimentally studies the complex phase behavior of $NiSO_4$, which exhibits two hydrates, one of them crystallizing in two different lattices. The appearance of the phases is linked enantiotropically and can be influenced by an additive.
5. **Thermal Deformations of Crystal Structures in the L-Aspartic Acid/L-Glutamic Acid System and DL-Aspartic Acid** exploits detailed temperature-resolved PXRD studies to derive temperature-dependent crystal lattice shifts of single phases, a solid solution and a racemate of two similar amino acids.
6. **Comparison of the Nucleation Parameters of Aqueous L-Glycine Solutions in the Presence of L-Arginine from Induction Time and Metastable-Zone-Width Data** applies the classical nucleation theory to derive interfacial free energies and pre-exponential nucleation factors from extensive experimental results, which clearly prove the significant impact of the impurity concentration on the MSZW.

7. **Contact-Mediated Nucleation of Subcooled Droplets in Melt Emulsions: A Microfluidic Approach** reports fundamental data and model approaches for this specific secondary nucleation mechanism together with experimental evidence of its dependency on different operational parameters.
8. **Reactive Crystallization Kinetics of K_2SO_4 from Picromerite-Based $MgSO_4$ and KCl** investigates an important fertilizer component and its formation kinetics, i.e., growth and nucleation, during reactive crystallization-based reciprocal conversion.
9. **Structural Properties of Bacterial Cellulose Film Obtained on a Substrate Containing Sweet Potato Waste** describes an opportunity to produce tailor-made, microcrystalline cellulose, which is usually created from wood using heat and/or chemicals via aerobic microorganisms using food waste as the source material.
10. **Gypsum Seeding to Prevent Scaling** deals with eutectic freeze concentration, a promising process to generate and separate two solid phases from aqueous solutions. Solid-free heat exchanger surfaces are a key requirement for process performance, and it is shown that seeding can greatly reduce the risk of the crystal layer formation, i.e., scaling in this context.
11. **Continuous Isolation of Particles with Varying Aspect Ratios up to Thin Needles Achieving Free-Flowing Products** tackles basic challenges in continuous solid/liquid separation processes. It is shown that in a specific, small-scale vacuum screw filter prototype, crystal size alteration can be excluded and even the crystal size distribution of needle-forming systems can be preserved.
12. **Flow Map for Hydrodynamics and Suspension Behavior in a Continuous Archimedes Tube Crystallizer** combines intensive CFD studies with detailed images as well as experimental data to provide a systematic approach for estimating flow regimes in a novel crystallizer-type in order to influence the produced crystal size distributions.
13. **Crystallization in Fluidized Bed Reactors: From Fundamental Knowledge to Full-Scale Applications** presents a comprehensive review of the past 50 years of fluidized bed reactor developments and their application to various separation and purification tasks.

Heike Lorenz, Alison Emslie Lewis, and Erik Temmel
Editors

Obituary

In Memoriam—Gerda van Rosmalen †

Slobodan Jančić

SJJ Management Consulting, Baggastiel 34, 9475 Sevelen, Switzerland; boban.jancic@catv.rol.ch
† 27 May 1936–18 January 2021.

Citation: Jančić, S. In Memoriam—Gerda van Rosmalen. *Crystals* **2022**, *12*, 177. https://doi.org/10.3390/cryst12020177

Academic Editors: Heike Lorenz, Alison Emslie Lewis, Erik Temmel and Jens-Petter Andreassen

Received: 19 January 2022
Accepted: 19 January 2022
Published: 26 January 2022

Publisher's Note: MDPI stays neutral with regard to jurisdictional claims in published maps and institutional affiliations.

Copyright: © 2022 by the author. Licensee MDPI, Basel, Switzerland. This article is an open access article distributed under the terms and conditions of the Creative Commons Attribution (CC BY) license (https://creativecommons.org/licenses/by/4.0/).

Dear colleagues,

This Special Issue is in memory and in honor of Professor Gerda van Rosmalen, who saddened the crystallization community with her departure and left us with the wide and deep heritage of her work and most pleasant personal memories. If those who knew her were asked what thoughts were in her mind at the time of the early nineteen seventies when this picture was taken and she was still at the start of her long and fruitful career in crystallization, the likely answers would be: "Oh, oh . . . Crystallization as a unit operation is not a subject at any university, only some universities perform research on crystallization, and many industrial companies use crystallization but experience substantial problems. More researchers should be aware of what is truly happening in large crystallizers and more users in the industry should be aware of what researchers are truly discovering. What can I do to contribute?"

She started in Delft in 1970, became professor in 1987 and supervised 26 Ph.D. students until she retired in 2001. Some 150 publications and 3 patents crown her work in the area of industrial crystallization. After her retirement, she continued to guide Ph.D. students and maintained contact with her ex-colleagues. In 2015, fourteen years after her retirement as professor, she co-authored a book on industrial crystallization with authors from three continents. This says a lot about her diversity of thinking and doing.

Gerda received other thoughts on crystallization well and offered her own ideas very openly in all discussions during her courses, research and consultation. It is not surprising that she became a renowned member of the crystallization community and a Board Member of the Working Party on Industrial Crystallization of the European Federation of Chemical Engineers.

We in the crystallization community shall hold warm memories of her work and her person.

- Boban Jančić, Sevelen, August 2021

The two pictures (Figures 1 and 2) below show the crystallization community that participated to one of the symposia in Japan and a cordial meeting with the author three years after he left Delft University.

Figure 1. International Symposium on Separation Process Engineering, Tokyo, Japan, 26–27 September 1986.

Figure 2. Who knew her work—respected her. Who knew her as well—also liked her.

Funding: This research received no external funding.

Conflicts of Interest: The authors declare no conflict of interest.

Article

Solubility and Crystallization of Piroxicam from Different Solvents in Evaporative and Cooling Crystallization

Iben Ostergaard and Haiyan Qu *

Department of Green Technology, University of Southern Denmark, Campusvej 55, 5230 Odense, Denmark; IBOS@lundbeck.com
* Correspondence: haq@igt.sdu.dk

Abstract: In this work, the solubility of a non-steroidal anti-inflammatory drug (NSAID), piroxicam, is investigated. The polymorphic form II, which is the most stable form at room temperature, was investigated in seven different solvents with various polarities. It has been found that the solubility of piroxicam in the solvents is in the following order: chloroform > dichloromethane > acetone > ethyl acetate > acetonitrile > acetic acid > methanol > hexane. Crystallization of piroxicam from different solvents has been performed with evaporative crystallization and cooling crystallization; the effects of solvent evaporation rate and solute concentration have also been studied. Both form I and form II could be produced in cooling and evaporative crystallization, and no simple link can be identified between the operating parameters and the polymorphic outcome. Results obtained in the present work showed the stochastic nature of the nucleation of different polymorphs as well as the complexity of the crystallization of a polymorphic system.

Keywords: solubility; polymorphism; nucleation; crystallization

1. Introduction

Crystallization often serves as the final separation and purification step in the production of active pharmaceutical ingredients (APIs) [1]. Generally, cooling, solvent evaporation or the addition of an antisolvent are used to create supersaturation in the solution, which will induce nucleation and hence crystallization. A large percent of APIs can exist in different solid states as polymorphs, hydrates, solvates or cocrystals. Different solid forms of an API have different physical and chemical properties, which will influence the further processing and formulation of the API as well as product effectiveness, such as bioavailability [2,3]. Although it is common that the thermodynamically most stable form is selected for formulating the final dosage, metastable forms may be occasionally selected due to their higher solubility and dissolution rate in water. In the latter case, the polymorphic purity of the crystallized product is very important as a trace amount of the stable form can induce and facilitate the transformation of the metastable form to the stable form [4]. Consequently, it is of paramount importance to control the crystallization process so that the desired solid form is produced without any minor contamination of other forms [5].

The formation of a specific crystalline state of the API is influenced by the operating conditions during crystallization, such as solvent, temperature, supersaturation, crystallization method and so on. However, nucleation of different solid forms from a supersaturated solution represents a process with a very complex nature, and it is difficult to identify the underlying mechanism behind the formation of the different solid forms. A link between the solvent and polymorphic form of isonicotinamide (INA) being nucleated was reported by Kulkarni et al. [6]. In their work, the hydrogen bonding capabilities of given solvents were investigated and observed to have an influence on the polymorphic form yielded from the crystallization process. Nevertheless, other process parameters should be considered as these will also influence the polymorphic form. It was observed in our previous study that the formation of INA polymorphs also depended on the solute concentration as well as the temperature at which nucleation was onset [7,8]. The complex nature of the nucleation of polymorphs has been demonstrated in our previous work by the cooling crystallization of piroxicam from acetone–water solutions [9]. A solid form landscape has been established to show the dependence of the nucleated polymorph on the solute concentration. However, a further study revealed the specific challenge for upscaling the cooling crystallization from 100 mL to 2 liters. Batch cooling crystallization with the same operating conditions yielded different polymorphs of piroxicam in the small and large-scale systems [10].

Piroxicam is a non-steroidal anti-inflammatory drug (NSAID), forming four anhydrous polymorphs and one monohydrate [11–14]. In the present work, the solubility of piroxicam in different organic solvents has been investigated at temperatures ranging from 10 °C to 40 °C. The nucleation of the different solid forms of piroxicam was examined in evaporative and cooling crystallization. The effects of evaporation rate as well as the solute concentration and temperature on the polymorphic outcome were studied. It was observed that both form I (BIYSEH01, 03, 04, 10, 13, 14, 16) and form II (BIYSEH02, 08) of piroxicam could be yielded; the frequency of occurrence of the polymorphs did not show any clear dependence on the characteristics of the solvents and the other operating parameters. This is in agreement with recently published literature [15] in which the stochastic nucleation of INA polymorphs was demonstrated by both experimental and modeling approaches.

2. Experimental Method

2.1. Materials and Equipment

Piroxicam was purchased from Hyper Chemicals Limited, Zhejiang, China, and identified as the polymorphic form II. Analytical grade acetic acid (AcOH), chloroform (TCM), dichloromethane (DCM) and methanol (MEOH) were purchased from Sigma Aldrich (St. Louis, MO, USA). The other solvents, acetonitrile (CAN), ethyl acetate (EtOAc) and hexane, are HPLC standards purchased from VWR Chemicals (Radnor, PA, USA).

Cooling crystallization experiments were conducted using a Mettler Toledo EasyMax 102 Advanced Synthesis Workstation with two 100 mL glass reactors (Mettler Toledo, Columbus, OH, USA), an overhead stirrer and a solid-state thermostat cooling/heating jacket. The setup was controlled using N_2 as the purge gas and using iControl software. Evaporative crystallization experiments were conducted using a Buchi Rotavapor R-210 rotating evaporator (BÜCHI Labortechnik AG, Flawil, Switzerland) and using a temperature-controlled water basin.

Solvates were characterized using a simultaneous thermal analyzer (STA 449 F3 Jupiter®) from NETZSCH (Erich NETZSCH GmbH & Co. Holding KG, Selb, Germany). Samples were heated from room temperature up to 250 °C in a ceramic crucible at a heating rate of 5 °C/min. It has been confirmed in our previous work that the polymorphic forms of piroxicam have very distinguishable Raman spectra [9,10,16]. Raman spectroscopy was used in this work to identify the polymorphic forms obtained from the crystallization processes. A Bruker Senterra Dispersive Raman microscope (Bruker, Billerica, MA, USA) with a 785 nm laser operating at 100 mW with a 5 s integration time and two scans was used to collect the spectra.

2.2. Solubility Measurement

The gravimetric method was used to measure the solubility of piroxicam in seven different solvents: dichloromethane, chloroform, ethyl acetate, acetonitrile, acetic acid, methanol and hexane. The solubility of piroxicam form II was measured at temperatures 10, 20, 30 and 40 °C, except for dichloromethane, where the solubility was measured up to 30 °C as further temperature increase would exceed the boiling point. A suspension with an excess amount of solute (piroxicam) was prepared in 8 mL glass vials with 5 mL solvent. The vials were sealed and maintained under stirring for 24 h at a constant temperature to ensure that solid–liquid equilibrium was reached. The clear solution was sampled with a syringe filter and weighed. After the solvent was evaporated, the mass of the dried solid was measured to obtain the solubility of piroxicam. The experiments were reproduced in triplets for all solvents at the different temperatures and solubility profiles made.

2.3. Cooling Crystallization and Nucleation Kinetics of Piroxicam by Metastable Zone Width Measurement

The outcome of the solubility measurement showed the solubility of piroxicam increased with increasing temperature in five solvents: dichloromethane, ethyl acetate, acetonitrile, chloroform and acetic acid. These were then chosen as the solvents for performing cooling crystallization. The effects of the five different solvents on the nucleation of piroxicam have been investigated by measuring the Metastable Zone Width (MSZW) in the given solvents. Saturated solutions at 10 °C or 30 °C in the five solvents were prepared by dissolving an appropriate amount of piroxicam (form II) in approximately 90 g of solvent in the reactor. The solutions were heated to 40 °C and kept at this temperature for 30 min to ensure the complete dissolution of piroxicam. Then the solution was cooled linearly at 0.5 °C/min. The experiments were performed as duplicates. The MSZW was characterized by the visually observed sudden increase in turbidity of the solution. Nucleation was also verified from the temperature profile of the thermostat as the exothermic nucleation caused a sudden deviation of the reactor temperature from the cooling profile. After complete crystallization, the suspension was filtered with a 250 mL Büchner funnel, the crystals were dried off at room temperature and analyzed with Raman spectroscopy to identify the polymorphic form.

2.4. Fast Evaporative Crystallization

The evaporative crystallization of piroxicam was performed from six solvents: dichloromethane, chloroform, ethyl acetate, acetonitrile, acetic acid and methanol. Saturated solutions at both 20 °C and 30 °C in the six solvents were prepared by dissolving an appropriate amount of piroxicam (form II) in approximately 90 g of solvent. To ensure complete dissolution, the temperature was elevated during the solution preparation. When complete dissolution occurred, the solution was transferred to the rotating evaporator, and the pressure was set to the saturation vapor pressure of the corresponding solvent. The temperature of the solution was controlled by a water basin at 30 °C. The conditions were kept until all of the solvent had evaporated. These experiments were conducted as duplicates. The crystal form was investigated by Raman spectroscopy to determine the polymorphic form.

2.5. Slow Evaporative Crystallization

The effect of solvent evaporation rate was investigated by performing slow evaporative crystallization. Saturated solutions prepared from piroxicam (form II) at room temperature with the abovementioned six solvents were prepared by adding 5 mL of solvent and the appropriate amount of piroxicam to an 8 mL vial. The samples were filtered with 2 µm cellulose membranes to new vials to ensure the removal of any non-dissolved particles. The filtered samples were left uncapped, wrapped in aluminum foil to protect the samples from light. The crystal form was obtained after all solvents had evaporated and was analyzed with Raman spectroscopy.

3. Results

3.1. Solubility of Piroxicam

The solubility of piroxicam form II measured in the seven solvents, including dichloromethane, chloroform, ethyl acetate, acetonitrile, acetic acid, methanol and hexane are shown in Table 1 and Figure 1a,b. From the figure, it is obvious that the highest solubility of piroxicam was found in dichloromethane and chloroform, while the lowest solubility was found in hexane and methanol. An investigation regarding the solubility of piroxicam in acetone was found from a previous study [9] and shown in Figure 1a. The solubility was measured with the same method but at temperatures 25, 35, and 45 °C. Moreover, it was observed that the solubility of piroxicam in all solvents, except methanol and hexane, increased with increasing temperature. In cooling crystallization, the yield depended on the slope of the solubility curve. It is desirable to perform cooling crystallization using solvents where the solubility increases significantly with increasing temperature. Consequently, the solvents dichloromethane, ethyl acetate, acetonitrile, chloroform and acetic acid were chosen as solvents when conducting the cooling crystallization of piroxicam. Chloroform and acetone showed the largest solubility increase with approximately 20 mg/g solvent from 10–40 °C and 25–45 °C, respectively.

Table 1. Solubility of piroxicam form II in different solvents at varied temperatures (in mg/g solvent).

Solvents	10 °C	20 °C	30 °C	40 °C
Chloroform	67.73 ± 9.54	68.79 ± 0.67	78.40 ± 1.22	91.38 ± 2.38
Dichloromethane	61.00 ± 2.71	66.76 ± 0.76	73.80 ± 10.76	
Ethyl acetate	12.42 ± 0.24	13.82 ± 0.48	15.55 ± 0.47	17.61 ± 0.67
Acetonitrile	10.73 ± 0.78	10.87 ± 0.34	13.30 ± 0.23	16.61 ± 0.91
Acetic acid	6.82	9.43 ± 0.32	11.97 ± 0.24	15.69 ± 0.04
Methanol	3.19 ± 0.08	4.44 ± 1.53	4.91 ± 0.05	5.08 ± 0.38
n-Hexane	0.68 ± 0.05	0.14 ± 0.25	0.16 ± 0.11	0.29 ± 0.01

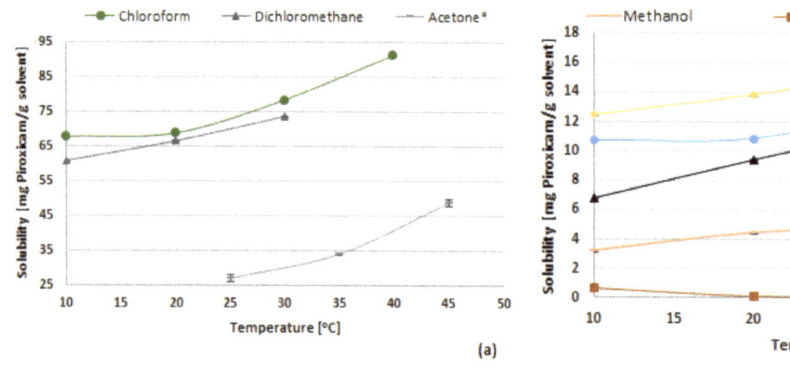
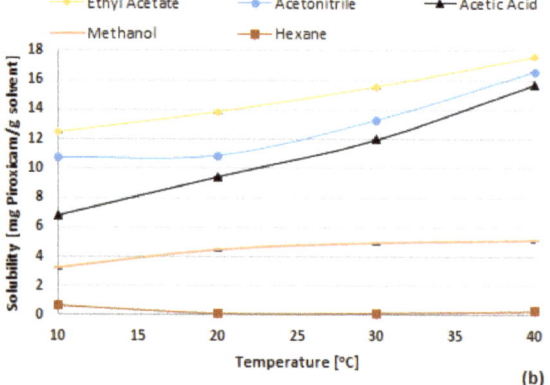

Figure 1. (2-column fitting) Solubility of piroxicam form II in different solvents. Solid lines are drawn for visual guidance. (**a**) Solubility of piroxicam in chloroform, dichloromethane and acetone. (**b**) Solubility of piroxicam in ethyl acetate, acetonitrile, methanol, hexane and acetic acid. * Acetone solubility was obtained in an earlier study [9].

The solubility of a solute in a solvent is determined by how well these materials interact. A way of evaluating these interactions is by the Hansen solubility parameters (HSPs). The HSP propose that the total force of the various interactions between the molecules can be divided into partial solubility parameters, i.e., dispersion forces (δ_d), polar bonding (δ_p) and hydrogen bonding (δ_h) [17]:

$$\delta_T^2 = \delta_d^2 + \delta_P^2 + \delta_h^2 \tag{1}$$

A larger similarity between the HSPs of the solvent and solute implies a high degree of similarity of the molecular polarities and hence could imply a higher solubility. The HSPs of piroxicam and the solvents are shown in Figure 2a,b with a plot of δ_h versus δ_T, and δ_d versus δ_p, respectively. It can be seen from Figure 2 that chloroform and dichloromethane are the most similar to piroxicam in terms of the total solubility parameter and the three partial solubility parameters, which are also in agreement with these solvents giving the highest solubility. The solvents including ethyl acetate, acetone, acetonitrile and acetic acid possess a similar total solubility parameter as piroxicam; however, one or two of their partial solubility parameters are significantly different from that for piroxicam. Finally, hexane and methanol have very different HSPs from piroxicam, which was also in agreement with the lowest solubility of piroxicam in these two solvents.

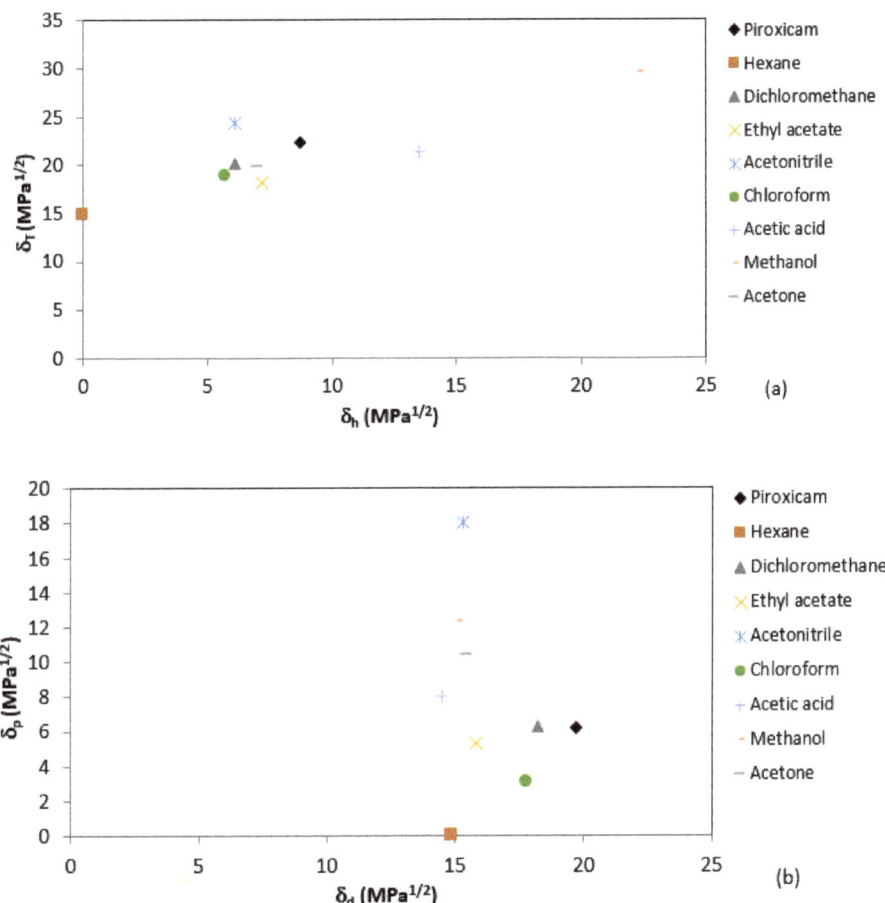

Figure 2. (2-column fitting) Hansen solubility parameters for piroxicam and solvents. (**a**) δ_T versus δ_h; (**b**) δ_p versus δ_d (Hansen solubility parameters for solvents and piroxicam are found in [17] and [18], respectively).

3.2. Piroxicam Polymorphism and Characterization

Piroxicam is a polymorphic compound; it can form at least four anhydrous polymorphs and one monohydrate [11–14,19]. The focus in this work is on the polymorphic forms I and II. It has been reported that solution crystallization at moderate temperatures yielded either form I or form II, while the preparation of the unstable form III and IV requires more extreme conditions [11,14]. The relative stability of form I and form II has been investigated using suspension conversion and melting temperature [16]. It has been reported that form I and form II are enantiotropically related. Form II is the most thermodynamically stable polymorph at temperatures up to 60 °C, which was confirmed by the suspension conversion method. Form I should be more stable at elevated temperatures as form I has a higher melting temperature than form II [11,16]. The transition temperature of the two forms was found between 60 and 196 °C by [16]. A redetermination of this transition temperature would be preferred to obtain a better insight into the parameters that govern the selectivity between the two polymorphic forms. The difference between the two polymorphs arises from the different intermolecular hydrogen bonding. The orientation of the molecules of the two polymorphic forms are shown in Figure 3. Form I show a head-to-head configuration, while form II show a head-to-tail configuration. Cruz-Cabeza et al. [20] reported that being able to form different intermolecular hydrogen bonding does not lead to any significant higher propensity for the molecule to form polymorphism. However, it could be expected that for the polymorphs with different intermolecular hydrogen bonding, the influence of solvents used in the crystallization process may be more significant on the formation of polymorphic forms because the solvent can affect the self-association of the solute molecules in a supersaturated solution. It has been observed in our previous work [9,16] that the two polymorphs of piroxicam can be identified and characterized using Raman spectroscopy. As shown in Figure 4, several specific Raman shifts for forms I and II are marked with the arrows.

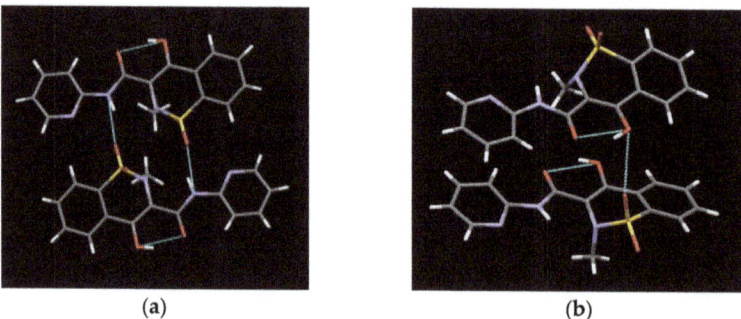

Figure 3. (1.5-column fitting) Illustration of the difference in orientation between the two forms: Piroxicam form I (**a**) (Cambridge Crystallographic Datacenter, reference BIYSEH04) and form II (**b**) (Cambridge Crystallographic Datacenter, reference BIYSEH08). The light blue lines illustrate the hydrogen bonds between the piroxicam molecules in each form, and hydrogen atoms are omitted for clarity.

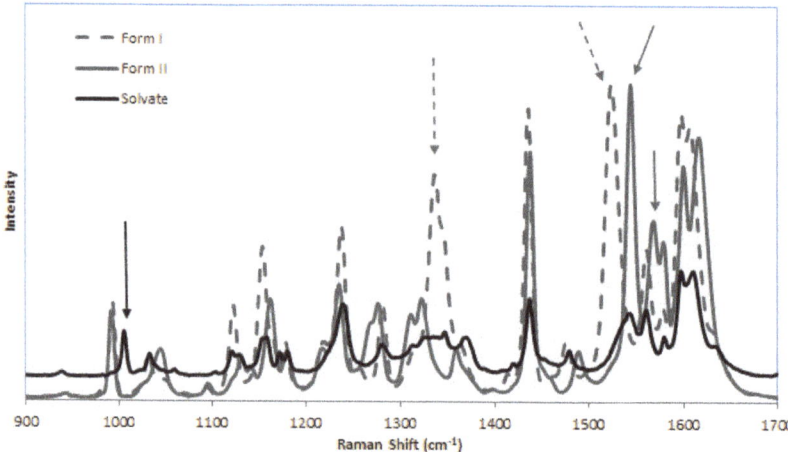

Figure 4. (1-column fitting) Raman spectra of piroxicam form I, form II and the solvate of acetic acid. Arrows are inserted to show characteristic peaks.

3.3. Cooling Crystallization and the Effect of Solvent on Nucleation Kinetics of Piroxicam

The effect of a solvent on the nucleation kinetics of piroxicam has been investigated via the measurement of MSZW, the results are shown in Figure 5. The MSZW of piroxicam in acetic acid is relatively narrow compared with the MSZWs in other solvents. The MSZW in solutions with dichloromethane, ethyl acetate and acetonitrile are between 25 °C and 35 °C. Furthermore, piroxicam showed a very wide MSZW in chloroform solution and could not be determined due to the cooling limit of the EasyMax system (approximately −30 °C). The piroxicam–chloroform solution saturated at 30 °C was cooled down to −30 °C and remained as a clear solution. The highly supersaturated solution was left overnight, and the crystallized piroxicam was analyzed with Raman spectroscopy. These observations imply a high energy barrier for the primary nucleation of piroxicam in most of the solvents studied in this work, which could be attributed to the formation of the hydrogen bonds between piroxicam and the solvent. Hydrogen bond formation between the solute and solvent has been observed to have an effect on both the crystallization process and the polymorphic form yielded in other studies [21–23]. The relatively wide MSZW of piroxicam in most of the studied solvents could suggest seeding as a feasible strategy for controlling the polymorphism as well as the particle size distribution of piroxicam from a cooling crystallization. The wide MSZW provides a large operating space for selecting optimal seeding parameters, such as seed with the desired polymorph, seed loading and seeding time, to direct the crystallization process towards the desired properties of the crystalline product.

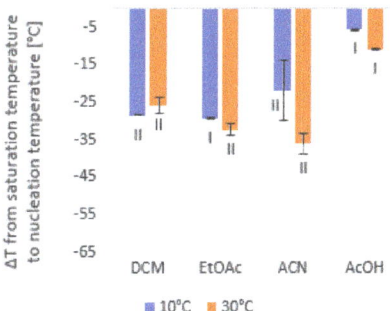

Figure 5. (1-column fitting) Metastable zone width (MSZW) of piroxicam in different solvents. Polymorphism of the obtained solid is also shown in the figure. Saturated solutions at 10 and 30 °C have been used, respectively. DCM (dichloromethane), EtOAc (Ethyl Acetate), ACN (Acetonitrile) and AcOH (Acetic Acid).

After the measurement of MSZW, the nucleated piroxicam crystals were recovered by filtration and dried at room temperature. Subsequently, the polymorphism of the crystals was analyzed with Raman spectroscopy. It can be seen from Figure 5 that both form I and form II were yielded from the cooling crystallization. It has been discovered in our previous work [9] that the solute concentration has a significant influence on the polymorphism of piroxicam in cooling crystallization from acetone–water mixtures. It was observed in our previous study that low piroxicam concentrations would yield form I, while at higher concentrations (e.g., saturated at 30 °C and 40 °C), form II was obtained [9]. A similar effect of the solute concentration has been observed in the present work with cooling crystallization from ethyl acetate. However, the solute concentration showed no effect on the polymorphism of piroxicam in crystallization from dichloromethane, acetonitrile, chloroform and acetic acid. Form II was solely crystallized out from solutions with dichloromethane and acetonitrile, while form I was solely produced from solutions with acetic acid and chloroform, regardless of the very different solute concentrations in the solutions. Combing the MSZW and the polymorphic forms obtained in the cooling experiments (shown in Table 2), it seems that there is no direct link between the nucleation kinetics and the polymorphic outcome. The MSZW in acetic acid is relatively narrow while the MSZW in chloroform is extremely wide (>30 °C, could not be detected), form I was produced from both solvents.

Table 2. Piroxicam polymorphic forms obtained from cooling and evaporative crystallizations.

Saturated Solution at Given Temperature (Solubility Rank)	Linear Cooling Crystallization		Fast Evaporative Crystallization		Slow Evaporative Crystallization
	10 °C	30 °C	20 °C	30 °C	20 °C
TCM (1)	-	Form I	-	Form II	Form II
DCM (2)	Form II	Form II	Form II	Form II	Form II
EtOAc (3)	Form I	Form II	Form II	Form II	Form II+I *
ACN (4)	Form II	Form II	Form II	Form II	Form I
AcOH (5)	Form I	Form I	Solvate	Solvate	-
MeOH (6)	-	-	Form II	Form II	Form II

* Form I was observed in one vial and form II in another vial. The dark grey background denotes piroxicam form II.

The crystallization of polymorphs and solvates from organic compounds is a complex process and represents a particular challenge in the production of APIs. It has been observed that the solute–solvent interaction has an effect on the formation of the self-associations of the solute molecules. Certain solvents can facilitate the formation of head-to-head dimers of the solute molecules; however, the exiting of these dimers does not necessarily promote the nucleation of the polymorphs with similar head-to-head molecular configurations [7]. The stochastic nature of the nucleation of different polymorphic forms in solution crystallization has been demonstrated in the crystallization of isonicotinamide and piroxicam, which has been reported in our previous work [7,9,10] as well as in studies from other groups [6,14].

3.4. Evaporative Crystallization with Fast and Slow Evaporation Rates

It has been observed in our previous work [9] that the crystallization method has a significant effect on the yielded polymorphism of piroxicam when cooling and anti-solvent crystallization was investigated in the solvent system of acetone and water. Piroxicam monohydrate crystallized dominantly from cooling crystallization of piroxicam from acetone–water solutions if the concentration of water was higher than 10 wt.%. However, the anhydrous form I of piroxicam was formed in anti-solvent crystallization from acetone solutions using water as the anti-solvent regardless of the water concentration exceeding 10 wt.%. In this study, the crystallization of piroxicam polymorphs in evaporative crystallization from different organic solvents was investigated. The effects of solvent and evaporating rates have been studied. The overview of polymorphic forms obtained in both cooling and evaporative crystallization of piroxicam from the various solvents is shown in Table 2. The dark grey defines piroxicam form II, and the parentheses represent the given solvent ranking according to the solubility measurement. The temperature listed in the table denotes the temperature at which the saturated solution is prepared.

It is shown in Table 2 (Raman spectra of the samples shown in Table 1 are included in Supplementary Materials Figure S1) that the polymorph formed from the different experiments changes with the crystallization technique (cooling or evaporative) and with the solvent utilized. The cooling crystallization of piroxicam from ethyl acetate and chloroform could produce either form I or form II; evaporative crystallization of piroxicam from all studied solvents (except acetic acid) yielded form II when fast solvent evaporation was applied. Slowing down the evaporation rate led to the formation of form I in solutions with acetonitrile and ethyl acetate. Interestingly, form I was obtained from the cooling crystallization of acetic acid solutions regardless of the different initial solute concentrations, and evaporative crystallization produced a solvate (Raman spectra shown in Figure 4). As shown in Figure 6, the solid form was analyzed by thermogravimetric analysis, confirming the acetic acid solvate with the ratio of 1:1. This observation is in agreement with the literature [11], where a mono-solvate of piroxicam with acetic acid was discovered. The solubility of the solvent is shown in parentheses in Table 2 where one corresponds to the solvent where piroxicam was most soluble, and six to the least soluble. No correlation can be drawn between the solubility of piroxicam in the solvents with the polymorphic form obtained from cooling and evaporative crystallization.

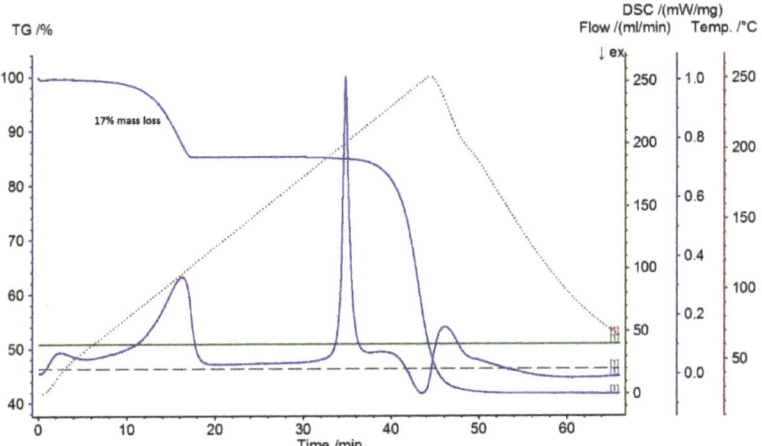

Figure 6. (1-column fitting) Thermogravimetric analysis of acetic acid solid-state form obtained from evaporative crystallization with a saturated concentration at 20 °C. The loss of 17% mass at the temperature 18–120 °C corresponding to one mole of acetic acid pr. mole of piroxicam, confirming the solvate.

The polymorphic form overview in Table 2 clearly demonstrates the stochastic nature of the nucleation of different polymorphic forms during solution crystallization and the complex interplay of the operation parameters that affect the outcome of a crystallization process. It has been hypothesized that the solvent–solute interaction may have a more significant effect on determining the polymorphism of the crystallized piroxicam, which have different intermolecular hydrogen bonding and configurations of molecules. However, the results obtained in the present work do not verify this hypothesis, as both form I and form II can be crystallized out of several solvents (see Table 2).

4. Conclusions

The solubility of piroxicam form II was measured in seven different solvents of varying Hansen's solubility parameters at different temperatures (10, 20, 30 and 40 °C). The highest solubility of piroxicam form II was found in dichloromethane and chloroform, while the lowest solubility was found in methanol and hexane. These observations of solubility are also in accordance with the HSP$_E$, where the similarity of parameters between piroxicam and the solvent resulted in higher solubility and vice versa.

The energy barrier for the nucleation of piroxicam depends on the solvent. Solutions with dichloromethane, ethyl acetate and acetonitrile showed a metastable zone width (MSZW) of approximately 30 °C. A narrow MSZW was observed when using the solvent acetic acid, while a very broad MSZW was encountered for chloroform. Polymorphism of piroxicam from cooling and evaporative crystallization could be affected by the operation parameters. Comparing linear cooling crystallization with fast and slow evaporative crystallization indicated some tendencies; form II was favored from both fast and slow evaporative crystallization techniques. Regardless of the crystallization method used, form II was yielded from dichloromethane and methanol. Form II was also yielded by all studied solvents (except acetic acid yielding a solvate) when applying fast evaporative crystallization. This indicates that fast evaporative crystallization is a method that can be used for the production of Form II. However, the obtained results showed the complex and stochastic nature of the nucleation of different polymorphic forms in solution crystallization. No simple link can be suggested to correlate the polymorphism outcome with any operation parameter, such as solvent, solute concentration, or how fast the supersaturation was generated.

Supplementary Materials: The following are available online at https://www.mdpi.com/article/10.3390/cryst11121552/s1, Figure S1: Raman spectra for samples obtained from cooling and evaporative crystallizations of piroxicam from chloroform (**a**), dichloromethane (**b**), ethyl acetate (**c**), acetonitrile (**d**), acetic acid (**e**) and methanol (**f**).

Author Contributions: Conceptualization, H.Q.; methodology, H.Q.; validation, H.Q. and I.O.; investigation, H.Q. and I.O.; writing – original draft writing, I.O.; writing – review and editing, H.Q.; visualization, I.O.; supervision, H.Q.; project administration, H.Q.; funding acquisition, H.Q. All authors have read and agreed to the published version of the manuscript.

Funding: This research was partially funded by the Danish Council for Independent Research (DFF) grant ID DFF-6111-00077B.

Data Availability Statement: The data presented in this study are available on request from the corresponding author. The data are not publicly available due to further continuing research.

Acknowledgments: The Danish Council for Independent Research (DFF) is acknowledged for the financial support (grant ID: DFF-6111-00077B). The authors would also like to thank Joakim Tobias Schack and Morten Ørndrup Nielsen for carrying out the experiments throughout their bachelor thesis in the Department of Chemical Engineering, Biotechnology and Environmental Technology, University of Southern Denmark.

Conflicts of Interest: The authors declare no conflict of interest.

References

1. Braatz, R.D. Advanced control of crystallization processes. *Annu. Rev. Control.* **2002**, *26*, 87–99. [CrossRef]
2. Datta, S.; Grant, D.J.W. Crystal structures of drugs: Advances in determination, prediction and engineering. *Nat. Rev. Drug Discov.* **2004**, *3*, 42–57. [CrossRef]
3. Chen, J.; Sarma, B.; Evans, J.M.B.; Myerson, A.S. Pharmaceutical Crystallization. *Cryst. Growth Des.* **2011**, *11*, 887–895. [CrossRef]
4. Qu, H.; Kohonen, J.; Louhi-Kultanen, M.; Reinikainen, S.P.; Kallas, J. Spectroscopic Monitoring of Carbamazepine Crystallization and Phase Transformation in Ethanol-Water Solution. *Ind. Eng. Chem. Res.* **2008**, *47*, 6991–6998. [CrossRef]
5. Nagy, Z.K.; Braatz, R.D. Advances and New Directions in Crystallization Control. *Annu. Rev. Chem. Biomol. Eng.* **2012**, *3*, 55–75. [CrossRef]
6. Kulkarni, S.A.; McGarrity, E.S.; Meekes, H.; ter Horst, J.H. Isonicotinamide self-association: The link between solvent and polymorph nucleation. *Chem. Commun.* **2012**, *48*, 4983–4985. [CrossRef] [PubMed]
7. Hansen, T.B.; Taris, A.; Rong, B.G.; Grosso, M.; Qu, H. Polymorphic behavior of isonicotinamide in cooling crystallization from various solvents. *J. Cryst. Growth* **2016**, *450*, 81–90. [CrossRef]
8. Taris, A.; Hansen, T.B.; Rong, B.G.; Grosso, M.; Qu, H. Detection of Nucleation during Cooling Crystallization through Moving Window PCA Applied to in Situ Infrared Data. *Org. Process. Res. Dev.* **2017**, *21*, 966–975. [CrossRef]
9. Hansen, T.B.; Qu, H. Formation of Piroxicam Polymorphism in Solution Crystallization: Effect and Interplay of Operation Parameters. *Cryst. Growth Des.* **2015**, *15*, 4694–4700. [CrossRef]
10. Hansen, T.B.; Simone, E.; Nagy, Z.; Qu, H. Process Analytical Tools to Control Polymorphism and Particle Size in Batch Crystallization Processes. *Org. Process Res. Dev.* **2017**, *21*, 855–865. [CrossRef]
11. Vrecer, F.; Srcic, S.; Smid-Korbar, J. Investigation of piroxicam polymorphism. *Int. J. Pharm.* **1991**, *68*, 35–41. [CrossRef]
12. Mishnev, A.; Kiselovs, G. New Crystalline Forms of Piroxicam. *Zeitschrift für Naturforschung B* **2013**, *68*, 168–174. [CrossRef]
13. Sheth, A.R.; Bates, S.; Muller, F.X.; Grant, D.J.W. Polymorphism in Piroxicam. *Cryst. Growth Des.* **2004**, *4*, 1091–1098. [CrossRef]
14. Vrečer, F.; Vrbinc, M.; Meden, A. Characterization of piroxicam crystal modifications. *Int. J. Pharm.* **2003**, *256*, 3–15. [CrossRef]
15. Maggioni, G.M.; Bezinge, L.; Mazzotti, M. Stochastic Nucleation of Polymorphs: Experimental Evidence and Mathematical Modeling. *Cryst. Growth Des.* **2017**, *17*, 6703–6711. [CrossRef]
16. Hansen, C.M. *Hansen Solubility Parameters a User's Handbook*, 2nd ed.; CRC Press: Boca Raton, FL, USA, 2007.

17. Bustamante, P.; Peña, M.A.; Barra, J. Partial solubility parameters of piroxicam and niflumic acid. *Int. J. Pharm.* **1998**, *174*, 141–150. [CrossRef]
18. Naelapää, K.; van de Streek, J.; Rantanen, J.; Bond, A.D. Complementing High-Throughput X-ray Powder Diffraction Data with Quantum–Chemical Calculations: Application to Piroxicam Form III. *J. Pharm. Sci.* **2012**, *101*, 4214–4219. [CrossRef]
19. Liu, G.; Hansen, T.B.; Qu, H.; Yang, M.; Pajander, J.P.; Rantanen, J.; Christensen, L.P. Crystallization of Piroxicam Solid Forms and the Effects of Additives. *Chem. Eng. Technol.* **2014**, *37*, 1297–1304. [CrossRef]
20. Cruz-Cabeza, A.J.; Reutzel-Edens, S.M.; Bernstein, J. Facts and fictions about polymorphism. *Chem. Soc. Rev.* **2015**, *44*, 8619–8635. [CrossRef]
21. Sanz, A.; Jiménez-Ruiz, M.; Nogales, A.; Martín y Marero, D.; Ezquerra, T.A. Hydrogen-Bond Network Breakage as a First Step to Isopropanol Crystallization. *Phys. Rev. Lett.* **2004**, *93*, 015503. [CrossRef]
22. Cui, P.; Zhang, X.; Yin, Q.; Gong, J. Evidence of Hydrogen-Bond Formation during Crystallization of Cefodizime Sodium from Induction-Time Measurements and in Situ Raman Spectroscopy. *Ind. Eng. Chem. Res.* **2012**, *51*, 13663–13669. [CrossRef]
23. Kelly, R.C.; Naír, R.H. Solvent Effects on the Crystallization and Preferential Nucleation of Carbamazepine Anhydrous Polymorphs: A Molecular Recognition Perspective Abstract. *Org. Process. Res. Dev.* **2009**, *13*, 1291–1300. [CrossRef]

Article

Phase Diagram Determination and Process Development for Continuous Antisolvent Crystallizations

Corin Mack [1], Johannes Hoffmann [1], Jan Sefcik [2] and Joop H. ter Horst [1,3,*]

[1] Future Manufacturing Research Hub in Continuous Manufacturing and Advanced Crystallisation (CMAC), Technology and Innovation Centre, Strathclyde Institute of Pharmacy and Biomedical Sciences, University of Strathclyde, Glasgow G1 1RD, UK

[2] Future Manufacturing Research Hub in Continuous Manufacturing and Advanced Crystallisation (CMAC), Department of Chemical and Process Engineering, University of Strathclyde, Glasgow G1 1XJ, UK

[3] Laboratoire Sciences et Méthodes Séparatives, SMS EA 3233, Université de Rouen Normandie, Place Emile Blondel, 76821 Mont Saint Aignan, France

* Correspondence: joop.terhorst@strath.ac.uk

Abstract: The development of an antisolvent crystallization process requires the construction of an accurate phase diagram for this ternary system of compound, solvent and antisolvent, preferably as a function of temperature. This study gives an efficient methodology to systematically determine such antisolvent phase diagrams, exemplified with four model compounds: Sodium bromate, DL-Asparagine Monohydrate, Mefenamic acid and Lovastatin. Using clear point temperature measurements, single solvent and mixed solvent-antisolvent solubilities are obtained, showing strongly non-linear solubility dependencies as well as more complex solubility behaviour as a function of antisolvent fraction. A semi-empirical model equation is used to describe the phase diagram of the antisolvent crystallization system as a function of both temperature and antisolvent fraction. The phase diagram model then allows for the identification of condition ranges for optimal productivity, yield, and suspension density in continuous antisolvent crystallization processes.

Keywords: solubility in mixed solvents; antisolvent crystallization; continuous crystallization; crystallization process development

1. Introduction

The access to accurate solubilities of pharmaceutical compounds sets the foundations towards the optimal design and optimization of crystallization processes [1,2]. Although there are methods developed to determine the solubility using computational methods, such as the SAFT-γ MIE approach [3], they are not accurate enough to rely on in the crystallization process design. It is thus required to experimentally determine the solubility to have access to an accurate phase diagram, enabling a reliable crystallization process development, but also to satisfy regulatory approval conditions [4–6]. While solubility data of pure compounds in single solvents is relatively easily accessible nowadays using commercially available equipment and standardized methods [7], solid solubilities in more complex systems such as binary solvent mixtures are much less easy to obtain [8–11]. Solid solubilities in binary solvent mixtures are needed to build phase diagrams to use in the antisolvent crystallization process development [12,13]. Antisolvent crystallization is based on the substantial and non-linear dependence of the solubility upon increasing the antisolvent fraction, a solubility decrease that should be substantially more than the concentration decreases due to the dilution because of the antisolvent addition, as shown in Figure 1 [1]. Antisolvent crystallization can only take place if, at a specific antisolvent fraction, the solubility (the solid line in Figure 1) drops below the diluted concentration (the dashed line in Figure 1). Continuous antisolvent crystallization can be performed in various process configurations using Plug Flow Tubes [14–18] or Continuously Stirred

Tanks (CST) [13,19–21]. One of the possible continuous antisolvent CST crystallization configurations is shown next to the phase diagram in Figure 1.

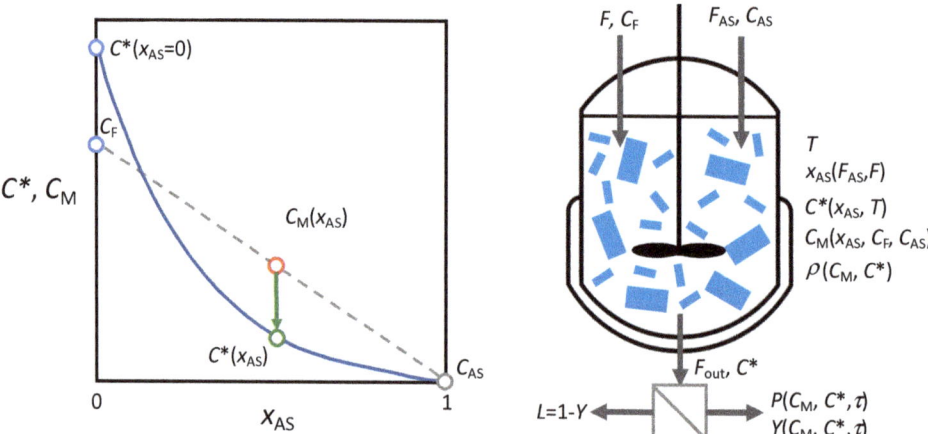

Figure 1. A schematic of the isothermal ternary phase diagram (**left**) for the development of a continuous antisolvent crystallization process (**right**). The feed solution with concentration C_F and pure antisolvent are continuously fed into the crystallization vessel at temperature T and at specific feed rates for the solution F and antisolvent F_{AS} with continuous product removal stream F_{out} to provide a specific residence time τ and antisolvent fraction x_{AS}. The overall concentration C_M and the solubility C^* determine the attainable suspension density r, the productivity P, the yield Y and the loss L.

In order to establish the phase diagram in antisolvent crystallization process development, a common method used is the gravimetric method, in which a suspension is equilibrated and the solution concentration is determined at a specific temperature [11,16,22,23]. Work by Reus et al. [7] showed a variation of this equilibrium method, utilizing solvent addition to a suspension in a mixed solvent antisolvent system at constant temperature in order to dissolve the suspended crystals and obtain the antisolvent fraction at which the overall concentration equals the saturation concentration. Alongside these methods, spectroscopic methods [24] are used to determine the solubility in equilibrated mixed solvent suspensions such as in the case of sulfadiazine [25]. Although effective, the constant temperature method commonly used takes substantial laboratory effort and time.

The temperature variation method, in which the temperature of a suspension is slowly increased until the temperature dependent solubility matches the overall sample composition and all crystals dissolve [26], has become more widely used. In this method, the saturation temperature is approximated by the clear point temperature, the temperature at which, upon increasing the temperature of a suspension, the suspension turns into a clear solution. This method has been utilized in many single solvent systems to establish single solvent solubility data. However, its applicability has not yet been widely used for the process of establishing mixed solvent solubility data, as temperature is introduced as an additional variable, next to compound concentration and antisolvent fraction, increasing the complexity of the analysis. One previous example of the use of the temperature variation method for such a complex system related to antisolvent crystallization is reported by Vellema et al. [27] describing the solubility of lorazepam at varying levels of glucose solution, in order to determine the correct operating window to prevent recrystallization of lorazepam mixed solution during infusion in intensive care units.

The aim of this work is to outline a systematic approach to accurately obtain temperature-dependent phase diagrams for the development of antisolvent crystallization processes. For four model systems, the phase diagram for antisolvent crystallization is determined as a function of temperature and antisolvent fraction. From the solubility data, a solubility model is established. The model is then used to propose operation conditions for a continuous antisolvent crystallization process identifying the most optimal region in terms of antisolvent fraction and temperature to achieve a productivity P, yield Y and suspension density r within specifications.

2. Materials and Methods

The molecular structures of the model compounds are shows in Figure 2. Lovastatin (LOV) was supplied from Molekula (>98%, Darlington), DL-Asparagine Monohydrate (ASN), Sodium Bromate (NaBrO$_3$) and Mefenamic Acid (MFA) were obtained from Sigma Aldrich (>99%). All compounds were used as received. For solution preparation, acetone (99%) and ethanol (99%) were obtained from VWR International. The distilled and filtered water was obtained from an in-house Millipore Systems setup.

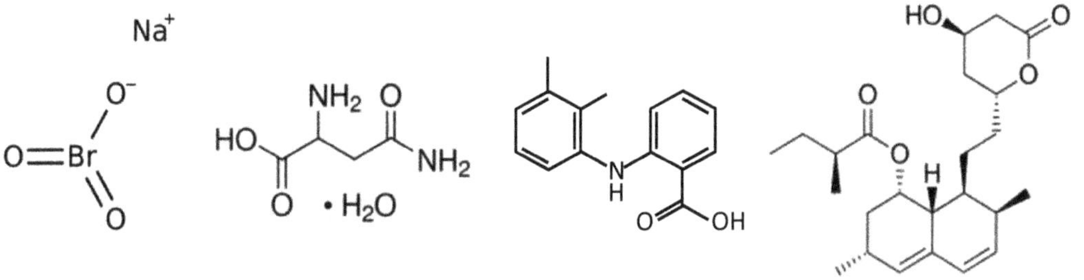

Figure 2. Molecular and ionic structure of the four model compounds. From left to right: Sodium Bromate (NaBrO$_3$), DL-Asparagine Monohydrate (ASN), Mefenamic Acid (MFA) and Lovastatin (LOV).

For the clear point measurements a known amount of the crystalline compound was added to a standard HPLC vial. Then, about 1 mL of a solvent/antisolvent mixture with known antisolvent mass fraction x_{AS} was pipetted into the vial from a larger volume of prepared solvent/antisolvent mixture stock solution. These vials were weighed before and after addition of solvent mixture to exactly determine the mass of solvent mixture added. The composition of the sample is denoted by the (solute-free) antisolvent mass fraction x_{AS} and the solute concentration C in units of g/g-solvent mixture. Due to its low solubility, we chose to neglect the influence on the sample antisolvent fraction of the amount of water in the DL-Asn.H$_2$O crystals.

The Crystal16 Multiple Reactor Setup (Technobis Crystallization Systems) was used to determine the clear point temperature of the prepared samples. The clear point temperature is the temperature at which, upon heating, a suspension turns into a clear solution. The clear point temperature of the samples was determined in triplicate using automated temperature profiles. First, the suspension was dissolved by keeping the sample at a temperature well above the saturation temperature for 30 min. Then, the sample was recrystallized by cooling down to a temperature well below the saturation temperature and keeping the sample at that temperature for at least 30 min. Finally, three temperature cycles were performed to determine 3 clear point temperatures in the heating part of the temperature cycle. A single temperature cycle consisted of a heating part in which the suspension was heated with a heating rate of 0.2 °C/min, a constant high temperature part of 30 min, a cooling part in which the clear solution was cooled with a cooling rate of 0.4 °C/minute, and a constant low temperature part of 30 min. The average of the 3 clear point measurements was assumed to be equal to the saturation temperature of the sample concentration C at the antisolvent fraction x_{AS} in the vial.

Occasionally, samples displayed larger than 1 °C difference between clear point temperatures measured in subsequent cycles and these measurements were then discarded, and fresh samples were used in a re-run. This deviation in clear points was usually coinciding with crowning of crystals just above the liquid level in the sample vials. The occurrence of crowning substantially decreased when the stirring rate in the constant high temperature region of the temperature profile was increased.

Due to known issues of lovastatin degrading over time in solutions at higher water content, the total experimental time experienced by each sample containing lovastatin was kept smaller than 24 h. There were no observations of large decreases in subsequent clear point temperatures of the same sample, indicating that the degradation of lovastatin is negligible.

The procedure was slightly adapted for MFA measurements due to the polymorphic nature of MFA to remove the impact of potential nucleation of the undesired form during the dissolution–recrystallization cycling. Instead of 3, only a single measurement was obtained for a sample, using the initial suspension in the vial without the initial dissolution step. Vials filled with pure ethanol were used to calibrate the transmission of light in the Crystal16 equipment for a clear solution. Subsequently, the prepared vials containing MFA slurries with the original raw material (Form 1) in ethanol–water mixtures were added to the machine and a single temperature cycle was performed. The clear point temperature at which 100% light transmission was reached was taken as the saturation temperature of the sample. A larger number of measurement samples were prepared for this model compound.

Fitting of experimental data was done using the MatLab Curve Fitting tool and by using the scipy.optimize curve_fit routine within Python.

3. Results

3.1. Single Solvent Solubility

Figure 3a displays the measured temperature-dependent solubilities of $NaBrO_3$ in water, DL-Asn.H_2O in water, MFA in ethanol and LOV in acetone. Each system shows a strong temperature dependence of the solubility.

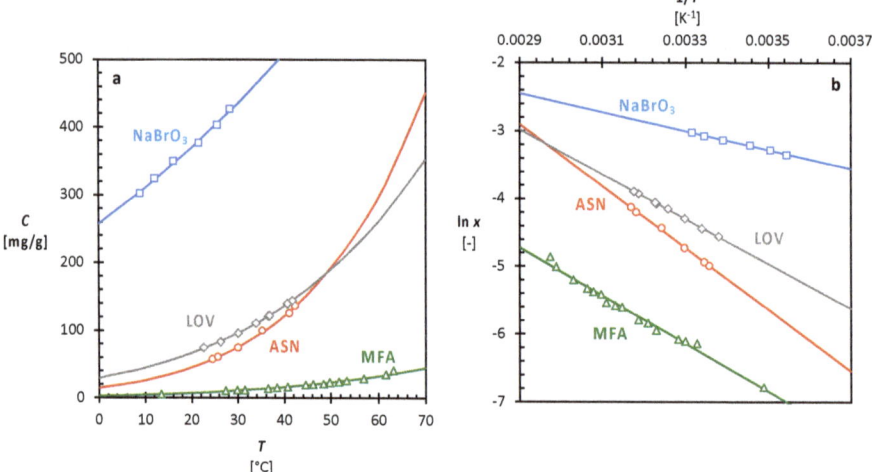

Figure 3. (**a**): temperature dependent solubilities of $NaBrO_3$ in water (blue, □), ASN in water (red, ○), MFA in ethanol (green, △), and LOV in acetone (grey, ◇). (**b**): The same data with the fits to Equation (2). The lines through the points in both graphs are best fits to Equation (2).

Sodium bromate ($NaBrO_3$, Figure 2) is a water-soluble salt. Dissolved in water, it is achiral but it crystallizes in the chiral space group $P2_13$ [28]. Similar to some other sodium salts [29] the solubility of $NaBrO_3$ in water is high, 367 mg/g at 25 °C, and it increases with temperature to 505 mg/g at 40 °C (Figure 3a). As $NaBrO_3$ is a salt with a high solubility in water and a low solubility in ethanol, it can be produced by antisolvent crystallization from solutions in water using ethanol as an antisolvent [21].

Lovastatin (LOV) from the statin family has a molecular structure (Figure 2) with several chiral centres and it is used in the treatment of high cholesterol and cardiovascular disease. It does not appear to have any known polymorphs. The solubility of LOV in acetone is ranging from 80 to 180 mg/g between 20 and 40 °C (Figure 3a). The very low solubility of LOV in water in comparison to acetone makes it an interesting model compound for antisolvent crystallization [30].

The racemic asparagine monohydrate (ASN, Figure 2) is a non-essential amino acid and is known to crystallize as a conglomerate forming system. Asparagine monohydrate has a solubility in water of around 80 to 180 mg/g between 20 and 40 °C (Figure 3a). The solubility of ASN is known to decrease with the antisolvent ethanol fraction while ASN recrystallizes as the anhydrous form at sufficiently large ethanol fractions [7].

Mefenamic acid (MFA, Figure 2) is a member of the anthranilic acid derivatives of non-steroidal anti-inflammatory drugs (NSAIDs), used for the treatment of mild and moderate pains. It is not widely used compared to other NSAID due to its higher costs. MFA is known to crystallize in three polymorphic forms [31,32]. The solubility of MFA in ethanol is the lowest of the measured pure solvent systems, with solubilities between 10 and 42 mg/g at temperatures ranging from 25 to 60 °C (Figure 3a). MFA has a very low solubility in water, similar to other NSAID on the market, making water a potential antisolvent.

The ideal solubility x_{id} is usually described as a function of temperature T using only two solid state properties: melting temperature T_m and heat of fusion ΔH (kJ mol^{-1}) [33]:

$$x_{id} = \exp\left(-\frac{\Delta H}{R}\left(\frac{1}{T} - \frac{1}{T_m}\right)\right) \quad (1)$$

with the molar gas constant R. Due to non-ideality in the solution, the ideal solubility can substantially deviate from the measured solubilities. Often, Equation (2), the linearized form of Equation (1), gives a good fit to experimental data within a sufficiently narrow temperature region using the parameters A and B as fitting parameters:

$$\ln x = \frac{A}{T} + B \quad (2)$$

The parameters A and B can be determined from a linear fit of Equation (2) to the experimental data in the plot of $\ln x$ versus $1/T$. Equation (2) describes well the measured solubilities in (Figure 3b) within the measured temperature region.

3.2. Solubility in Solvent/Antisolvent Mixtures

Figure 4a,c,e,g display the temperature dependent solubility for all model compounds obtained at specific anti-solvent fractions x_{AS}. In all instances, at a constant antisolvent fraction x_{AS}, the solubility increases with temperature. The solubility at a particular antisolvent fraction x_{AS} can well be described by Equation (2) as shown by the dashed lines in the right graphs of Figure 4b,d,f,h.

For NaBrO$_3$, at a specific temperature the solubility reduces non-linearly as a function of the antisolvent fraction x_{AS} (Figure 4a,b). When going from $x_{AS} = 0$ to 0.1 at 30 °C the solubility drops 51% from 366 to 181 mg/g. The solubility drops further to 94 mg/g going to $x_{AS} = 0.2$, and at $x_{AS} = 0.5$ it has decreased by 95% from its solubility in pure water. This trend of a strong non-linear solubility decrease as a function of antisolvent is consistent with the preferred antisolvent phase diagram behaviour for systems with optimal antisolvent crystallization potential [34,35]. ASN and MFA show similar behaviour with the solubilities decreasing as a function of antisolvent fraction (Figure 4c–f). At 30 °C the solubility shows, respectively, a 40% and 37% drop, going from $x_{AS} = 0$ to 0.1. The solubility of ASN, for instance, drops from 75 mg/g at $x_{AS} = 0$ to 45 mg/g at $x_{AS} = 0.1$ at 30 °C for DL-Asn.H$_2$O.

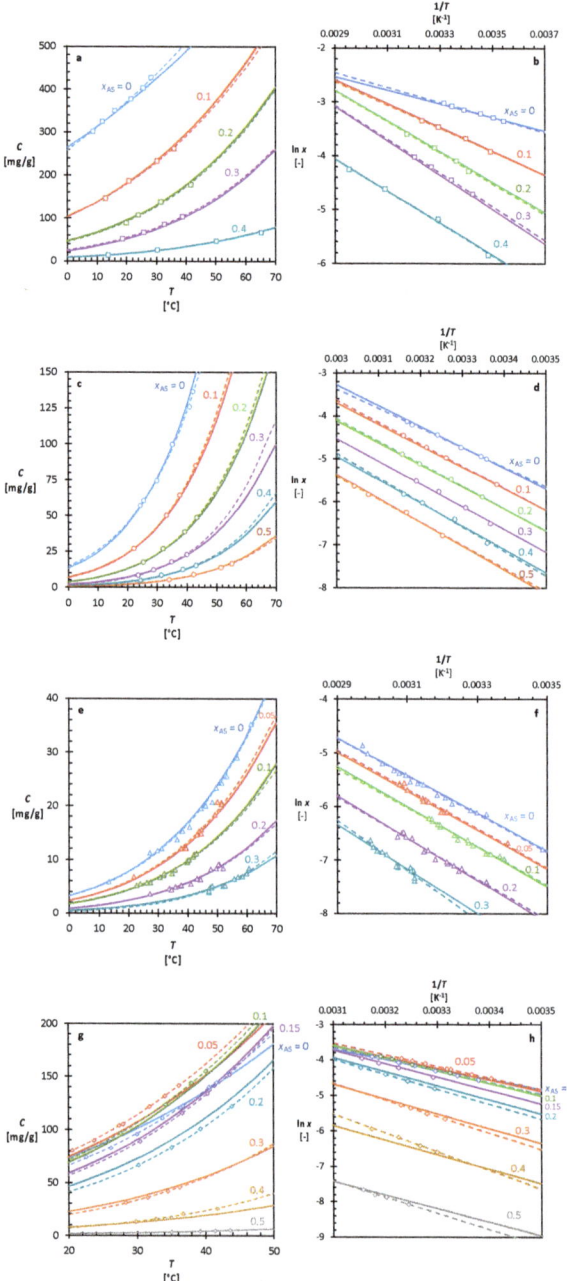

Figure 4. Concentration–temperature (left) and lnx—1/T (right) diagrams of: (**a,b**). (□) NaBrO$_3$ in water–ethanol, (**c,d**). (○) ASN in water–ethanol, (**e,f**). (△) MFA in ethanol/water, and (**g,h**). (◇) LOV in acetone/water at various antisolvent fractions. The experimental data are represented by the markers, with the colours and labels representing the antisolvent fraction. The dashed lines give the best fit of the data at a specific antisolvent fraction to Equation (2). The solid lines represent the predicted solubility at each antisolvent fraction using the best fit of all data to Equation (4).

Conversely, LOV in ethanol–water shows, at small antisolvent fractions, an increase in solubility compared to the solubility in absence of antisolvent (Figure 4g,h). Going from $x_{AS} = 0$ to 0.05 at 30 °C, the solubility increases from 98 to 122 mg/g, a 26% increase. As the antisolvent fraction increases further, the solubility decreases. At $x_{AS} = 0.1$ the solubility reduces to 98 mg/g and it reduces to below its pure system concentration at $x_{AS} = 0.2$. At $x_{AS} = 0.5$ the solubility has further decreased to 4.6 mg/g. Such a solubility maximum at a specific antisolvent fraction may be a complex combination of molecular, thermodynamic, and entropic effects. For instance, antisolvent molecules of water may shield unfavourable polar parts of the complex molecule LOV to enable a better interaction of the solvent molecules of acetone with LOV, increasing the solubility. Several systems are reported to exhibit solubility behaviour similar to LOV in solvent–antisolvent mixtures [24,25,31].

3.3. Antisolvent Crystallization Phase Diagrams

For the system of NaBrO$_3$ in water/ethanol, the fitted parameters A and B to Equation (2) at each antisolvent fraction are shown in Figure 5. The parameters A and B change substantially with the antisolvent fraction x_{AS}. The other systems show a similar behaviour. Utilizing the observed behaviour of the solubility as a function of temperature and antisolvent fraction a single empirical equation is proposed based on Equation (2), in which the antisolvent fraction dependence is of parameters A and B is captured by simple polynomials:

$$\ln x = \left(\sum_{i=0}^{n} a_i x_{AS}^i\right) \frac{1}{T} + \sum_{i=0}^{n} b_i x_{AS}^i \qquad (3)$$

Table 1. Fitting parameters and their standard errors obtained from the model Equation (3) using the entire dataset for a combination of compound, solvent and antisolvent. The relative standard deviations $\sigma_{\ln x}$ and σ_x from Equation (4) for each system are also provided.

Compound	NaBrO$_3$	ASN	MFA	LOV
Solvent	Water	Water	Ethanol (EtOH)	Acetone (AcO)
Antisolvent	Ethanol (EtOH)	Ethanol (EtOH)	Water	Water
N	22	27	65	39
n	2	1	1	2
$a_0 \times 10^{-3}$	-1.26 ± 0.2	-4.84 ± 0.16	-3.50 ± 0.09	-2.80 ± 0.36
$a_1 \times 10^{-3}$	-10.8 ± 1.6	-1.40 ± 0.47	-2.12 ± 0.64	-8.30 ± 5.10
$a_2 \times 10^{-3}$	14.8 ± 2.8			12.6 ± 11.2
b_0	1.12 ± 0.55	11.5 ± 0.5	5.44 ± 0.27	5.0 ± 1.2
b_1	31.4 ± 5.4	0.0 ± 1.5	0.75 ± 1.97	28.8 ± 16.5
b_2	-48.9 ± 9.5			-60.0 ± 36.2
$\sigma_{\ln x}$	0.8%	0.7%	0.8%	1.7%
σ_x	3.3%	3.7%	5.1%	8.9%

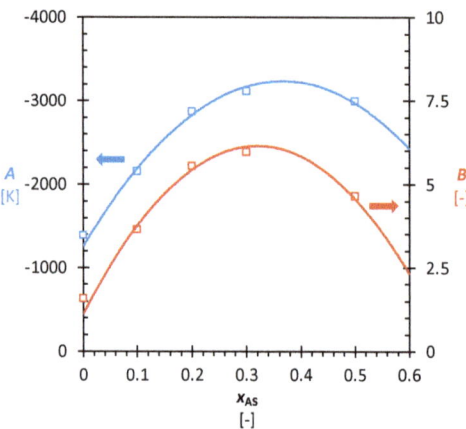

Figure 5. The points are the parameters A (blue) and B (red) for the Na_2BrO_3 system from the fits of the datasets at each antisolvent fraction x_{AS} to Equation (2). The lines are the antisolvent faction dependent parameters A (blue) and B (red) determined from a fit of the entire dataset of the Na_2BrO_3 system to Equation (3) with $n = 2$. The parameters can be found in Table 1.

The goodness of the fit for the model is determined for each system using Equation (4) for the relative standard deviation σ_z in % with N the total number of experimental points used, with $z = x$ or $\ln x$:

$$\sigma_z = 100 \times \sqrt{\frac{\sum_{i=1}^{N}\left(\frac{z_i - z_{pred}}{z_i}\right)^2}{N-1}} \qquad (4)$$

An $\sigma_{\ln x} = 1\%$ means that the resulting parameters describe the $\ln x$ data with on average 68% of the datapoints having less than a 1% deviation from the model $\ln x$ value.

The simple model of Equation (3) allows us to fit all experimental data for a specific system and the fitted parameters are shown in Table 1. The order n of the polynomial is chosen to be the $0 \leq n \leq 2$ for which the relative standard deviation $\sigma_{\ln x}$ is below 1% or else the smallest. We chose to fit $\ln x$ rather than x, but also give the value for σ_x in Table 1 to show how well the parameters describe solubility fraction x.

The model with $n = 2$ describes the $NaBrO_3$ system well, with a standard deviation $\sigma_{\ln x} = 0.8\%$, which means that on average 68% of the experimental $\ln x$ values deviate less than 0.8% from those predicted by the model. The standard deviation in the mole fraction x is slightly larger, $\sigma_x = 3.3\%$, as the fit was performed on $\ln x$ rather than on x. While the $NaBrO_3$ system needs six parameters to be described well, the systems of ASN and MFA can do with four, using $n = 1$ in Equation (3). The standard deviations of the ASN and MFA systems are, respectively, $\sigma_{\ln x} = 0.7\%$ and 0.8%. Looking at the large relative standard error in parameter b_1 in Table 1, both systems can also be described using three rather than four parameters, with an antisolvent fraction independent B-parameter ($b_1 = 0$). The standard deviation using three parameters is not substantially increasing for both systems. For the LOV system, with the more complex antisolvent fraction dependent solubility behaviour, the model, using $n = 2$, performs reasonably with a $\sigma_{\ln x} = 1.7\%$. The ability of the model to capture the change in antisolvent and temperature is shown by the solid lines in the $\ln x$—$1/T$ diagrams in Figure 4.

The parameters in Table 1 can be used to construct an antisolvent crystallization phase diagram at constant temperature, where the solubility is shown as a function of the antisolvent fraction x_{AS}. Figure 6a,c,e,g show the antisolvent crystallization phase diagrams for all the systems constructed using the model at a specific temperature. The solubility decreases strongly linearly with increasing antisolvent fraction for the systems of $NaBrO_3$, ASN and MFA. For the LOV system, the increase of the solubility at low antisolvent fractions is clearly seen, while at a higher antisolvent fraction, the solubility sharply decreases with the antisolvent fraction. It is interesting to note that the solubility continues to decrease towards pure antisolvent (x_{AS} = 1), even for antisolvent fractions larger than the ones at which the solubilities were measured.

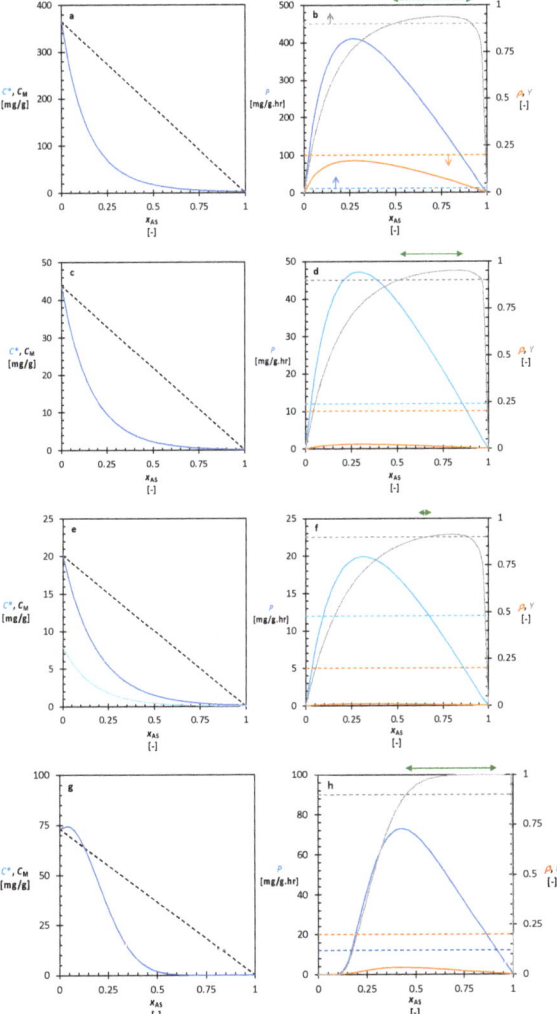

Figure 6. The isothermal phase diagram of solubility C^* against antisolvent fraction x_{AS} (left, **a,c,e,g**) and productivity P, yield Y and suspension density r (right, **b,d,f,h**) of: (**a,b**). NaBrO$_3$ in water with antisolvent EtOH, (**c,d**). ASN in water with antisolvent EtOH, (**e,f**). MFA in EtOH with antisolvent water, and (**g,h**). LOV in AcO with antisolvent water. All phase diagrams are at temperature T = 20 °C while in the case of the MFA system also the one at T = 45 °C is shown. The dashed line represents the dilution line with total concentration C_M as a function of the antisolvent fraction x_{AS} that results from the mixing of a saturated solution feed and a pure antisolvent feed. On the right, the predicted productivity P (—), yield Y (—) and slurry density r (—) in a continuous antisolvent crystallization process with a saturated feed and a pure antisolvent feed as a function of antisolvent fraction x_{AS} are shown. The residence time for each system is 0.5 h. The dashed horizontal lines on the right indicate the process specifications: $Y \geq 90\%$, $P \geq 0.012$ g/g.hr and $\rho \leq 20\%$ while the optimal region of the antisolvent fraction is indicated with a green horizontal arrow above the graphs on the right.

The determined phase diagrams are very helpful in the development of continuous antisolvent crystallization processes such as that in Figure 1.

3.4. Continuous Antisolvent Crystallization

The pharmaceutical industry has been increasingly adopting continuous manufacturing for both primary (drug substance) and secondary (drug product) processing. The use of continuous crystallization rather than batch-wise crystallization allows higher productivities and yields to be obtained, while allowing better control of critical quality attributes and reducing product variations, often seen from batch to batch [36]. A continuously stirred tank antisolvent crystallization process has continuous inlets of concentrated solution and of antisolvent as well as a continuous suspension outlet stream and is operated in a steady state in which, in principle, the suspension properties do not change over operation time (Figure 1).

In continuous antisolvent crystallization processes, the mixing of the pure antisolvent and the solution results in two effects, shown in the schematic in Figure 1. On the one hand, by adding an antisolvent the solubility $C^*(x_{AS}, T)$ decreases in the resulting mixture. This antisolvent fraction-dependent solubility at constant temperature is shown for all systems in Figure 6, calculated using the constructed phase diagram model. On the other hand, the addition of antisolvent dilutes the solution, decreasing the overall concentration in the mixed solution compared to the feed concentration C_F. The overall concentration C_M at a specific antisolvent x_{AS} fraction is a function of the feed concentration C_F and solute concentration C_{AS} in the antisolvent.

$$C_M(x_{AS}) = C_F - (C_F - C_{AS}) \frac{x_{AS} - x_{AS}^F}{x_{AS}^{AS} - x_{AS}^F} \tag{5}$$

Here we also account for the antisolvent fractions in the solution feed and antisolvent feed which might deviate from $x_{AS}^F = 0$ and $x_{AS}^{AS} = 1$, respectively. The dashed straight line in Figure 6 represents the dilution line $C_M(x_{AS}) = C_F(1 - x_{AS})$ for a process with a feed of solute in pure solvent, saturated at the process temperature ($C_F = C^*(x_{AS}^F = 0, T)$), and a feed of pure antisolvent ($C_{AS} = 0$, $x_{AS}^{AS} = 0$) at the same temperature. Only at an antisolvent fraction at which the overall concentration C_M is higher than the solubility C^* (the dilution line is above the solubility line), overall supersaturation is created and antisolvent crystallization can take place [37].

A continuous antisolvent crystallization at ambient pressure is determined by the operating conditions of temperate T, residence time τ and antisolvent fraction x_{AS} [38]. The operating temperature T in the crystallizer determines the solubility $C^*(x_{AS}, T)$ at the prevailing antisolvent fraction. In steady state, the antisolvent fraction x_{AS} in the crystallizer is constant and determined by the size of the antisolvent feed flow rate with respect to the sum of the feed flow rates. The residence time $\tau = v/\dot{v}$ is defined as the average time molecules spend in the crystallizer and is determined by the crystallizer volume V and the flow rate v of the suspension leaving the crystallizer.

In continuous antisolvent crystallization process development, we can define three limiting process requirements such as in Figure 1, for yield Y, productivity P, and suspension density r. Long residence times allow the complete consumption of the supersaturation by the growing crystals and therefore a maximum product yield Y to be obtained. Then, we can use the equilibrium stage concept often used in the early stages of separation technology process development [39]. The maximum yield Y at a specific antisolvent fraction x_{AS} is then defined by the difference between overall concentration $C_M(x_{AS})$ and solubility $C^*(x_{AS})$ at that antisolvent fraction, relative to the overall concentration:

$$Y(x_{AS}) = \frac{C_M(x_{AS}) - C^*(x_{AS})}{C_M(x_{AS})} \tag{6}$$

However, as the solubility $C^*(x_{AS})$ is not zero, some solute remains in solution and creates a loss of potential product. The yield Y is related to the loss $L = 1 - Y$ of product that remains dissolved in the solvent mixture. The loss of product needs to be as small as possible, and a maximum acceptable product loss can be defined, for example, as $L_{max} = 0.1$, in which case 10% of the incoming solute is not crystallized. The target minimum yield therefore is set to $Y_{min} = 0.9$.

If the crystallization kinetics are relatively fast, a comparatively short residence time may suffice to achieve a high productivity, the amount of product produced per unit of time and feed mass. The maximum productivity P at a specific antisolvent fraction x_{AS} is defined by the difference in overall concentration C_M and solubility C^* at that antisolvent fraction and the residence time τ:

$$P(x_{AS}) = \frac{C_M(x_{AS}) - C^*(x_{AS})}{\tau} \quad (7)$$

At a specific residence time, the maximum productivity therefore can be achieved at the antisolvent fraction at which the difference between overall concentration and solubility is the largest. A target minimum productivity of an economically viable chemical process leading can be regarded, for example, to be about $P_{min} = 12$ mg/g.hr. Since information on crystallization kinetics is usually not available during early stages of the process development, a lower limit of the residence time is not a priori known. However, if the residence time becomes too short, the maximum productivity will not be achieved as the solution concentration in the crystallizer will significantly deviate from the saturation concentration since the crystallization kinetics will not be fast enough to consume all supersaturation. We can assume that a reasonable residence time may be at $\tau = 0.5$ hr, for which the crystals would need a linear growth rate of crystal faces of about 28 nm/s or an overall crystal growth rate of 3.3 μm/minute to arrive at a product size of about 100 μm, however, a suitable residence time should be based on experimental validation for a particular system.

The third limiting process requirement is set on the suspension density ρ, which is defined as the weight fraction of the crystallizer suspension occupied by the solid phase. This is calculated from the following equation, with the units for concentration in mg/g-solution:

$$\rho = \frac{C_M(x_{AS}) - C^*(x_{AS})}{1000 + C_M(x_{AS}) - C^*(x_{AS})} \quad (8)$$

The more product crystallizes from the solution, the larger the suspension density. If the suspension density becomes too high, the suspension of crystals is hampered which would decrease the product quality. We assume that a suspension density of $\rho_{max} = 0.2$ is the upper limiting process requirement of a continuous antisolvent crystallization to maintain proper mixing and with it a good product quality.

The predicted yield Y, productivity P, and slurry density ρ as a function of the antisolvent fraction x_{AS} are presented in Figure 6b,d,f,h for each system using a residence time $\tau = 0.5$ h and a saturated feed at the process temperature T. The productivity of $NaBrO_3$ increases significantly from $x_{AS} = 0$ up to $x_{AS} = 0.25$ with a maximum productivity of 409 mg/g.hr (Figure 6a). At higher antisolvent fractions the productivity decreases again. The productivity is well above its minimum value $P_{min} = 12$ mg/g.hr for almost the entire x_{AS} region. The yield Y increases up to an antisolvent fraction of $x_{AS} = 0.75$ and a yield $Y = 0.94$, after which the yield drops to zero. The yield $Y > Y_{min} = 0.9$ in the antisolvent fraction region $0.5 < x_{AS} < 0.9$. Even for this high drop in solubility the suspension density $r < r_{max}$ remains low enough for all $0 < x_{AS} < 1$. Considering productivity, yield, and suspension density the optimal antisolvent fraction region to perform this process is therefore $0.5 < x_{AS} < 0.9$, indicated by the green arrow in Figure 6b.

The ASN system shows a similar behaviour for productivity P, yield Y and suspension density r, although productivity and suspension density are much lower than in the case of $NaBrO_3$ as the ASN solubility is much lower (Figure 6d). In this case the optimal antisolvent fraction is $0.5 < x_{AS} < 0.95$. However, at these high antisolvent fractions it might well be that the anhydrate rather than the hydrate of ASN is formed within the process.

For the MFA system the solubility is even lower and therefore, at 20 °C, does not result in a productivity $P > P_{min}$ in the entire range of antisolvent fraction (Figure 6f). By increasing the process temperature to 45 °C (Figure 6e, f), the productivity increases and $P > P_{min}$ from $x_{AS} = 0.09$ to an antisolvent fraction slightly higher than $x_{AS} = 0.65$. At 45 °C, the yield $Y > Y_{min}$ in the region $0.65 < x_{AS} < 0.85$. The optimal antisolvent fraction is therefore around $x_{AS} = 0.65$ at which the yield is $Y = 0.9$ and productivity $P = 12.67$ mg/g.hr.

Because of the increasing solubility at smaller antisolvent fractions, the productivity and yield are zero below $x_{AS} = 0.15$ for the LOV system (Figure 6h). However, beyond that antisolvent fraction the solubility shows a huge drop to close to zero resulting in high productivities ($0.2 < x_{AS} < 0.9$) and yields ($0.45 < x_{AS} < 0.99$). This makes the optimal antisolvent fraction for continuous antisolvent crystallization to be $0.45 < x_{AS} < 0.9$ as shown by the green arrow in Figure 6h.

4. Discussion

By applying the experimental methodology introduced here, antisolvent phase diagrams can be accurately determined across a range of solvent-antisolvent mixture compositions and temperatures. The methodology is based on measuring a number of clear point temperatures at several antisolvent fractions and uses the data to construct a phase diagram model. The methodology not only works for typical antisolvent phase diagrams such as for the $NaBrO_3$, ASN and MFA systems, but also for a system such as that of LOV, where low amounts of antisolvent induce solubility increases before establishing a more standard antisolvent behaviour at much higher antisolvent fractions. The determined phase diagram can then be used for the development of a continuous antisolvent crystallization process. However, the process development would make use of some prior knowledge while appropriate limiting process requirements will have to be set for a particular process.

At the point of the determination of the phase diagram, some prior knowledge should already be available. First, the chosen solvent and antisolvent are miscible in all ratios. Second, the solubility in the solvent is relatively high while that in the antisolvent is relatively low. The solubility in the antisolvent is connected to the minimum yield. In the case of $Y_{min} > 0.9$, at $x_{AS} = 0.5$ and a saturated feed, the solubility in the mixture needs to be at most 5% of that in a saturated feed solution. A good estimate of the maximum solubility in the antisolvent then is about 1% of that in the pure solvent. Further, the prior knowledge indicates that antisolvent crystallization for the selected system is possible, which can be validated by simple small-scale batch-wise laboratory experiments. These simple batch experiments would show crystal formation upon adding small volumes of antisolvent to a solution and indicate a strongly non-linear decrease of the solubility with the antisolvent fraction. Finally, account should be given to hazard and safety issues concerning the solvent, antisolvent and the compound.

The envisaged process must fulfil relevant limiting process requirements, which are process specifications that are either due to upstream processes or equipment limitations or that are set limiting process requirements for the process. The feed concentration and solvent may have, for instance, a fixed value and composition due to the process stream from an upstream synthesis unit operation. Another possibility may be that the solution feed comes from a continuous cooling crystallization where the continuous antisolvent crystallization is used to recover a recyclable product from the remaining solution of the cooling crystallization process. Here, we chose to work with a feed solution, saturated at the crystallization process temperature, which could represent the latter possibility. Figure 1 shows a situation where $C_F < C^*(x_{AS} = 0)$ to note that the feed concentration C_F does not have to be saturated. Additionally, we chose a pure antisolvent. However, the antisolvent could be recycled when it is separated from the remaining solution in a downstream process. The recycled antisolvent then might contain solvent ($x_{AS} \neq 1$) or solute ($C_{AS} \neq 0$), which influences the route through the phase diagram of the dilution line describing overall concentration C_M in the crystallizer.

Other limiting process requirements, such as the minimal productivity P_{min}, minimal yield Y_{min} and maximal suspension density r_{max} used in this paper, involve the product or the process operation. By assuming equilibrium between the crystal and solution phases productivity P, yield Y and suspension density r can be determined as a function of temperature and antisolvent fraction such as in Figure 6 and compared to the set limiting process requirements conditions P_{min}, Y_{min} and r_{max}. However, the phase diagram does not give any information on the crystallization kinetics. The kinetics of crystallization are strongly influencing the residence time τ needed to achieve the required productivity. In our analysis, we chose a residence time of $\tau = 0.5$ hr but a slow growing compound might need much more time to reach close-to-equilibrium conditions. The lack of information on the crystallization kinetics also reduces the possibility to predict product quality aspects such as crystal size distribution of the product, which can depend on the chosen mixing configuration, the antisolvent crystallization equipment and the scale of the crystallization process.

The knowledge of the phase diagram opens the route towards the rational development of a continuous antisolvent crystallization process and suggests process conditions to consider in the subsequent experimental optimization of the continuous antisolvent crystallization process.

5. Conclusions

The accurate measurement of solubilities as a function of antisolvent fraction and temperature was enabled by clear point measurements, using a small number of experiments per antisolvent fraction. A simple empirical equation was proposed to describe the antisolvent crystallization phase diagrams for four systems using the experimental data. The phase diagrams of $NaBrO_3$ and ASN in water with antisolvent ethanol and MFA in ethanol with antisolvent water all resemble the typical antisolvent crystallization phase diagram with a significant reduction in solubility at small antisolvent fractions. LOV in acetone with antisolvent water shows a solubility increases with antisolvent fraction before a strong decrease. The proposed approach to determine the phase diagram enables the development of continuous antisolvent crystallization processes to determine optimal temperature and antisolvent fraction for the specified productivity, yield, and suspension density.

Author Contributions: Conceptualization, C.M., J.S. and J.H.t.H.; methodology, C.M., J.H.t.H.; formal analysis, C.M., J.H.t.H.; investigation, C.M., J.H.; writing—original draft preparation, C.M.; writing—review and editing, C.M., J.S., J.H.t.H.; supervision, J.S., J.H.t.H.; funding acquisition, J.H.t.H. All authors have read and agreed to the published version of the manuscript.

Funding: Part of this research has received funding as part of the CORE project (October 2016–September 2020) from the European Union's Horizon 2020 research and innovation programme under the Marie Sklodowska-Curie grant agreement No 722456 CORE ITN. We thank the EPSRC Centre for Innovative Manufacturing in Continuous Manufacturing and Crystallisation for support (EPSRC funding under grant reference: EP/I033459/1).

Institutional Review Board Statement: Not applicable.

Informed Consent Statement: Not applicable.

Data Availability Statement: The solubility data is available as ESI with this article.

Conflicts of Interest: The authors declare no conflict of interest.

References

1. Ter Horst, J.H.; Schmidt, C.; Ulrich, J. Fundamentals of Industrial Crystallization. In *Handbook of Crystal Growth*; Elsevier: Amsterdam, The Netherlands, 2015; pp. 1317–1346.
2. Wichianphong, N.; Charoenchaitrakool, M. Statistical optimization for production of mefenamic acid-nicotinamide cocrystals using gas anti-solvent (GAS) process. *J. Ind. Eng. Chem.* **2018**, *62*, 375–382. [CrossRef]
3. Febra, S.A.; Bernet, T.; Mack, C.; McGinty, J.; Onyemelukwe, I.I.; Urwin, S.J.; Sefcik, J.; Ter Horst, J.H.; Adjiman, C.S.; Jackson, G.; et al. Extending the SAFT-γ Mie approach to model benzoic acid, diphenylamine, and mefenamic acid: Solubility prediction and experimental measurement. *Fluid Phase Equilibria* **2021**, *540*, 113002. [CrossRef]
4. Sheikholeslamzadeh, E.; Rohani, S. Solubility prediction of pharmaceutical and chemical compounds in pure and mixed solvents using predictive models. *Ind. Eng. Chem. Res.* **2012**, *51*, 464–473. [CrossRef]
5. Nti-Gyabaah, J.; Chmielowski, R.; Chan, V.; Chiew, Y.C. Solubility of lovastatin in a family of six alcohols: Ethanol, 1-propanol, 1-butanol, 1-pentanol, 1-hexanol, and 1-octanol. *Int. J. Pharm.* **2008**, *359*, 111–117. [CrossRef] [PubMed]
6. Mudalip, S.K.A.; Bakar, M.R.A.; Jamal, P.; Adam, F. Prediction of Mefenamic Acid Solubility and Molecular Interaction Energies in Different Classes of Organic Solvents and Water. *Ind. Eng. Chem. Res.* **2018**, *58*, 762–770. [CrossRef]
7. Reus, M.A.; Van Der Heijden, A.E.D.M.; Ter Horst, J.H. Solubility Determination from Clear Points upon Solvent Addition. *Org. Process Res. Dev.* **2015**, *19*, 1004–1011. [CrossRef]
8. Macedo, E.A. Solubility of amino acids, sugars, and proteins. *Pure Appl. Chem.* **2005**, *77*, 559–568. [CrossRef]
9. Sun, H.; Gong, J.B.; Wang, J.K. Solubility of Lovastatin in acetone, methanol, ethanol, ethyl acetate, and butyl acetate between 283 K and 323 K. *J. Chem. Eng. Data* **2005**, *50*, 1389–1391. [CrossRef]
10. Lorimer, J.W. Thermodynamics of solubility in mixed solvent systems. *Pure Appl. Chem.* **1993**, *65*, 183–191. [CrossRef]
11. Sun, H.; Wang, J. Solubility of lovastatin in acetone + water solvent mixtures. *J. Chem. Eng. Data* **2008**, *53*, 1335–1337. [CrossRef]
12. Zijlema, T.G.; Geertman, R.M.; Witkamp, G.-J.; van Rosmalen, G.M.; de Graauw, J. Antisolvent Crystallization as an Alternative to Evaporative Crystallization for the Production of Sodium Chloride. *Ind. Eng. Chem. Res.* **2000**, *39*, 1330–1337. [CrossRef]
13. Oosterhof, H.; Witkamp, G.-J.; van Rosmalen, G.M. Antisolvent crystallization of anhydrous sodium carbonate at atmospherical conditions. *AIChE J.* **2001**, *47*, 602–608. [CrossRef]
14. Johnson, M.D.; Burcham, C.L.; May, S.A.; Calvin, J.R.; McClary Groh, J.; Myers, S.S.; Webster, L.P.; Roberts, J.C.; Reddy, V.R.; Luciani, C.V.; et al. API Continuous Cooling and Antisolvent Crystallization for Kinetic Impurity Rejection in cGMP Manufacturing. *Org. Process Res. Dev.* **2021**, *25*, 1284–1351. [CrossRef]
15. Hussain, M.N.; Jordens, J.; John, J.J.; Braeken, L.; Van Gerven, T. Enhancing pharmaceutical crystallization in a flow crystallizer with ultrasound: Anti-solvent crystallization. *Ultrason. Sonochem.* **2019**, *59*, 104743. [CrossRef]
16. McGinty, J.; Chong, M.W.S.; Manson, A.; Brown, C.J.; Nordon, A.; Sefcik, J. Effect of Process Conditions on Particle Size and Shape in Continuous Antisolvent Crystallisation of Lovastatin. *Crystals* **2020**, *10*, 925. [CrossRef]
17. Raza, S.A.; Schacht, U.; Svoboda, V.; Edwards, D.P.; Florence, A.J.; Pulham, C.R.; Sefcik, J.; Oswald, I.D.H. Rapid Continuous Antisolvent Crystallization of Multicomponent Systems. *Cryst. Growth Des.* **2018**, *18*, 210–218. [CrossRef]
18. Svoboda, V.; MacFhionnghaile, P.; McGinty, J.; Connor, L.E.; Oswald, I.D.H.; Sefcik, J. Continuous Cocrystallization of Benzoic Acid and Isonicotinamide by Mixing-Induced Supersaturation: Exploring Opportunities between Reactive and Antisolvent Crystallization Concepts. *Cryst. Growth Des.* **2017**, *17*, 1902–1909. [CrossRef]
19. Ferguson, S.; Morris, G.; Hao, H.; Barrett, M.; Glennon, B. In-situ monitoring and characterization of plug flow crystallizers. *Chem. Eng. Sci.* **2012**, *77*, 105–111. [CrossRef]
20. Ostergaard, I.; de Diego, H.L.; Qu, H.; Nagy, Z.K. Risk-Based Operation of a Continuous Mixed-Suspension-Mixed-Product-Removal Antisolvent Crystallization Process for Polymorphic Control. *Org. Process Res. Dev.* **2020**, *24*, 2840–2852. [CrossRef]
21. Hoffmann, J.; Flannigan, J.; Cashmore, A.; Briuglia, M.L.; Steendam, R.R.E.; Gerard, C.J.J.; Haw, M.D.; Sefcik, J.; ter Horst, J.H. The unexpected dominance of secondary over primary nucleation. *Faraday Discuss.* **2022**, *235*, 109–131. [CrossRef]
22. Acree, W.E. Comments on "Solubility and Dissolution Thermodynamic Data of Cefpiramide in Pure Solvents and Binary Solvents". *J. Solut. Chem.* **2018**, *47*, 198–200. [CrossRef]
23. Svärd, M.; Nordström, F.L.; Jasnobulka, T.; Rasmuson, Å.C. Thermodynamics and nucleation kinetics of m-aminobenzoic acid polymorphs. *Cryst. Growth Des.* **2010**, *10*, 195–204. [CrossRef]
24. Pramanik, R.; Bagchi, S. Studies on solvation interaction: Solubility of a betaine dye and a ketocyanine dye in homogeneous and heterogeneous media. *Indian J. Chem. Sect. A Inorg. Phys. Theor. Anal. Chem.* **2002**, *41*, 1580–1587.
25. Ruidiaz, M.A.; Delgado, D.R.; Martínez, F.; Marcus, Y. Solubility and preferential solvation of sulfadiazine in 1,4-dioxane+water solvent mixtures. *Fluid Phase Equilibria* **2010**, *299*, 259–265. [CrossRef]
26. Ter Horst, J.H.; Deij, M.A.; Cains, P.W. Discovering New Co-Crystals. *Cryst. Growth Des.* **2009**, *9*, 1531–1537. [CrossRef]
27. Vellema, J.; Hunfeld, N.G.M.; Van Den Akker, H.E.A.; Ter Horst, J.H. Avoiding crystallization of lorazepam during infusion. *Eur. J. Pharm. Sci.* **2011**, *44*, 621–626. [CrossRef]
28. Abrahams, S.C.; Bernstein, J.L. Remeasurement of optically active $NaClO_3$ and $NaBrO_3$. *Acta Crystallogr. Sect. B* **1977**, *33*, 3601–3604. [CrossRef]
29. Pinho, S.P.; Macedo, E.A. Solubility of NaCl, NaBr, and KCl in water, methanol, ethanol, and their mixed solvents. *J. Chem. Eng. Data* **2005**, *50*, 29–32. [CrossRef]

30. Nagy, Z.K.; Fujiwara, M.; Braatz, R.D. Modelling and control of combined cooling and antisolvent crystallization processes. *J. Process Control* **2008**, *18*, 856–864. [CrossRef]
31. Romero, S.; Escalera, B.; Bustamante, P. Solubility behavior of polymorphs I and II of mefenamic acid in solvent mixtures. *Int. J. Pharm.* **1999**, *178*, 193–202. [CrossRef]
32. SeethaLekshmi, S.; Row, T.N.G. Conformational Polymorphism in a Non-steroidal Anti-inflammatory Drug, Mefenamic Acid. *Cryst. Growth Des.* **2012**, *12*, 4283–4289. [CrossRef]
33. Poling, B.E.; Prausnitz, J.M.; O'Connell, J.P. *The Properties of Gases and Liquids*, 5th ed.; McGraw-Hill: New York, NY, USA, 2001; ISBN 978-0-07-011682-5.
34. Shayanfar, A.; Fakhree, M.A.A.; Acree, W.E.; Jouyban, A. Solubility of lamotrigine, diazepam, and clonazepam in ethanol + water mixtures at 298.15 K. *J. Chem. Eng. Data* **2009**, *54*, 1107–1109. [CrossRef]
35. Yang, Y.; Tang, W.; Li, X.; Han, D.; Liu, Y.; Du, S.; Zhang, T.; Liu, S.; Gong, J. Solubility of Benzoin in Six Monosolvents and in Some Binary Solvent Mixtures at Various Temperatures. *J. Chem. Eng. Data* **2017**, *62*, 3071–3083. [CrossRef]
36. Pawar, N.; Agrawal, S.; Methekar, R. Continuous Antisolvent Crystallization of α-Lactose Monohydrate: Impact of Process Parameters, Kinetic Estimation, and Dynamic Analysis. *Org. Process Res. Dev.* **2019**, *23*, 2394–2404. [CrossRef]
37. Giulietti, M.; Bernardo, A. Crystallization by Antisolvent Addition and Cooling. *Cryst. Sci. Technol.* **2012**. [CrossRef]
38. Zhang, D.; Xu, S.; Du, S.; Wang, J.; Gong, J. Progress of Pharmaceutical Continuous Crystallization. *Engineering* **2017**, *3*, 354–364. [CrossRef]
39. Seader, J.D.; Henley, E.J.; Roper, D.K. *Separation Process Principles: Chemical and Biochemical Operations*, 3rd ed.; Wiley: Hoboken, NJ, USA, 2011; ISBN 978-0-470-48183-7.

Article

The Effect of Reaction Conditions and Presence of Magnesium on the Crystallization of Nickel Sulfate

Ina Beate Jenssen [1], Oluf Bøckman [2], Jens-Petter Andreassen [1] and Seniz Ucar [1,*]

[1] Department of Chemical Engineering, Norwegian University of Science and Technology, 7491 Trondheim, Norway; ina.b.jenssen@ntnu.no (I.B.J.); jens-petter.andreassen@ntnu.no (J.-P.A.)
[2] Glencore Nikkelverk AS, 4613 Kristiansand, Norway; Oluf.Bockman@glencore.no
* Correspondence: seniz.ucar@ntnu.no

Citation: Jenssen, I.B.; Bøckman, O.; Andreassen, J.-P.; Ucar, S. The Effect of Reaction Conditions and Presence of Magnesium on the Crystallization of Nickel Sulfate. *Crystals* **2021**, *11*, 1485. https://doi.org/10.3390/cryst11121485

Academic Editor: Brahim Benyahia

Received: 30 October 2021
Accepted: 29 November 2021
Published: 30 November 2021

Publisher's Note: MDPI stays neutral with regard to jurisdictional claims in published maps and institutional affiliations.

Copyright: © 2021 by the authors. Licensee MDPI, Basel, Switzerland. This article is an open access article distributed under the terms and conditions of the Creative Commons Attribution (CC BY) license (https://creativecommons.org/licenses/by/4.0/).

Abstract: Recycling of valuable metals such as nickel is instrumental to meet the need from the dramatic increase in electric vehicle battery production and to improve its sustainability. Nickel required in the battery manufacture can be recovered from the hydrometallurgical industrial process streams by crystallization of nickel sulfate. Here, crystallization of nickel sulfate is studied from an industrial point of view, investigating the effects of temperature, seeding and presence of magnesium on the formation of various solid phases for the evaluation of their potential influence on the process design. Results showed that the precipitating phase was dictated both by seed amount and reaction temperature. Transformation of metastable phases both in suspension and in a dry state was observed over time. Presence of magnesium was shown to promote formation of $NiSO_4 \cdot 7H_2O$ in solution and increased its stability in a dry form. In their dry state, nickel sulfate that was formed in the absence of magnesium transformed towards α-$NiSO_4 \cdot 6H_2O$, whereas those precipitated in the presence of high magnesium concentrations transformed towards β-$NiSO_4 \cdot 6H_2O$, indicating that magnesium inhibited the phase transformation towards α-$NiSO_4 \cdot 6H_2O$. Knowledge about various solid phases of varying crystal morphology and stability can be used as input to decisions for the best suited solid product type and how this relates to the initial conditions of the sidestreams.

Keywords: crystallization; nickel sulfate; solid phases; phase transformation

1. Introduction

Nickel is an important component in a wide range of materials such as stainless steel, nickel- or copper-based alloys and catalysts, crystal UV-filters, and the cathode of ion batteries [1–4]. Among these, nickel containing electric vehicle batteries (EVBs) has seen a dramatic increase in production over the last decade as a result of the environmental targets set by, amongst others, the EU [5,6]. This situation not only creates a high demand for the production of metals used in batteries, but also necessitates their efficient recycling [7,8]. In addition to its market value, recycling of nickel reduces the environmental pollution caused by nickel-containing waste and the CO_2 footprint of the produced metal [9].

Nickel can be recycled from various sources such as spent batteries, spent catalysts, or from process streams in the hydrometallurgical industry [4,9–11]. In industrial hydrometallurgical production by electrowinning, nickel metal is produced from an electrolyte solution of nickel sulfate. Yet, in this process, significant amounts of nickel are still present in various sidestreams, such as from the purification steps and the spent electrolyte from electrowinning, from which nickel can potentially be recovered in the form of nickel sulfate salts. Due to their primary use in Li-ion batteries, production of nickel sulfate salts is expected to become the key growth area for nickel in the coming years with the rapidly growing battery sector and develop into the second-largest application for nickel after stainless steel by 2030 [12].

One method for achieving high purity nickel sulfate products from process or waste streams is crystallization [10,11]. A variety of techniques can be used for this purpose such as evaporative, antisolvent, eutectic freeze and cooling crystallization and for each technique in-depth analyses are needed to optimize yield, energy efficiency, purity and particle characteristics of the final product [13]. In this work, seeded cooling crystallization of nickel sulfate from industrially relevant solutions are investigated.

Nickel sulfate can precipitate as hydrated salts with different hydration levels ranging from mono to heptahydrates, depending on the reaction temperature [14]. Crystallization at temperatures between 5 °C to 100 °C at neutral conditions results in the precipitation of the hexa- and heptahydrates, and thus makes them the most studied nickel sulfate phases [14]. The hexahydrate form of nickel sulfate can crystallize into two polymorphs—α-NiSO$_4$·6H$_2$O (retgersite) and β-NiSO$_4$·6H$_2$O—with different stability regions [15,16]. The solubility curves of nickel sulfate hexahydrates and heptahydrate are presented in Figure 1, showing the most stable hydrate form in solution as a function of temperature. The high dependency of phase stability on reaction temperature forecasts significant implications on the cooling crystallization of nickel sulfates. Both the nucleation temperature and the time frame crystals spend in mother liquor at different temperature ranges will be commanding on the crystallization pathway and the final product population. In such a complex system, where multiple possible phases can crystallize, seeding of a supersaturated solution is a good method for inducing crystallization of a desired form, and is commonly used in industrial crystallization to control the crystallized product and product size. However, seeding does not guarantee phase purity in crystal systems with high solubility and close solubility values, such as nickel sulfate salts, due to strong contributions of the kinetic factors of crystallization [17,18]. Moreover, phase transformation of nickel sulfate takes place also in its dry state. Crystals of the NiSO$_4$·7H$_2$O phase decompose in air, losing water, even at room temperature [19]. At ambient conditions, nickel sulfate heptahydrate will lose one crystal water and transform into the hexahydrate form. The transition temperatures and stability order of phases in a dry state differ from the ones reported in solution [20,21]. At temperatures above the stability range of β-NiSO$_4$·6H$_2$O, the nickel sulfate can lose several crystal waters and convert into phases such as NiSO$_4$·4H$_2$O, NiSO$_4$·2H$_2$O or anhydrous NiSO$_4$.

The crystallization studies of nickel sulfate from industrial streams with the aim of producing battery-grade material have shown that common impurities found in such streams such as sodium, chloride and magnesium can be adsorbed or incorporated into the crystal structure. Their incorporation not only affects the purity of the product, but also can have consequences on the pathways of crystallization. Impurities present in a growth medium can influence the solubility of the crystallizing phases, and hence have significant effects on the course of reaction progression, and considerable effects on the properties of the crystal solid [18]. In the case of transformation between polymorphs or different phases, the rate of dissolution of a metastable form may decrease by the addition of specific impurities, and hence slows down the transformation towards the stable phase [17]. When the additives that will promote or inhibit certain forms of a compound are determined, it is possible to direct the system towards the desired product. Examples of this strategy can be found with L-glutamic acid, where the transformation from the metastable to the stable form can be inhibited by specific additives and thus allow kinetic factors to dominate the system [22,23]. Hence, from an industrial point of view, it is important to know how impurities in the process stream will influence the final crystallization product, in terms of properties such as purity and stability.

Having an understanding of different polymorphs or phases of nickel sulfates and transformation between these are of great importance for industrial production. Nickel sulfate hexahydrates are the preferred products over the heptahydrate form due to their lower water content. Additionally, stable crystals are favored during storage and transport, which reduces the risk of free water production and changes in crystal properties over time. Particle properties such as size, size distribution and shape effect their filterability, and thus should be controlled if possible [25].

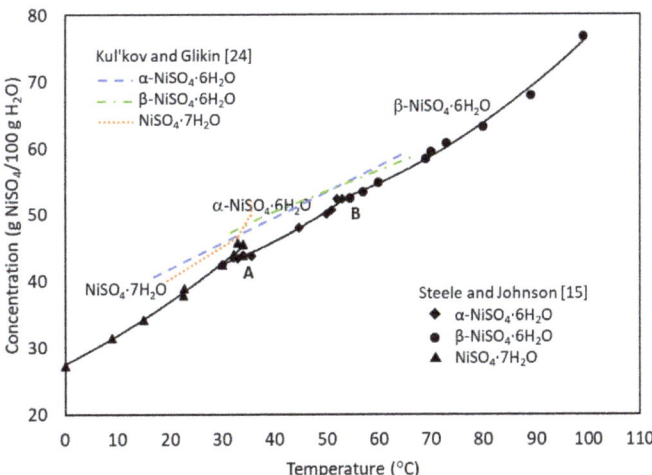

Figure 1. Solubility of nickel sulfate hydrates as a function of temperature, adapted from Refs [15,24]. Solubility curves in both studies show two eutonic points, A and B, representing the transition points for the thermodynamically most stable phase. Below 31.5–32 °C, the pale green heptahydrate, NiSO$_4$·7H$_2$O (morenosite), is the most stable phase in acid-free solutions and above this temperature nickel sulfate hexahydrate will form in solution. The transition point between the two hexahydrate polymorphs has been observed to be between 53–55 °C [15,24]. Near eutonic points precipitation of nickel sulfate salts out of their stability regions and reversible phase transformation between phases are probable. Metastable equilibria here were shown to be comparable to thermodynamic equilibria.

On the basis of the described knowledge needs, this work aims to study the cooling crystallization of nickel sulfate from an industrial point of view, at relevant temperatures for process streams, containing nickel, which could potentially be used for extracting nickel by crystallization of nickel sulfate. For this purpose, crystallization experiments seeded with α-NSH6 are employed as a strategy to control the precipitation product. Among the nickel sulfate phases shown in Figure 1 that can be precipitated in the determined experimental range, α-NSH6 is preferred due to its lower water content and its higher stability in storage conditions. The effects of magnesium on the crystallization and stability of nickel sulfate is investigated due to earlier work showing high levels of uptake and possible incorporation of magnesium in the crystal bulk [4,26]. Finally, stability of precipitated crystals both in solution and in dry state is followed, which can have practical implications on the recovery and storage of nickel salts.

2. Materials and Methods

2.1. Experimental Set-Up for Crystallization Experiments

Crystallization experiments were performed using a 100 mL temperature-controlled double-walled batch reactor (Figure 2), as described in detail in our previous work [26]. Proper mixing was ensured by a magnetic stirrer operating at 600 rpm and two baffles were attached to the lid. Temperature and pH were logged throughout the experiments, using a pH probe connected to a Mettler Toledo SevenExcellence multiparameter module. The temperature of the reactor was controlled by a Julabo refrigerated/heating circulator. The temperature logging was used for observing the point of nucleation, which could be seen as a small temperature increase, due to the heat of crystallization liberated when a solute crystallizes out of a solution [17].

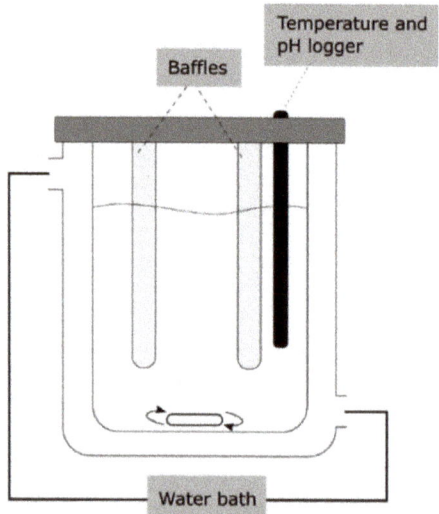

Figure 2. Sketch of the experimental set-up used for crystallization of nickel sulfate. Batch reactor, temperature-controlled by circulating water bath. Proper mixing was ensured by magnetic stirring and baffles attached to the lid.

2.2. Experimental Methods and Characterization

2.2.1. Preparation of Nickel Sulfate Seeds

In order to produce pure α-NiSO$_4$·6H$_2$O seeds, commercial nickel sulfate hexahydrate (reagent grade ≥ 98%, Sigma-Aldrich, Darmstadt, Germany) was dissolved in distilled water, and was left for evaporation at 40 °C overnight [27,28]. The phase purity of the seeds was verified by powder X-ray diffraction (XRD) analysis and comparing the spectra with theoretical diffraction patterns of nickel sulfate hydrates (see Supplementary Information, Section A). Seeds were stored in their dry state at room temperature until further use. Their stability over storage time was confirmed by XRD analyses (Figure 3), and seeds from the same batch were used for all the seeded crystallization experiments in this work.

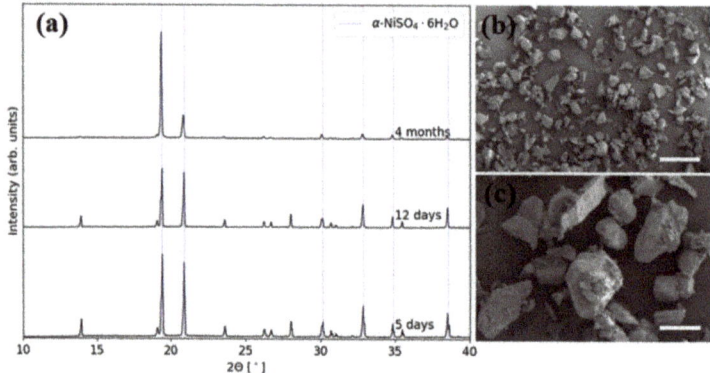

Figure 3. (**a**) XRD spectra of phase pure α-NiSO$_4$·6H$_2$O seeds, synthesized by evaporative crystallization at 40 °C. Spectra shows pure α-NSH at all times, with blue lines indicating the most prominent characteristic peaks. Changes in the relative intensities of the peaks were interpreted as a result of crystal alignment. (**b**,**c**) SEM micrographs of the seeds are shown at various magnifications. Scale bar: (**b**) 400 μm, (**c**) 100 μm.

2.2.2. Crystallization Experiments

Seeded and unseeded crystallization experiments were conducted at two temperature values: 25 °C and 70 °C. These values were chosen based on the temperatures of nickel-containing process streams that are potential candidates for the extraction of nickel. Initial nickel sulfate concentrations that allow seeding without spontaneous crystallization at the chosen temperatures were determined at preliminary experiments without impurities. For all the experiments performed at a low temperature (25 °C), the concentration of nickel sulfate was 0.53 g $NiSO_4 \cdot 6H_2O$/g solution, and at high temperature (70 °C) the concentration was 0.70 g $NiSO_4 \cdot 6H_2O$/g solution. The chosen concentrations corresponded to the same supersaturation level with respect to the most stable $NiSO_4$ phase at the respective temperatures. The solutions were prepared by dissolving the corresponding amounts of the nickel sulfate salt by heating, followed by cooling down to the chosen reaction temperatures at a cooling rate of 0.29 °C min^{-1} (Figure 4). An overview of the experimental conditions is given in Table 1.

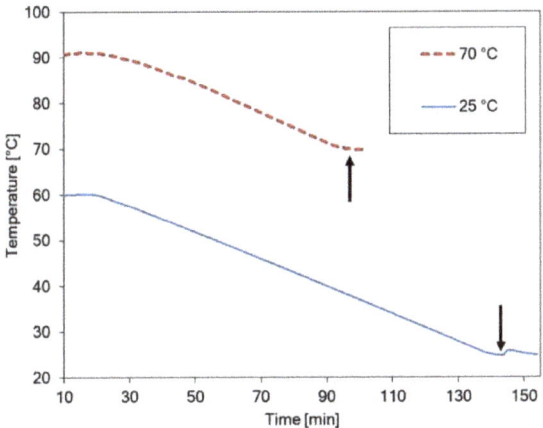

Figure 4. Temperature evolution during crystallization of nickel sulfate at temperatures 25 and 70 °C. Seeds were added once the desired temperature was reached, where the arrows represent the seeding time points. The cooling rate for both cases is 0.29 °C min^{-1}. Note the temperature increase in the blue curve, showing the nucleation of nickel sulfate after seed addition.

Table 1. Overview of experiments and their respective temperatures, seed amounts, and initial magnesium concentrations (mg Mg/g total solution) in the reactor solution.

Seed Amount (g)	Mg Content (mg/g tot)
T = 25 °C	
0	0
0	18
0.2	0
0.2	5
0.2	9
0.2	18
2	0
2	9
T = 70 °C	
0.2	0
2	0

2.2.3. Seeded Experiments

Once the chosen temperature was reached, nickel sulfate seeds were added to the reactor, and the suspension was left for 10 min before the samples were taken out for characterization to ensure sufficient crystal growth while minimizing phase transformation. Low and high seed amounts were tested, with 0.2 and 2 g respectively, in order to study the effect of seed amount on the final nickel sulfate product. Seeded experiments were performed in solutions both without Mg and with different amounts of Mg.

2.2.4. Unseeded Experiments

Unseeded experiments were performed to study the behavior of the system at 25 °C without seeding. The experiments were allowed to run until nucleation was observed via an increase in temperature, and precipitates were left for 10 min before sampling. Unseeded experiments were not repeated at 70 °C since the nucleation point could not be detected at this temperature.

2.2.5. Magnesium Uptake during Crystallization

Based on our previous results [26], magnesium was selected as the impurity of interest in this study, due to its capability to be incorporated into the nickel sulfate crystal structure. Experiments in the presence of magnesium was conducted by the addition of $MgSO_4$ (anhydrous, ReagentPlus® $\geq$ 99.5%, Sigma-Aldrich) at varying concentrations to the reaction medium, using the same setup as described above.

2.2.6. Characterization

At the end of the experiments, small samples of the suspension were taken out from the reactor, in order to characterize the precipitated nickel sulfate. Directly after sampling, the samples were filtered, washed with ethanol and dried at room temperature in air overnight. Filtration of samples from experiments conducted at 70 °C were performed with a preheated funnel to avoid additional crystallization on the filtration equipment. The crystal phase was determined by powder X-ray diffraction (XRD) with a Bruker D8 A25 DaVinci X-ray Diffractometer with CuKα radiation. All XRD analyses were performed within 24 h after the experiment in order to avoid phase transformations, since solid-state transition was observed when dry samples were stored for longer times. Due to the crowded XRD spectra of precipitating phases, two characteristic peaks, representing the highest intensity peaks of each nickel sulfate phase, were selected and marked in all provided data to make it easier for the readers to distinguish the produced nickel sulfate phases. The complete XRD spectra of all the relevant phases are given in the Supplementary Information, Section A. Table 2 shows the assigned peaks and the abbreviations for each nickel sulfate hydrate phase in this work. The morphology and size of the crystals was studied by scanning electron microscopy (SEM) (Hitachi S-3400 N). Light microscopy images of nickel sulfate products can be found in the Supplementary Information, Section C.

Table 2. Overview of the nickel sulfate hydrates, and their abbreviations used in this work, and selected characteristic peaks with highest intensities. The selected peaks were used for identification and comparison of products from the experiments in this work.

Nickel Sulfate Hydrate	Abbreviations	Characteristic Peaks (2θ)
α-NiSO$_4$·6H$_2$O	α-NSH6	19.4° and 20.9°
β-NiSO$_4$·6H$_2$O	β-NSH6	20.4° and 22.3°
NiSO$_4$·7H$_2$O	NSH7	16.7° and 21.1°

2.2.7. Phase Transformation in Solution and in Solid-State

The phase transformation of nickel sulfate hydrate salts was studied both in solution and in solid-state. The crystallization and solution-mediated phase transformation of nickel sulfates at chosen temperatures were followed by light microscopy (Axio Imager A1m, Zeiss, Oberkochen, Germany) via monitoring the change in crystal morphology. For this purpose, nickel sulfate solution was first cooled down to either 40 °C or 70 °C, and a drop of this solution was placed on a microscope glass slide. The glass slide was then placed under a microscope without further control of temperature and was monitored for 60 min.

Phase transformation in solid-state was investigated by using washed and dried samples that were obtained from experiments depicted in Table 1. Collected samples were stored in closed glass vials at room temperature and their phase transformation was followed by analyzing selected samples with XRD after 24 h and up to 10 days.

3. Results

3.1. Crystallization of Nickel Sulfate

3.1.1. Effect of Temperature

Seeded experiments in additive-free solutions were conducted at 25 °C and 70 °C by cooling the nickel sulfate solutions to establish supersaturation, then adding the seeds once the desired reaction temperature was reached.

Crystallization from solutions at 25 °C and seeded with 0.2 g of α-NiSO$_4$·6H$_2$O, resulted in a mixture consisting of α-NiSO$_4$·6H$_2$O (α-NSH6), β-NiSO$_4$·6H$_2$O (β-NSH6) and NiSO$_4$·7H$_2$O (NSH7), while at 70 °C only β-NSH6 was produced, as observed from XRD analyses (Figure 5a). SEM images showed a distinct difference in crystal shape of the two products, in good accordance with the phases observed from the XRD analyses (Figure 5b,c).

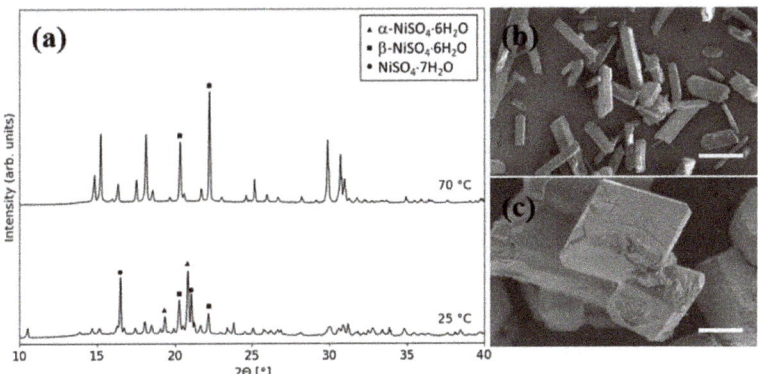

Figure 5. (a) XRD spectra of nickel sulfate crystals produced with 0.2 g α-NiSO$_4$.6H$_2$O seeding and without impurities at 25 °C and 70 °C, showing the effect of crystallization temperature. (b,c) SEM images of crystals precipitated at 25 °C and 70 °C, respectively. Scale bar: 100 µm.

The typical morphologies of nickel sulfate phases produced in our work correspond well with the previous observations of Kul'kov and Glikin [24]. They reported that NSH7 precipitates typically have the shape of fibrous or short-prismatic crystals, α-NSH6 are thick plates or short-prismatic crystals, and the β-NSH6 crystals are observed as basal plates.

3.1.2. Effect of Seed Amount

By increasing the seed amount by a tenfold, from 0.2 g to 2 g, it was possible to produce pure α-NSH6 at 25 °C, as verified by XRD analysis (Figure 6a). SEM images showed particles of a short-prismatic shape, typical for α-NSH6, and the long fibrous particles associated with NSH7 were absent (Figure 6a,b). On the contrary, the precipitating phase could not be altered at 70 °C even with high seed amount, and only β-NSH6 was detected in the XRD spectrum. The plate-like crystal morphology was also maintained at higher seed amount (Figure 6c).

Unseeded experiments were also performed following the same procedure as for the seeded experiments at 25 °C. The reaction was left to proceed for 48 h, but no crystal formation was observed in the system during this time interval.

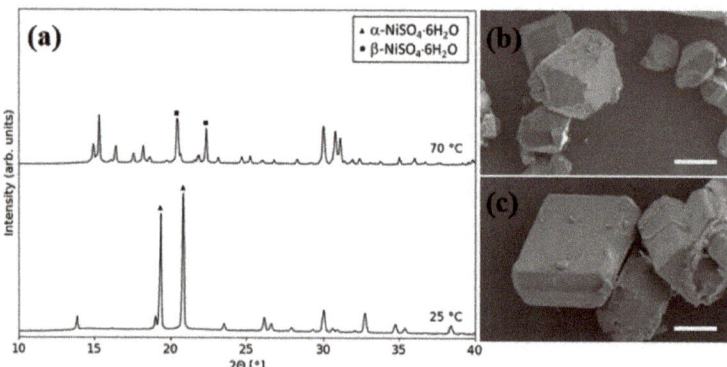

Figure 6. (a) XRD spectra of nickel sulfate phases produced at 25 °C and 70 °C, with addition of 2 g seeds and without impurities. (b,c) SEM images of crystals precipitated at 25 °C and 70 °C, respectively. Scale bar: 100 μm.

3.2. Crystallization of Nickel Sulfate in the Presence of Magnesium

3.2.1. Effect of Initial Mg Content

In order to study the effect of Mg uptake on the crystallization behavior of nickel sulfate, different initial concentrations of Mg were added in the reaction solution in experiments seeded with 0.2 g of α-NSH6 at 25 °C (Table 1). From XRD analyses (Figure 7a) it was seen that nickel sulfate precipitated with Mg concentrations of 0 and 5 mg g^{-1} contained a mixture of α-NSH6, β-NSH6 and NSH7. At an initial Mg concentration of 9 mg g^{-1}, the product consisted of β-NSH6 and NSH7, and with 18 mg g^{-1} only NSH7 was produced. As the initial Mg concentration increased in the reaction solution, the diffraction peaks of the α-NSH6 got weaker, while the peaks of the heptahydrate phase were more pronounced. As can be seen from SEM images in Figure 7, there was no significant difference in particle shape or size for the three lower concentrations of Mg, while the crystals were significantly bigger in the presence of high Mg concentration.

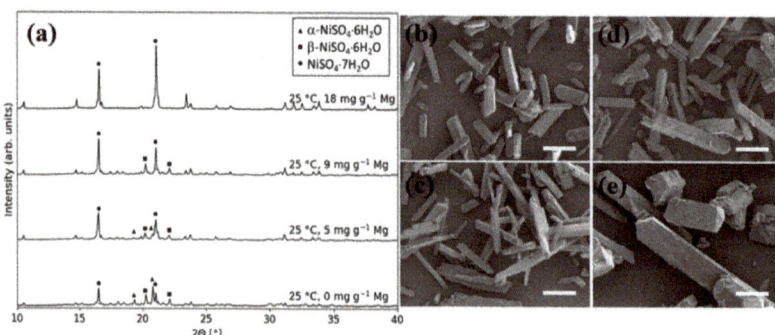

Figure 7. (a) XRD spectra of nickel sulfate phases precipitated with 0.2 g α-NSH6 seeds at 25 °C in the absence and presence of Mg at indicated concentration levels. SEM images of crystals collected from corresponding experiments; (b) no magnesium, (c) 5 mg g^{-1} Mg, (d) 9 mg g^{-1} Mg, and (e) 18 mg g^{-1} Mg. Scale bar: 100 μm.

3.2.2. Effect of Seed Amount

The effect of seed amount on crystallization in the presence of magnesium was tested only at a Mg concentration of 9 mg g^{-1} and at 25 °C. The precipitating phase and morphology of the particles did not show any changes when the seed amount was increased from 0.2 g to 2 g, as seen from XRD analyses (Figures 7a and 8a, respectively) and SEM images (Figures 7d and 8a inset, respectively), showing a mixture of β-NSH6 and NSH7 crystals.

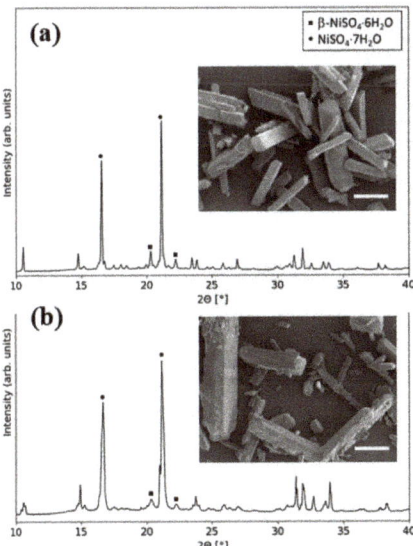

Figure 8. XRD spectra and (insets) SEM images of crystals precipitated at 25 °C (**a**) in the presence of 9 mg g^{-1} Mg with addition of 2 g seeds, and (**b**) in the presence of 18 mg g^{-1} Mg, without the addition of seeds. Scale bar: 100 μm.

Unseeded experiments were performed at parallel reaction conditions without Mg and with an initial Mg concentration of 18 mg g^{-1}. In contrast to Mg free experiments, where no precipitation was observed for 48 h, nucleation was detected during initial cooling in the presence of Mg by an increase of temperature at 28 °C (see Supplementary Information, Section B), and the product was shown to consist of mainly NSH7, with weak peaks of β-NSH6 (Figure 8b). The characteristic long fibrous crystal form for NSH7 was observed by SEM (Figure 8b inset).

3.3. Phase Transformation of Nickel Sulfate Crystals
3.3.1. Phase Transformation in Suspension

The phase stability of nickel sulfate in solution was investigated at low and high temperatures, without seeding. For this purpose, a small amount of solution was taken out from the reactor, placed on a microscopy glass slide and imaged at predetermined time points during 60 min. For low and high temperature, the nickel sulfate solution was cooled to 40 °C and 70 °C, respectively, before sampling. After sampling, the glass slide was observed under the microscope for 60 min, without further control on temperature. At low temperatures, crystals with morphologies resembling α-NSH6 were present after 5 min of sampling, and no significant changes could be observed from the light microscopy images during 60 min (Figure 9a,b). It should be noted that unseeded batch experiments at 25 °C, which were also cooled down from 40 °C, did not yield any crystal formation. This discrepancy can be explained by the faster cooling rate when the sample is placed on the glass slide, and the large surface provided by the glass slide, allowing nucleation to take place. At high temperatures, many crystals with varying morphologies were observed from the initial time point, and phase transformation was observed in time (Figure 9c,d). Further growth of certain crystals accompanied by the consumption of neighboring ones showed that a solution-mediated phase transformation was active.

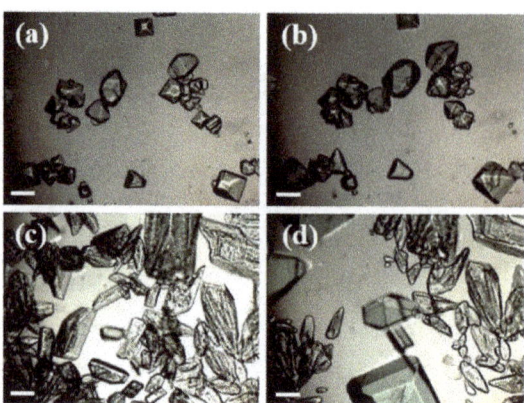

Figure 9. Nickel sulfate crystallized on microscope glass slide and imaged by optical microscope. Samples taken out: (**a**,**b**) at 40 °C and (**c**,**d**) at 70 °C, and imaged after 5 min and 60 min, respectively. Scale bar: 100 µm.

3.3.2. Phase Transformation in the Dry State

From preliminary experiments, it was observed that the nickel sulfate in its dry state transforms over time when stored at room temperature. Therefore, XRD analyses were performed to follow the transformation of chosen dry samples over several days (see Supplementary Information, Section D).

For nickel sulfate produced at 25 °C with addition of 0.2 g seeds and without Mg, the β-NSH6 peaks disappeared within 7 days of storage time, the NSH7 peaks got weaker, while the α-NSH6 peaks got stronger and new α-NSH6 peaks were observed.

Nickel sulfate produced at 70 °C with addition of 0.2 g seeds and without Mg, showed β-NSH6 peaks disappearing over the same time period, the NSH7 peaks got weaker, while the α-NSH6 peaks got stronger and new α-NSH6 peaks were observed. The same trend was observed for nickel sulfate produced at 25 °C with an initial Mg concentration of 5 mg g^{-1}. However, at 25 °C with initial Mg concentration of 18 mg g^{-1}, α-NSH6 peaks were observed after storage for 10 days, and the heptahydrate peaks were stronger. In addition, some peaks of the β-polymorph appeared. The same trend was observed for the nickel sulfate produced in the unseeded experiment in the presence of 18 mg g^{-1} Mg. For nickel sulfate crystallized at 70 °C, the peaks of β-NSH6 got weaker with time, and after 10 days there were also new α-NSH6 peaks.

In summary, the dry nickel sulfate samples precipitated without Mg tend to transform towards α-NSH6, while the ones with Mg content showed stable NSH7 phase and emergence of β-NSH6.

4. Discussion

4.1. Effect of Temperature and Seed Amount

In nickel sulfate crystallization, temperature plays a pivotal role in determining the final product. Within the temperature range of 0–100 °C, the thermodynamic stability regions of different nickel sulfate phases are well-established. The crystalline phases of two hexahydrate polymorphs, α-NSH6 and β-NSH6, and the heptahydrate form, NSH7, are the most stable precipitation products at different temperature ranges marked by the eutonic points, according to their solubility curves as shown in Figure 1. Yet, near the eutonic points, these highly soluble salts can precipitate and persist in solution out of their stability zones as a result of comparable effects of kinetic factors to thermodynamic drivers.

At 25 and 70 °C, the most stable phases in solution are NSH7 and β-NSH6, respectively. Experiments conducted at the corresponding temperatures and with 0.2 g of α-NSH6 seeds showed that the presence of the seeds was not sufficient to manipulate the system towards precipitation of the metastable α polymorph. XRD analyses (Figure 5) showed that for crystallization of nickel sulfate at 70 °C, seeded with 0.2 g of α-NSH6, the product consisted of pure β-NSH6. Although seeds of the α-polymorph were added in the reaction medium, formation of the β-NSH6 at 70 °C was in good accordance with earlier findings showing that it is the most stable phase, above 53 °C [15,24]. The α-seeds must undergo dissolution and recrystallization reactions due to the higher solubility of α-NSH6 than β-NSH6 at this temperature. Even at a higher seed amount of 2 g, pure β-NSH6 was produced at 70 °C, verifying the high driving force in the system towards the formation of β-polymorph, which induces the phase transformation of the seed material (Figure 6). At 25 °C, our results showed that when seeded with 0.2 g of α-NSH6, the crystallization product contained both the α- and β-polymorph of the hexahydrate, in addition to the heptahydrate form which has been reported to be the most stable phase of nickel sulfate below 35 °C. For compounds that can undergo enantiotropic transformations, metastable phases can be crystallized, before transforming to the stable phase, following Ostwald's rule of stages [17]. This might explain the presence of the two NSH6 polymorphs even though the NSH7 is the most thermodynamically stable phase at 25 °C since reaction times were limited to 10 min. The observations of α-NSH6 in this case can be explained by the presence of the seeds; yet, the formation of other phases must be due to additional nucleation, possibly caused by insufficient seed amounts to ensure only growth in the system [17,29]. When the seed amount was increased by a ten-fold, only α-NSH6 was obtained at 25 °C. Higher seed amounts provide an increased surface area for growth [30], hence promoting growth over nucleation. It should also be noted that with both seed amounts, the total mass of final precipitates was measured to be approximately 10 g, showing that the supersaturation consumption by relative rates of nucleation and growth was influenced by the seed amount, while the yield was not drastically affected. Some secondary nucleation was also observed for this experiment, as seen by the presence of smaller crystals in SEM images (Figure 6a,b) and from light microscopy (see Supplementary Information, Section C, Figure C.3b). These results showed that in addition to the interplay between the supersaturation value and the seed amount, the kinetics of precipitation was of paramount importance in determining the metastability regions of the three phases.

For unseeded experiments at 25 °C, without Mg, the reaction was left for 48 h, without any signs of nucleation taking place, neither visually nor by an increase in temperature. This indicates that the system is very stable with a broad metastable zone where spontaneous crystallization is unlikely to occur [17].

4.2. Effect of Magnesium

Additives can cause changes in the stability zones of crystalline phases by effecting kinetic and/or thermodynamic factors. Magnesium was chosen as the additive of interest due to its relevance in industrial streams and previous reports signaling for incorporation of Mg in the crystalline structure of nickel sulfate crystals [26]. Incorporation of additives can drastically affect the crystal solubility, thus the thermodynamic stability, as well as kinetic stability of metastable phases [17,18].

The effect of magnesium on precipitation of nickel sulfate crystals was investigated at 25 °C, in the absence and presence of varying amounts of α-NSH6 seeds. With no magnesium present and in the absence of seeds, no crystal formation was observed even after 48 h in the system, whereas Mg additives at a concentration of 18 mg g^{-1} induced spontaneous formation of NSH7 even before reaching 25 °C by cooling. Promotion of NSH7 formation in the presence of Mg was further demonstrated in the seeded experiments, where increasing amounts of Mg shifted the product towards higher contents of NSH7. Experiments conducted at 25 °C without Mg and with 0.2 g of seeds resulted in a mixture of nickel sulfate hydrates as shown in Figure 5. With increasing Mg content under same experimental conditions, we observed promotion of the diffraction peaks corresponding to NSH7 and at the highest initial concentration of Mg (18 mg g^{-1}), the product consisted of phase pure NSH7 (Figure 7). It also must be noted that, in this experimental condition nucleation occurred at 33 °C (observed by a slight increase in temperature), hence prior to addition of the seeds. The primary nucleation of NSH7 during cooling can explain the phase pure product and the larger crystal size (Figure 7e). Since samples were taken out 10 min after reaching 25 °C for all experiments, the crystals nucleated during the cooling period were allowed to grow for a longer time period.

Seeded experiments with varying concentrations of Mg demonstrated that Mg promotes the formation of NSH7. Additionally, we cannot eliminate possible inhibition of the formation of hexahydrate polymorphs by Mg presence. Changing stability in the crystal product in the presence of additives is known. For crystallization of $CaCO_3$ as calcite and aragonite, similar effects of the presence of Mg has been observed, where Mg is favorable towards crystallization of aragonite, due to calcite becoming more soluble [31,32]. Moreover, for L-glutamic acid, certain additives have been shown to inhibit transformation from the metastable to the thermodynamically stable phase [22,23].

4.3. Transformation of Nickel Sulfate Hydrates in Solution and Dry State

Different polymorphs and phases of crystalline compounds have different physical properties that can be of importance in industrial applications [17,33]. Thus, phase transformations of nickel sulfate particles are investigated both in solution and in the dry state, which can be significant for the production and storage stages, respectively. By taking out a small sample of the nickel sulfate solution from the reactor at 40 °C and 70 °C, and immediately placing a few drops of the sample on a glass microscope slide, it was possible to follow the development of crystals by light microscopy over time. This could be a good indication of how the nickel sulfate crystals behave in solution. For these experiments, nucleation was observed within 5 min for both temperatures, whereas unseeded experiments at 25 °C did not produce any crystals. As mentioned in the previous section, this could be explained by the faster cooling rate when the sample was placed on the glass slide, and the large substrate surface. At 40 °C (Figure 9a,b), α-NSH6 was observed, based on the short-prismatic shape of the product compared with the morphologies of the α-NSH6 seeds (Figure 3c), α-NSH6 produced in our previous work [26], and previous findings by Kul'kov and Glikin [24]. During the time period of observation, it could be seen that the crystals were slightly growing while maintaining their shape. At 70 °C (Figure 9c,d), the particle shapes did not appear as well defined as for the sample taken out at 40 °C, but they showed resemblance to the typical plate like shape of β-NSH6 [24]. However, with time, the particles either had dissolved or the shape had changed significantly. After 60 min, short-prismatic crystals of α-NSH6 could be observed.

Nickel sulfate crystals in suspension have been observed to undergo transformations between the three mentioned phases during heating or cooling in the system [15]. Corresponding to the points A and B in Figure 1, the transition point between NSH7 and α-NSH6 is around 31.5 °C, and between α-NSH6 and β-NSH6 at 53.3 °C. From light microscopy images (Figure 9), phase transformation was observed at high temperature. Moreover, the crystallized product from experiments at 70 °C, seeded with α-NSH6, showed no indications of α-NSH6 in the XRD spectra. This could be explained by phase transformation from α-NSH6 to β-NSH6 in solution. Light microscopy images showed growth of crystals accompanied by consumption of neighboring crystals, indicating solution-mediated phase transformation.

In its dry state, nickel sulfate was observed to phase transform when stored as a dry powder. The phase transformations were followed by XRD analyses of chosen samples (see Supplementary Information, Section D). The general trend for samples not containing Mg was that with time, new α-NSH6 peaks appeared in the XRD spectra and α-NSH6 peaks already existing got more pronounced. Peaks of β-NSH6 got weaker with increasing storage time, and even disappeared completely in some cases. When precipitated in the presence of Mg, nickel sulfate crystals in their dry form showed a stable NSH7 phase and the emergence of β-NSH6. Phase transformations in the dry state at room temperature followed a different pathway to that in solution, which indicates a difference in the stability regions of crystal phases [34]. While NSH7 was the most stable phase in solution at 25 °C, in dry conditions α-NSH6 was favored. Earlier work shows that NSH7 loses one crystal water and transforms to α-NSH6, which is considered to be the most stable phase in dry state in air at room temperature [16,24,35]. Friesen et al. [20] observed that the NSH7 transforms spontaneously at room temperature via the pathway NSH7→β-NSH6→α-NSH6 [20,36]. This corresponds well with the transformations we observed for samples containing no or small amounts of Mg. On the other hand, this transformation was inhibited when crystals were precipitated in the presence of Mg, demonstrating stabilization of NSH7. However, suppression of α-NSH6 formation should also be considered as a possible outcome of Mg presence since at the highest concentration of the additive, NSH7 crystals were formed in solution transformed to β-NSH6.

5. Conclusions

Nickel sulfate salts show an increasing market share due to being a component of electric vehicle batteries, among other applications. Industrial hydrometallurgical nickel production by electrowinning offers a potential recovery route for nickel sulfates by precipitation since nickel metal is produced from an electrolyte solution of nickel sulfate and significant amounts of nickel are present in various sidestreams. Yet, the presence of multiple nickel sulfate hydrate phases, their metastability and significant effects of impurities necessitates detailed investigations to optimize the crystallization processes towards a desirable product. Accordingly, in this work, crystallization of nickel sulfate was investigated at 25 °C and 70 °C, with different seed amounts and in the absence and presence of various initial magnesium concentrations. The results showed that seeding alone is not always sufficient to control the crystallization of nickel sulfate phases from supersaturated solutions, and the reaction temperature as well as the reaction time may dominate the final precipitate. Following the phase transformations of nickel sulfate phases both in solution and in dry state further demonstrated the difficulty of manipulating this dynamic system towards a desired stable product. Experiments conducted with various initial concentrations of Mg showed that its presence promotes NSH7 formation and could have inhibitory effects on the formation of hexahydrate polymorphs in solution, as well as showing significant effects on the stability of different phases in the dry state. Although crystallization of nickel sulfate salts offer an attractive strategy to obtain high purity products and efficient recycling routes, our results show that the complexity of these systems should be recognized, where in-depth studies can provide the necessary input to optimize any chosen crystallization method for process design. Future work should include the designation of potential recycling routes for nickel sulfate salts from industrial streams and optimization of crystallization processes for their relevant conditions.

Supplementary Materials: The following are available online at https://www.mdpi.com/article/10.3390/cryst11121485/s1; (Figure SA) XRD spectra of nickel sulfate phases, (Figure SB) temperature profiles of crystallization reactions, (Figure SC) light microscopy images of particles, (Figure SD) XRD spectra and SEM images following phase transformation of nickel sulfate salts in dry state.

Author Contributions: Conceptualization, I.B.J., O.B., J.-P.A. and S.U.; Data curation, I.B.J. and S.U.; Formal analysis, I.B.J. and S.U.; Funding acquisition, J.-P.A.; Investigation, I.B.J. and S.U.; Methodology, I.B.J., J.-P.A. and S.U.; Project administration, J.-P.A.; Supervision, J.-P.A. and S.U.; Visualization, I.B.J.; Writing—Original draft, I.B.J. and S.U.; Writing—Review and editing, I.B.J., O.B., J.-P.A. and S.U. All authors have read and agreed to the published version of the manuscript.

Funding: This research was funded by the Norwegian Research Council, Yara, Glencore Nikkelverk AS, and Boliden Odda (Norwegian Research Council project number 236674).

Data Availability Statement: The data presented in this study are available on request from the corresponding author. The data are not publicly available due to confidentiality agreement with industrial partners.

Conflicts of Interest: The authors declare no conflict of interest.

Abbreviations

α-NSH6	α-NiSO$_4$·6H$_2$O
β-NSH6	β-NiSO$_4$·6H$_2$O
NSH7	NiSO$_4$·7H$_2$O
EVB	Electric vehicle battery
SEM	Scanning electron microscopy
XRD	X-ray diffraction

References

1. He, Y.; Su, G.; Yu, X.; Li, Z.; Huang, B.; Jang, R.; Zhao, Q. Growth of α-nickel sulphate hexahydrate for ultraviolet filters. *J. Cryst. Growth* **1996**, *169*, 193–195. [CrossRef]
2. Thirupathy, J.; Dhas, S.S.J.; Jose, M.; Dhas, S.M.B. An investigation on optical, mechanical and thermal properties of nickel sulphate hexahydrate single crystal—A UV band pass filter. *Mater. Res. Express* **2019**, *6*, 086206. [CrossRef]
3. Volkova, E.; Demidov, A. Production of nickel sulfate single crystals in processing of nickel oxide electrodes of alkaline batteries. *Russ. J. Appl. Chem.* **2010**, *83*, 1874–1876. [CrossRef]
4. Han, B.; Bøckman, O.; Wilson, B.P.; Lundström, M.; Louhi-Kultanen, M. Purification of Nickel Sulfate by Batch Cooling Crystallization. *Chem. Eng. Technol.* **2019**, *42*, 1475–1480. [CrossRef]
5. Berckmans, G.; Messagie, M.; Smekens, J.; Omar, N.; Vanhaverbeke, L.; Van Mierlo, J. Cost Projection of State of the Art Lithium-Ion Batteries for Electric Vehicles Up to 2030. *Energies* **2017**, *10*, 1314. [CrossRef]
6. The European Commission. *White Paper: European Transport Policy for 2010: Time to Decide*; The European Commission: Bruxelles, Belgium, 2001.
7. Bloomberg Finance. New Energy Finance, Electric Vehicle Outlook 2017—Executive Summary. 2017. Available online: http://data.bloomberglp.com/bnef/sites/14/2017/07/BNEF_EVO_2017_ExecutiveSummary.pdf (accessed on 16 July 2019).
8. Nitta, N.; Wu, F.; Lee, J.T.; Yushin, G. Li-ion battery materials: Present and future. *Mater. Today* **2015**, *18*, 252–264. [CrossRef]
9. Coman, V.; Robotin, B.; Ilea, P. Nickel recovery/removal from industrial wastes: A review. *Resour. Conserv. Recycl.* **2013**, *73*, 229–238. [CrossRef]
10. Nyirenda, R.; Phiri, W. The removal of nickel from copper electrorefining bleed-off electrolyte. *Miner. Eng.* **1998**, *11*, 23–37. [CrossRef]
11. Agrawal, A.; Manoj, M.K.; Kumari, S.; Bagchi, D.; Kumar, V.; Pandey, B.D. Extractive separation of copper and nickel from copper bleed stream by solvent extraction route. *Miner. Eng.* **2008**, *21*, 1126–1130. [CrossRef]
12. Roskill. Nickel Sulphate Outlook to 2030. 2021. Available online: https://roskill.com/market-report/nickel-sulphate/ (accessed on 25 October 2021).
13. Ma, Y.; Svärd, M.; Xiao, X.; Gardner, J.M.; Olsson, R.T.; Forsberg, K. Precipitation and Crystallization Used in the Production of Metal Salts for Li-Ion Battery Materials: A Review. *Metals* **2020**, *10*, 1609. [CrossRef]
14. Nicholls, D. *The Chemistry of Iron, Cobalt and Nickel: Comprehensive Inorganic Chemistry*; Elsevier Science: Amsterdam, The Netherlands, 2013; pp. 1109–1159.
15. Steele, B.D.; Johnson, F. XIII.—The solubility curves of the hydrates of nickel sulphate. *J. Chem. Soc. Trans.* **1904**, *85*, 113–120. [CrossRef]
16. Angel, R.; Finger, L. Polymorphism of nickel sulfate hexahydrate. *Acta Crystallogr. Sect. C Cryst. Struct. Commun.* **1988**, *44*, 1869–1873. [CrossRef]
17. Mullin, J.W. *Crystallization*; Elsevier Science: Amsterdam, The Netherlands, 2001; pp. 280–284.
18. Sangwal, K. *Additives and Crystallization Processes: From Fundamentals to Applications*; John Wiley & Sons: Hoboken, NJ, USA, 2007.
19. Manomenova, V.L.; Rudneva, E.B.; Voloshin, A.E.; Soboleva, L.V.; Vasil'ev, A.B.; Mchedlishvili, B.V. Growth of α-NiSO4·6H2O crystals at high rates. *Crystallogr. Rep.* **2005**, *50*, 877–882. [CrossRef]
20. Friesen, M.; Burt, H.; Mitchell, A. The dehydration of nickel sulfate. *Thermochim. Acta* **1980**, *41*, 167–174. [CrossRef]
21. Svarovsky, L. 2—Characterization of particles suspended in liquids. In *Solid-Liquid Separation*, 4th ed.; Svarovsky, L., Ed.; Butterworth-Heinemann: Oxford, UK, 2001; pp. 30–65.
22. Davey, R.; Blagden, N.; Potts, A.G.D.; Docherty, R. Polymorphism in Molecular Crystals: Stabilization of a Metastable Form by Conformational Mimicry. *J. Am. Chem. Soc.* **1997**, *119*, 1767–1772. [CrossRef]
23. Sakata, Y.; Takenouchi, K. Studies on the Behaviors of Impurities on the Crystallization of L-Glutamic Acid—Part III. Influences of Some Factors on the Solubility. *Agric. Biol. Chem.* **1962**, *26*, 824–830.
24. Kul'kov, A.M.; Glikin, A.E. Replacement of nickelhexahydrite with retgersite: Polymorphic-metasomatic structures. *Geol. Ore Depos.* **2007**, *49*, 821–826. [CrossRef]
25. Bourcier, D.; Féraud, J.; Colson, D.; Mandrick, K.; Ode, D.; Brackx, E.; Puel, F. Influence of particle size and shape properties on cake resistance and compressibility during pressure filtration. *Chem. Eng. Sci.* **2016**, *144*, 176–187. [CrossRef]
26. Jenssen, I.B.; Ucar, S.; Bøckman, O.; Dotterud, O.M.; Andreassen, J.-P. *Impurity Uptake During Cooling Crystallization of Nickel Sulfate*; Springer International Publishing: Cham, Switzerland, 2020; pp. 191–199.
27. Kathiravan, P.; Balakrishnan, T.; Srinath, C.; Ramamurthi, K.; Thamotharan, S. Growth and characterization of α-nickel sulphate hexahydrate single crystal. *Karbala Int. J. Mod. Sci.* **2016**, *2*, 226–238. [CrossRef]

28. Moldoveanu, G.A.; Demopoulos, G.P. Producing high-grade nickel sulfate with solvent displacement crystallization. *JOM* **2002**, *54*, 49–53. [CrossRef]
29. Kubota, N.; Doki, N.; Yokota, M.; Sato, A. Seeding policy in batch cooling crystallization. *Powder Technol.* **2001**, *121*, 31–38. [CrossRef]
30. Chung, S.H.; Ma, D.L.; Braatz, R.D. Optimal seeding in batch crystallization. *Can. J. Chem. Eng.* **1999**, *77*, 590–596. [CrossRef]
31. Berner, R.A. The role of magnesium in the crystal growth of calcite and aragonite from sea water. *Geochim. Cosmochim. Acta* **1975**, *39*, 489–504. [CrossRef]
32. Kitano, Y. The behavior of various inorganic ions in the separation of calcium carbonate from a bicarbonate solution. *Bull. Chem. Soc. Jpn.* **1962**, *35*, 1973–1980. [CrossRef]
33. Henck, J.O.; Kuhnert-Brandstatter, M. Demonstration of the terms enantiotropy and monotropy in polymorphism research exemplified by flurbiprofen. *J. Pharm. Sci.* **1999**, *88*, 103–108. [CrossRef] [PubMed]
34. Kitamura, M. Strategy for control of crystallization of polymorphs. *CrystEngComm* **2009**, *11*, 949–964. [CrossRef]
35. Frondel, C.; Palache, C. Retgersite, $NiSO_4 \cdot 6H_2O$, a new mineral. *Am. Mineral. J. Earth Planet. Mater.* **1949**, *34*, 188–194.
36. Sinha, S.; Deshpande, N.; Deshpande, D. Thermal dehydration of crystallin $NiSO_4 \cdot 6H_2O$. *Thermochim. Acta* **1989**, *144*, 83–93. [CrossRef]

Article

Thermal Deformations of Crystal Structures in the L-Aspartic Acid/L-Glutamic Acid System and DL-Aspartic Acid

Roman Sadovnichii [1], Elena Kotelnikova [1] and Heike Lorenz [2,*]

[1] Department of Crystallography, St. Petersburg State University, Universitetskaya emb. 7/9, 199034 St. Petersburg, Russia; rsadovnichii@gmail.com (R.S.); kotelnikova.45@mail.ru (E.K.)
[2] Max Planck Institute for Dynamics of Complex Technical Systems, Sandtorstrasse 1, 39106 Magdeburg, Germany
* Correspondence: lorenz@mpi-magdeburg.mpg.de

Abstract: The method of temperature-resolved powder X-ray diffraction (TRPXRD) was used to determine the elevated temperature behavior of L-aspartic acid (L-asp), DL-aspartic acid (DL-asp), L-glutamic acid (L-glu), and an L-asp$_{0.25}$,L-glu$_{0.75}$ solid solution. These amino acids were not found to undergo any solid-phase (polymorph) transformations. When heated, they all experienced only thermal deformations. The corresponding parameters of the monoclinic cells of L-asp and DL-asp, and the orthorhombic cells of L-glu and L-asp$_{0.25}$,L-glu$_{0.75}$, were calculated for the entire range of studied temperatures (up to 220 °C). The data obtained were used to calculate the parameters of the thermal deformation tensors, and to plot the figures of their thermal expansion coefficients. A correlation between the maximum and minimum values of thermal expansion coefficients and the length, type, direction, and number of hydrogen bonds in the crystal structures of the investigated amino acids was established. The observed negative thermal expansion (contraction) of crystal structures of L-asp and DL-asp along the *ac* plane can be explained as a result of shear deformations occurring in monoclinic crystals with a non-fixed angle β. The studies were related to the presence of amino acids in various natural and technological processes occurring at different temperatures.

Keywords: L-aspartic acid; DL-aspartic acid; L-glutamic acid; enantiomers; solid solution; thermal expansion; TRPXRD

1. Introduction

Amino acids of the general formula NH$_2$-CHR-COOH play an important role in natural biochemical processes. They are typical representatives of organic crystals composed of chiral molecules. Chirality is the ability of a substance to occur in two or more configurations of molecules, which are referred to as enantiomers and diastereomers. Molecules of enantiomers are mirror images of one another. Relative configurations of enantiomers are designated as L (levorotary) and D (dextrorotary). All of the naturally occurring protein molecules are composed of L-enantiomers of amino acids. The respective D-amino acids are less frequently found in nature; for example, they are present in the cellular walls of some bacteria, in some antibiotic drugs, etc. Amino acids are involved in the regulation of metabolic processes; they also stimulate oxidation–reduction reactions in the brain and perform neurotransmission functions in the central nervous system [1]. Moreover, amino acids are widely used in the pharmaceutical industry, and participate in many technological processes occurring at different temperatures [2–4]. The chirality of amino acids and their wide distribution in the geological media make these compounds useful tools for dating sediments and paleontological objects [5–7]. Considering all of the above-mentioned, investigation of amino acids' behavior at elevated temperatures is a subject of high relevance.

Thermal deformations in organic compounds [8,9] are considerably less studied than those in inorganic substances [10–12], while thermal deformations of chiral organic

compounds are hardly investigated at all. Surprisingly enough, there are only a few reported works discussing thermal deformations of amino acids and presenting figures of their thermal expansion coefficients (CTEs). Here, it is worth mentioning the studies of K. Nakata et al. [13], S.J. Coles et al. [14], and C.H. Görbitz et al. [15,16], which discuss the high-temperature polymorph transformations of L-and DL-methionine, L-norvaline, L- and DL-2-aminobutyric acid, and L- and DL-norleucine.

In our previous work, we investigated the thermal deformations of amino acid crystal structures and plotted the figures of CTEs for the following substances: (1) both components and two solid solutions formed in the L-threonine/L-*allo*-threonine diastereomer system [17], (2) both components and the non-equimolar discrete compound formed in the L-valine/L-isoleucine system [18], and (3) the components formed in the L-alanine/L-serine amino acids system [19].

The present work is seen as continuation of the previous investigations of thermal deformations in amino acid crystal structures, aimed at a better understanding of the structure–property relationships of this important class of compounds. In the system of levorotary enantiomers L-aspartic acid/β-L-glutamic acid, both components and the solid solution L-asp$_{0.25}$,L-glu$_{0.75}$ (L-asp/L-glu = 25/75 mol.%) were studied. In addition, racemic aspartic acid (DL-asp) was examined.

2. Materials and Methods

2.1. Materials

Aspartic acid ($C_4H_7NO_4$) and glutamic acid ($C_5H_9NO_4$) are proteinogenic aliphatic amino acids. They are characterized by having two carboxylic COOH groups in their molecules, and differ only in the absence or presence of a CH_2 group (Figure 1). Both acids have only one chiral center, i.e., they can exist in two different configurations, referred to as L and D enantiomers. In the event that the enantiomers form a binary equimolar compound, the D and L molecules contribute equally to the formation of the resulting DL crystalline structure. The crystals of L-asp and DL-asp show monoclinic syngony with the space groups $P2_1$ and $C2/c$, respectively [20,21]. The L-glu enantiomer possesses orthorhombic syngony with the space group $P2_12_12_1$ [22].

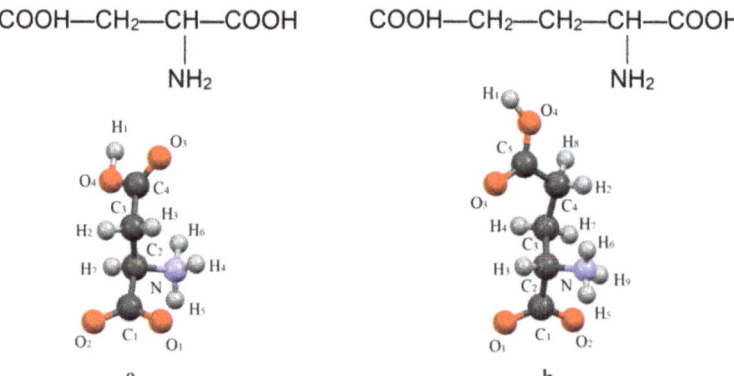

Figure 1. Structural formulae and spatial structures of the molecules of L-asp (**a**) and L-glu (**b**).

The reactants used were as follows: (1) L- and DL-aspartic acid of 99% purity, obtained from AppliChem GmbH, Germany, and (2) β-L-glutamic acid of 99% purity, obtained from Acros Organics–Thermo Fisher Scientific, Belgium. The samples of the L-asp and L-glu enantiomers, DL-asp, and the solid solution L-asp$_{0.25}$,L-glu$_{0.75}$, applied in the experiments, were obtained by dissolving the corresponding compounds in distilled water at ~70 °C, followed by crystallization via isothermal evaporation of the solvent at 55 °C for 48 h.

2.2. Methods

The above-mentioned samples were pre-investigated by means of the PXRD method (Rigaku MiniFlex II, Tokyo, Japan). The behavior of the samples at elevated temperatures was studied using the temperature-resolved powder X-ray diffraction (TRPXRD) technique. The experiments were performed in atmospheric air using a Rigaku Ultima IV (Tokyo, Japan) diffractometer equipped with a high-temperature accessory. The experimental conditions were as follows: $Co_{K\alpha}$ irradiation, 2θ range of 5–50°, and temperature range of 23–220 °C, with a temperature step of 10 °C. The unit cell parameters at each temperature were calculated via the Rietveld method using Topaz software (Bruker AXS GmbH, Karlsruhe, Germany).

Temperature dependencies of the unit cell parameters were applied to calculate the thermal expansion coefficients of the crystal structure (α) along the three mutually orthogonal axes of the thermal deformation tensor (α_{11}, α_{22}, α_{33}), and along the crystallographic axes a, b, and c. The resulting values were utilized to plot projections of the CTE figures onto the ab, ac, and bc planes of the orthorhombic (L-glu and L-asp$_{0.25}$,L-glu$_{0.75}$) and monoclinic (L-asp and DL-asp) unit cells. In the case of crystal structures with orthorhombic syngony, the tensor axes were found to be co-directional with the crystallographic axes. The tensor calculations and the plotting of the CTE figures were performed using TEV V1.0.1 software (Thomas Langreiter and Volker Kahlenberg, Institute of Mineralogy and Petrography, Innsbruck, Austria).

3. Results

3.1. Temperature-Resolved Powder X-ray Diffraction (TRPXRD) Data

Figures 2–5 show the TRPXRD patterns of L-asp, DL-asp, L-glu, and L-asp$_{0.25}$,L-glu$_{0.75}$ obtained at various temperatures. It can be seen that within the whole temperature range, all of the patterns maintain the fixed collection of peaks, and do not lose or acquire new ones. This means that the corresponding samples do not undergo any polymorph transformations up to the amorphization temperature of L- and DL-asp (amorphization occurs at 220–230 °C and 210–220 °C, respectively) (Figures 2 and 3), and up to the melting temperature of L-glu and L-asp$_{0.25}$,L-glu$_{0.75}$ (melting occurs between 170 and 180 °C) (Figures 4 and 5). Hence, it follows that, upon heating, the crystal structures of all of the studied samples experience only thermal deformations.

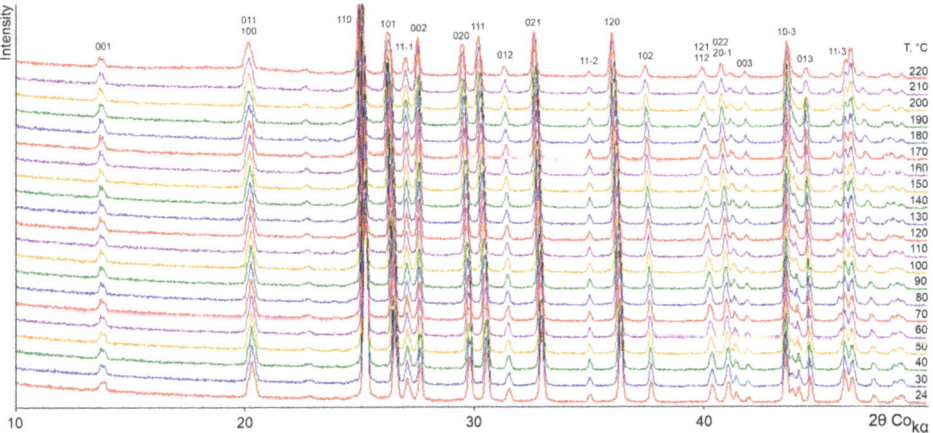

Figure 2. X-ray patterns of an L-asp sample obtained at various temperatures.

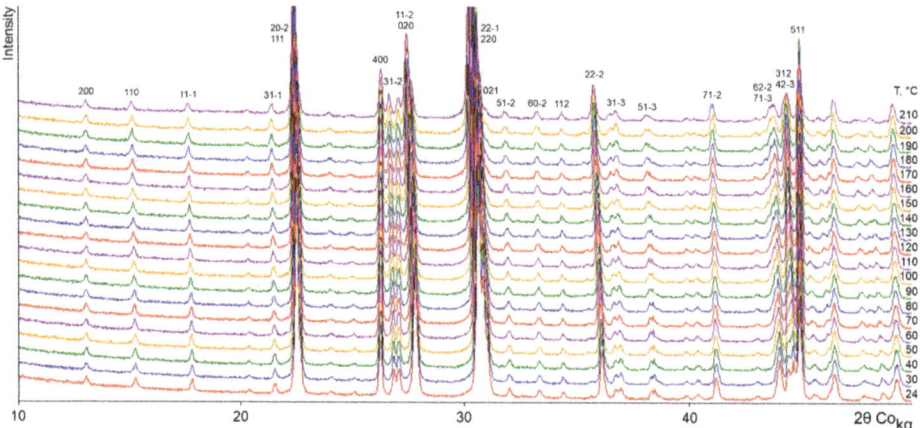

Figure 3. X-ray patterns of a DL-asp sample obtained at various temperatures.

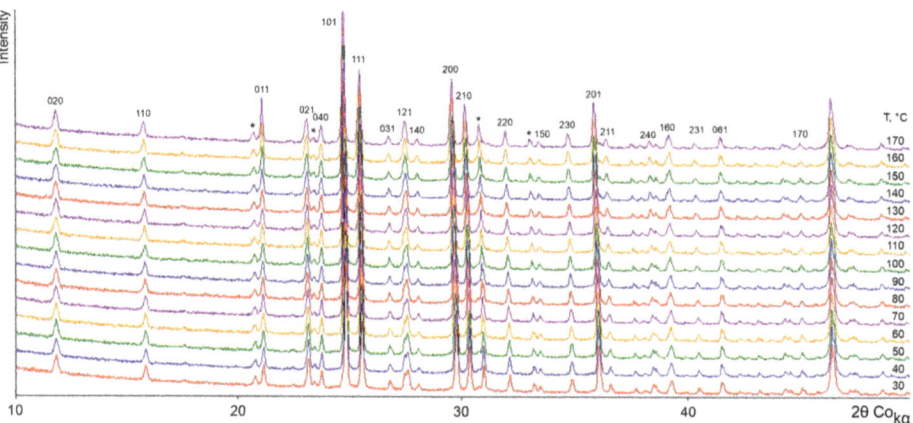

Figure 4. X-ray patterns of an L-glu sample obtained at various temperatures. The majority of the peaks correspond to the β-L-glu modification. The peaks belonging to the α-L-glu modification are marked with an asterisk.

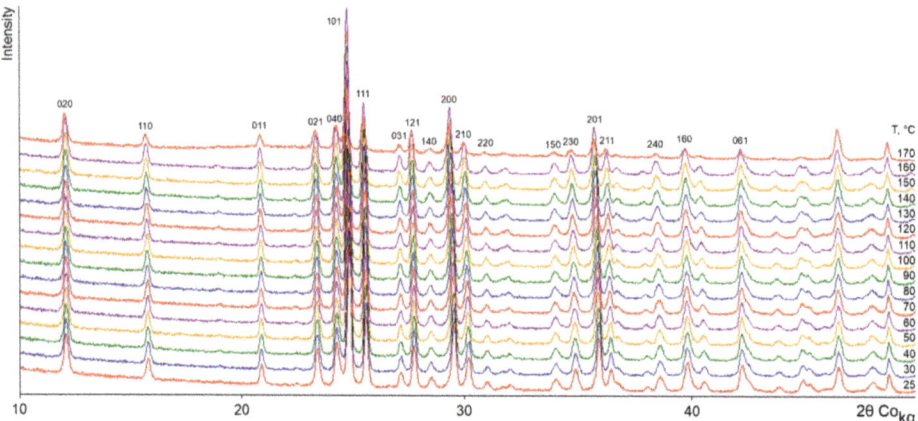

Figure 5. X-ray patterns of an L-asp$_{0.25}$,L-glu$_{0.75}$ solid solution obtained at various temperatures.

As the temperature increases, the peaks are shifted, usually towards the lower 2θ angles. Since the shift values of peaks with different *hkl* are not the same, some peaks merge.

The sample of the L-glu enantiomer (Figure 4) is already a mixture of two phases at the starting temperature. It is related to the polymorph modifications of β-L-glu (~86 wt.%) and α-L-glu (~14 wt.%). The intensity of the peaks corresponding to the α-L-glu component persists up to the melting point. This verifies that under the conditions of the TRPXRD experiment the β-L-glu and α-L-glu phases do not undergo solid-phase polymorph transformations to change into one another.

The X-ray patterns of the L-asp$_{0.25}$,L-glu$_{0.75}$ solid solution (Figure 5) correspond to those of the β-L-glu enantiomer. It is worth mentioning that the solid solution sample did not contain the α-L-glu admixture phase. This may be due to the fact that the conditions of preparing the solutions of L-glu and L-asp$_{0.25}$,L-glu$_{0.75}$ (the temperature, degree of supersaturation, and mode of stirring) were slightly different. Several attempts to crystallize a pure α-L-glu phase using the method proposed in [22–24] were unsuccessful.

The TRPXRD results were used to calculate the parameters and volume of monoclinic cells of L-asp and DL-asp and orthorhombic cells of L-glu and L-asp$_{0.25}$,L-glu$_{0.75}$ within the whole range of studied temperatures.

3.2. Temperature Dependencies and Thermal Expansion Coefficients

Figure 6 shows changes in the *a*, *b*, *c*, and β parameters, and in the volume *V* of the monoclinic cells of L- and DL-asp versus temperature. In all cases, the errors in calculating the parameters and volume did not exceed the marker size. For purposes of comparison, the corresponding dependencies of the parameters and the volume of L-asp (Figure 6a) and DL-asp (Figure 6b) are presented in the same scale. It can be seen that the linear parameters *a*, *b*, and *c*, as well as the volumes *V*, grow linearly with the increase in temperature for both L- and DL-asp. However, the monoclinic angle β behaves differently: it decreases linearly for L-asp and increases linearly for DL-asp. The CTE values α_V of the volumes calculated for the monoclinic cells of the said compounds are close to one another (Table 1).

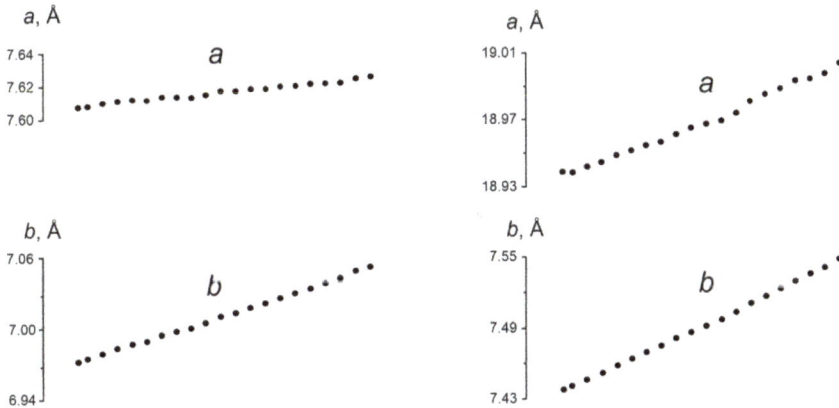

Figure 6. *Cont.*

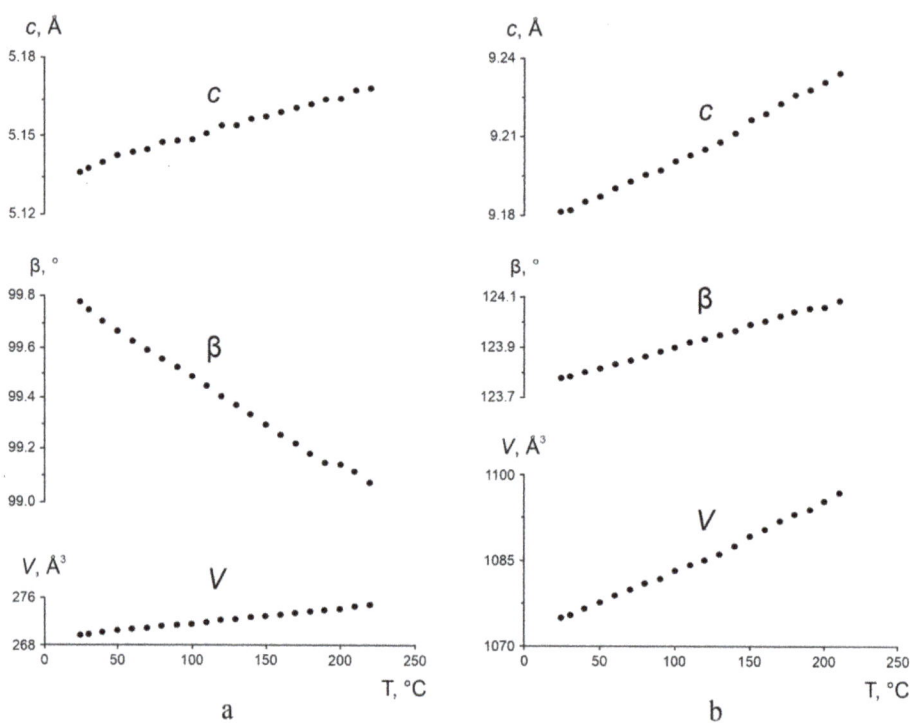

Figure 6. Temperature dependencies of the monoclinic cell parameters a, b, c (Å), β (°), and volume V (Å^3) of L-asp (**a**) and DL-asp (**b**). The errors in calculating the parameters and volume do not exceed the marker size.

Table 1. Thermal expansion coefficients ($\alpha_V * 10^{-6}$ °C^{-1}) for the monoclinic cell volume of L- and DL-asp.

Sample	α_V
L-asp	112(1)
DL-asp	107(1)

The temperature-derived changes in the orthorhombic cell parameters a, b, c, and β and volume V of L-glu and the L-asp$_{0.25}$,L-glu$_{0.75}$ solid solution are presented in Figure 7a,b, respectively—again at the same scale. For both the enantiomer and the solid solution, increasing the temperature caused a linear increase in all of the parameters and in the volume of the respective orthorhombic cells. This confirms that heating the samples results only in the thermal expansion of their crystal structures. As seen in Figure 7, elevating the temperature leads to a greater increase in the parameters and the volume of the unit cell for L-glu than for the L-asp$_{0.25}$,L-glu$_{0.75}$ solid solution. Accordingly, the CTE values α_V calculated for these samples appear to be rather different (Table 2).

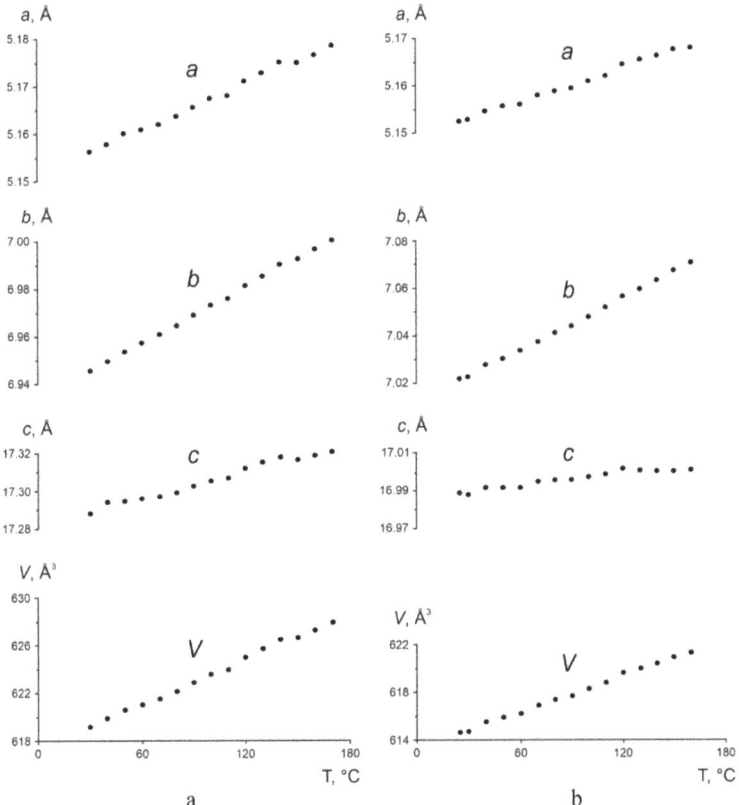

Figure 7. Temperature dependencies of the orthorhombic cell parameters a, b, c (Å), and volume V (Å^3) of L-glu (**a**) and the L-asp$_{0.25}$,L-glu$_{0.75}$ solid solution (**b**). The errors in calculating the parameters and volume do not exceed the marker size.

Table 2. Thermal expansion coefficients ($\alpha_V * 10^{-6}$ °C^{-1}) for the orthorhombic cell volume of L-glu and the L-asp$_{0.25}$,L-glu$_{0.75}$ solid solution.

Sample	α_V
L-glu	100(1)
L-asp$_{0.25}$,L-glu$_{0.75}$	79(2)

All of the temperature dependencies of the parameters and the volumes of the unit cells of L-asp, DL-asp, L-glu, and L-asp$_{0.25}$,L-glu$_{0.75}$ were approximated by polynomials of the first degree. The calculated data served as the basis for estimation of the CTE values of the crystal structures along the axes of the thermal deformation tensor and the crystallographic axes, given for L-asp and DL-asp in Table 3, and for L-glu and the L-asp$_{0.25}$,L-glu$_{0.75}$ solid solution in Table 4.

Table 3. Thermal expansion coefficients ($\alpha * 10^{-6}$ °C^{-1}) for the monoclinic crystal structures of L-asp and DL-asp along the axes of the thermal deformation tensor, and along the crystallographic axes.

Sample	α_{11}	α_{22}	α_{33}	α_a	$\alpha_b = \alpha_{22}$	α_c
L-asp	−7.2	57.9	59.8	10.9(5)	57.9(6)	30.8(9)
DL-asp	34.6	78.2	−4.1	18.7(6)	78.2(5)	31.5(7)

Table 4. Thermal expansion coefficients ($\alpha * 10^{-6}\ °C^{-1}$) for the orthorhombic crystal structures of L-glu and L-asp$_{0.25}$,L-glu$_{0.75}$ along the axes of the thermal deformation tensor, and the crystallographic axes.

Sample	$\alpha_a = \alpha_{11}$	$\alpha_b = \alpha_{22}$	$\alpha_c = \alpha_{33}$
L-glu	31.2(8)	56.6(6)	13.5(7)
L-asp$_{0.25}$,L-glu$_{0.75}$	23(1)	52(1)	6(1)

The data shown in Table 3 indicate the presence of negative (or anomalous) thermal expansion—that is, contraction—of the L-asp and DL-asp monoclinic crystal structures. In both substances, the contraction occurs in the crystallographic plane ac, but in L-asp it acts along the α_{11} axis of the tensor, while in DL-asp it acts along the α_{33} axis.

The data presented in Tables 3 and 4 were used to plot projections of the CTE figures onto the crystallographic planes ab, ac, and bc of L- and DL-asp (Figures 8 and 9, right), and L-glu and L-asp$_{0.25}$,L-glu$_{0.75}$ (Figure 10, right, solid black and red dashed lines, respectively). Figures 8 and 9 (left) also show projections of the monoclinic structures of L- and DL-asp onto the $a*b$, ac, and $bc*$ planes, which are orthogonal to the c, b, and a axes, respectively. The analogous Figure 10 (left) shows projections of the orthorhombic structure of L-glu onto the ab, ac, and bc planes. As the temperature dependencies of all of the unit cell parameters in all of the studied samples can be approximated by linear polynomials, the CTE figures do not change substantially when varying the temperature. For that reason, Figures 8–10 show the CTE figures obtained at only one temperature: 100 °C.

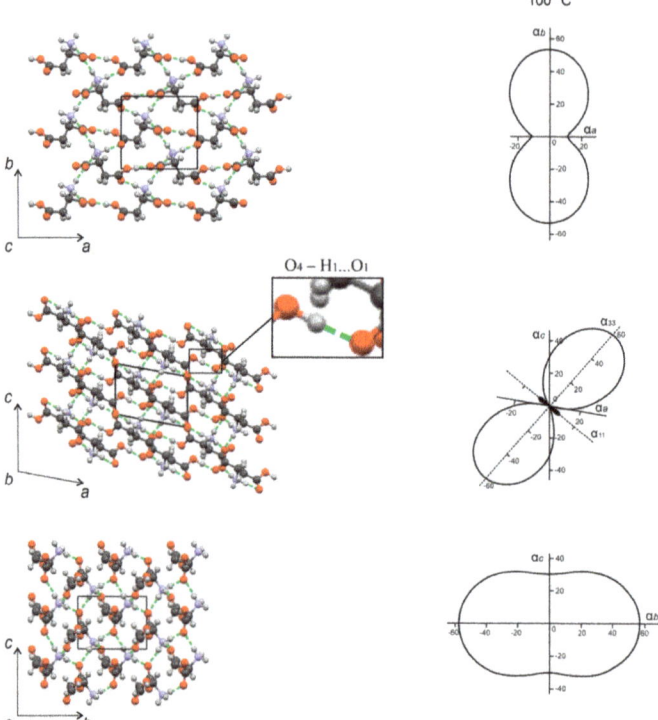

Figure 8. Projections of the CTE figures onto the ab, ac, and bc planes (**right**), and projections of the crystal structure onto the $a*b$, ac, and $bc*$ planes (**left**) in the monoclinic cell of L-aspartic acid. Black shade corresponds to the region of the negative thermal expansion. Hydrogen bonds are shown as green dashed lines. The crystal structure projections were plotted using the structural data from CSD (LASPRT) [25].

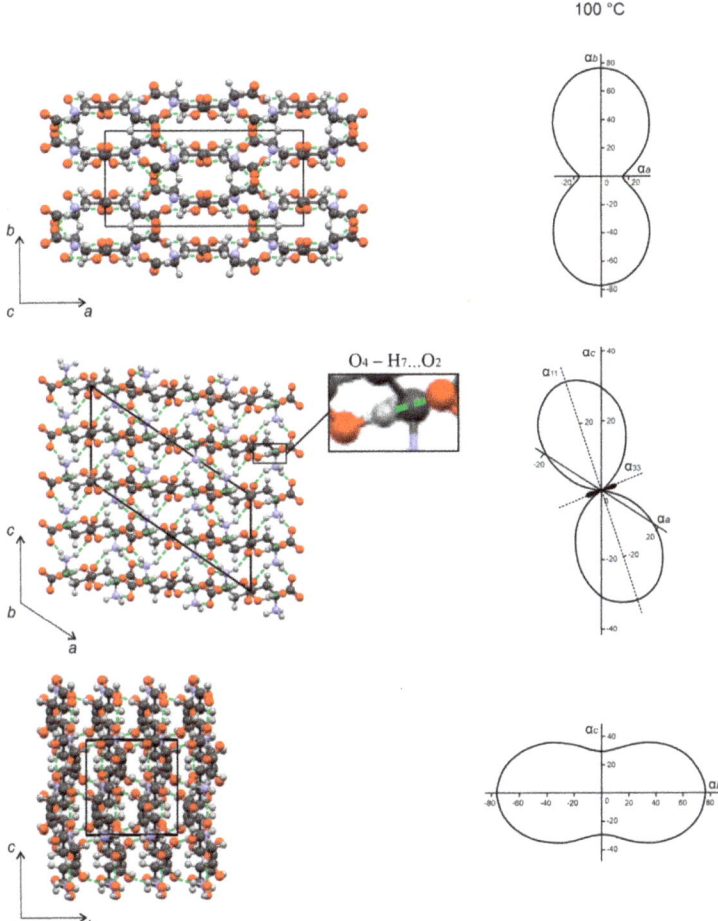

Figure 9. Projections of the CTE figures onto the *ab*, *ac*, and *bc* planes (**right**), and projections of the crystal structure onto the *a*b*, *ac*, and *bc** planes (**left**) in the monoclinic cell of DL-aspartic acid. Black shade corresponds to the region of the negative thermal expansion. Hydrogen bonds are shown as green dashed lines. The crystal structure projections were plotted using the structural data from CSD (DLASPA02) [25].

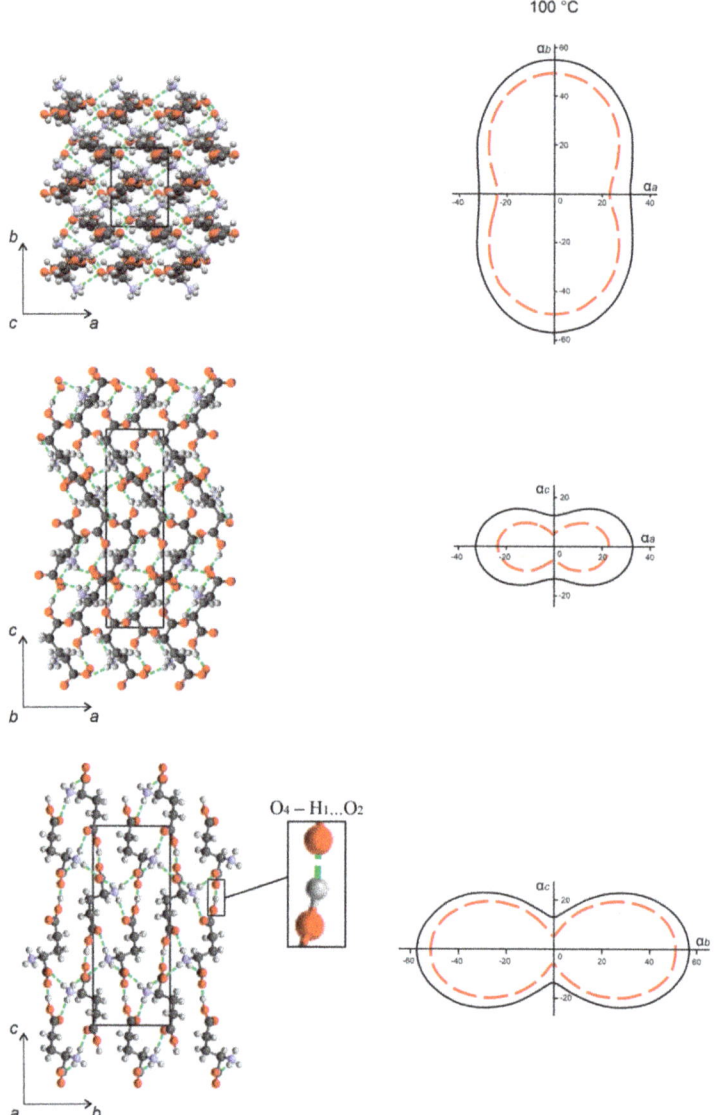

Figure 10. CTE figures of L-glutamic acid (solid black line) and the solid solution L-asp$_{0.25}$L-glu$_{0.75}$ (red dashed line) projected upon the *ab*, *ac*, and *bc* planes of the corresponding orthorhombic cells (**right**). Projections of the L-glu crystal structure onto the same planes (**left**). Hydrogen bonds are shown as green dashed lines. The crystal structure projections were plotted using the structural data from CSD (LGLUAC01) [25].

4. Discussion

In the following section, first, the results for L-asp and DL-asp crystallizing in a monoclinic lattice are discussed (Figures 8 and 9), followed by the findings for L-glu and the L-asp/L-glu solid solution crystallizing in an orthorhombic syngony (Figure 10).

4.1. Monoclinic Amino Acids L-asp and DL-asp

The molecules in the **L-asp** crystal structure (Figure 8) are arranged in zigzag chains extended along the direction close to that of the bisector of the *acute* angle $a\hat{}c$. The chain units are interconnected via the strongest hydrogen bond in the structure—namely, O_4–H_1 ... O_1 = 2577 Å—which is highlighted in the projection of the L-asp crystal structure onto the *ac* plane. This bond is essentially co-directional with the direction of the negative thermal expansion (α_{33}). There are two other hydrogen bonds—N–H_5 ... O_2 and N–H_6 ... O_3—formed in approximately the same direction and within the same plane. The maximum thermal expansion is directed along the α_{33} axis of the thermal deformation tensor, which approximates the bisector of the *obtuse* angle $a\hat{}c$, i.e., the β angle; only one hydrogen bond—namely, N–H_4 ... O_2—is extended in that direction (Figure 8, projection *ac*). A sufficiently great value of the CTE is observed along the α_{22} axis (which coincides with axis *b*), which can be explained by the relatively low concentration of hydrogen bonds along this direction compared to the crystallographic axis *a* (Figure 8, projection *ab*).

Therefore, at elevated temperatures, the L-asp crystal structure naturally contracts along the direction in which several hydrogen bonds are grouped, including the strongest hydrogen bond: O_4–H_1 ... O_1. This contraction is accompanied by a diminution in the β angle (see Figure 6a), and is characterized by maximum anisotropy of the thermal deformations within the *ac* plane (Figure 8).

The crystal structure of **DL-asp** is also characterized by negative thermal expansion in the *ac* plane (Figure 9). In contrast to L-asp, however, the racemate structure contracts along the direction of the bisector of the *obtuse* angle $a\hat{}c$, i.e., the β angle (Figure 9, projection *ac*). This may be due to the fact that this is also the direction of the strongest hydrogen bond: O_4–H_7 ... O_2 = 2558 Å. The closely located hydrogen bonds of the N–H ... O type may also facilitate the contraction. Therefore, in this case, contraction with increase in temperature is accompanied by enlargement of the β angle.

The greatest thermal expansion of the DL-asp crystal structure is observed in the direction of crystallographic axis *b* ($\alpha_b = \alpha_{22}$). This may be the result of molecular layers formed from interconnected chains of molecules that are arranged in the direction orthogonal to the *b* axis (within the *ac* plane). Each D- or L-asp molecule located within the layer is bonded to two molecules located in the previous and subsequent layers via N–H_2 ... O_1 and N–H_3 ... O_1 hydrogen bonds. This interlayer interaction is not very strong (Figure 9, left, projection *bc**) and, therefore, the maximum expansion of the DL-asp crystal structure is directed along the *b* axis.

In the crystal structures of the aspartic acid enantiomer and the racemate, the negative thermal expansion in the *ac* plane is co-directed with the direction of the strongest hydrogen bonds of the O–H ... O type: O_4–H_1 ... O_1 (L-asp) and O_4–H_7 ... O_2 (DL-asp). Individual particularities of thermal deformations in their crystal structures are most clearly visible in the changes in the monoclinic angle β, which for L-asp decreases, but for DL-asp increases. Accordingly, the maximum thermal expansion in both crystal structures occurs along the direction of relatively weak hydrogen bonds of the N–H ... O type. Therefore, for both structures, the hydrogen bonds of the O–H ... O type are stronger than the hydrogen bonds of the N–H ... O type. Additionally, the geometry and concertation of the bonds should be taken into account.

The effect of changing the angle β in monoclinic crystals with elevation of temperature is described in detail by S.K. Filatov et al. [26,27]. Such thermal behavior is typical of crystalline structures, where either one angular parameter (monoclinic cell) or three angular parameters (triclinic cell) are not symmetrically fixed. In addition to thermal expansion, such crystalline structures can also undergo shear deformations characterized by shifts in the molecular rows relative to one another.

Figure 11 illustrates simulated shear deformations in monoclinic crystal models projected for the *ac* plane. Increasing the β_1 angle of the initial orthorhombic structure (Figure 11a, solid line) to reach the β_2 value (dashed line) is accompanied by contraction of the crystal structure along the bisector of this angle, and by its simultaneous expansion in the orthogonal direction.

Similarly, decreasing the initial β_1 angle (Figure 11b, solid line) to the smaller value β_2 (dashed line) also leads to contraction of the crystal structure; however, this contraction occurs along the direction orthogonal to the β angle bisector, and at the same time the structure is expanded in the direction parallel to this bisector. Figure 11a,b also show the CTE figures and the regions corresponding to the positive (axis α_{11}) and negative (axis α_{33}) thermal expansion.

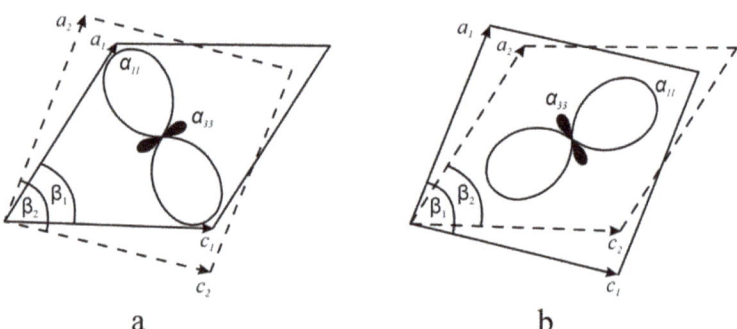

Figure 11. Schematic representation of shear deformation in relation to the figures of the thermal expansion coefficients (α_{11} and α_{33}) for increased (**a**) and decreased (**b**) angle β. Angles β_1 and β_2 are the monoclinic angles before and after the shear deformation, respectively. The image was plotted with partial use of the data reported in [26].

A decrease in the β angle with increase in temperature was also observed for enantiomers of two other monoclinic amino acids—namely, L-valine and L-isoleucine [9]. However, in these cases the changes in the crystal structures at elevated temperatures were not accompanied by negative thermal expansion, as revealed for L- and DL-aspartic acid.

4.2. Orthorhombic Amino Acids L-glu and L-asp$_{0.25}$,L-glu$_{0.75}$

The strongest hydrogen bond in the orthorhombic crystal structure of **L-glu** is O_4–H_1 ... O_2 = 2519 Å (Figure 10, left, highlighted), which bonds the molecules into zigzagged chains extended along the crystallographic axis *c*. This explains the minimal CTE value observed along the *c* axis compared to the other crystallographic directions (Figure 10, right, solid black line). The chains are joined to one another by means of a hydrogen bond (N–H_5 ... O_1) to form layers arranged orthogonally to the *b* axis. The layers, in turn, are bonded via two other hydrogen bonds, i.e., N–H_6 ... O_3 and N–H_9 ... O_1. The maximal thermal expansion observed along the *b* axis can be explained by the fact that the direction of the hydrogen bonds uniting the molecular layers deviates to a considerable extent from direction of the *b* axis. The most significant anisotropy of the thermal deformation was noticed within the *ac* and *bc* planes (Figure 10, right, solid black line).

Generally, projections of the CTE figures of the **L-asp$_{0.25}$,L-glu$_{0.75}$** solid solution onto the *ab*, *ac*, and *bc* planes (Figure 10, right, dashed red line) were of similar proportions—in fact, substantially alike—to the CTE figure of L-glu. However, the CTE values obtained for the solid solution were diminished along all three crystallographic directions in comparison to L-glu. The greatest decrease in the CTE values was observed in the projections upon the *ac* and *bc* planes.

The minimal CTE values in the crystal structures of L-glu and the L-asp$_{0.25}$,L-glu$_{0.75}$ solid solution are observed in the direction of the strongest hydrogen bond: O_4–H_1 ... O_2. Conversely, the maximal thermal expansion occurs in the direction of the weakest interaction between the molecules, which may be related to the geometry and low concentration of relatively weak hydrogen bonds of the N–H ... O type in the crystal structure of the L-glu enantiomer. The temperature dependencies of the orthorhombic cell parameters and the CTE figures of L-glu (see Figures 7a and 10, solid black line) and the L-asp$_{0.25}$,L-glu$_{0.75}$ solid solution (see Figures 7b and 10, dashed red line) have definite similarities. Discrepancies in the thermal changes to their orthorhombic cell parameters and, accordingly, different CTE values, can be caused by a different degree of imperfection of their crystal structure. Substitution of some of the L-glu molecules with smaller L-asp molecules results in the formation of defects, such as voids, in the crystal structure of the solid solution. Expansion of its crystal structure at elevated temperatures can be partially compensated for by the presence of these voids. For this reason, the intensity of changes in the L-asp$_{0.25}$,L-glu$_{0.75}$ orthorhombic parameters is considerably lower than that for the L-glu enantiomer.

The above results were compared with similar studies of thermal deformations for (1) the orthorhombic components in the L-threonine/L-*allo*-threonine (L-thr/L-*a*thr) diastereomer system, and two solid solutions of both—namely, L-thr$_{0.34}$,L-*a*thr$_{0.66}$ and L-thr$_{0.10}$,L-*a*thr$_{0.90}$ [17]—and (2) the orthorhombic components in the L-alanine (L-ala) and L-serine (L-ser) enantiomer system [19]. It appears that heating resulted in increases in all orthorhombic parameters of L-glu and those of the solid solutions L-asp$_{0.25}$,L-glu$_{0.75}$, L-thr$_{0.34}$,L-*a*thr$_{0.66}$, and L-thr$_{0.10}$,L-*a*thr$_{0.90}$, and in diminution of one of the parameters of the diastereomers L-thr and L-*a*thr and the enantiomers L-ala and L-ser. Increasing all of the parameters indicates the predominantly thermal expansion of the crystal structure. Decreasing one of the parameters allows, in addition, the use of the concept of the hinge mechanism of thermal deformations [26].

5. Conclusions

It was established that, when heated, the L-enantiomers of aspartic and glutamic acid, the aspartic acid racemate, and the L-asp$_{0.25}$,L-glu$_{0.75}$ solid solution do not undergo polymorphic transitions, but experience thermal deformations. The mode of the thermal deformations observed in all of the above crystal structures was consistent with the distance, type, geometry, and concentration of the intermolecular hydrogen bonds.

Thermal deformations of L-asp and DL-asp are characterized by the manifestation of negative thermal expansion—i.e., contraction—of their crystal structures in the monoclinicity plane *ac*. Explanation of this effect of contraction is based on the contribution of shear deformations to thermal deformations of the crystal structure.

Thermal deformations of L-glu and the L-asp$_{0.25}$,L-glu$_{0.75}$ solid solution are characterized by increases in all of the parameters of their orthorhombic cells. The CTE values obtained for the solid solution were lower than those calculated for the L-glu enantiomer. The reason for this difference seems to be imperfection of the solid solution crystal structure due to isomorphic substitution of some L-glu molecules by the smaller-sized L-asp molecules.

The results of the present study were compared with data for other amino acids—such as L-valine, L-isoleucine, L-threonine, L-*allo*-threonine, and solid solutions of the latter two—that we had investigated previously [9,17,19].

The results obtained made it possible to reveal the individual specific features of the amino acids' thermal behavior, which is important for their efficient application.

Author Contributions: Conceptualization, E.K. and H.L.; methodology, E.K., H.L. and R.S.; formal analysis, R.S.; investigation, R.S. and E.K.; writing—original draft preparation, R.S. and E.K.; writing—review and editing, R.S., E.K. and H.L.; visualization, E.K. and R.S.; supervision, H.L. All authors read and agreed to the published version of the manuscript.

Funding: This work was supported by the President of Russian Federation Grant to leading scientific schools (NSh-2526.2020.5).

Acknowledgments: The investigations were performed using equipment from the Resource Centre "Centre for X-ray Diffraction Studies" of St. Petersburg State University. The authors thank M.G. Krzhizhanovskaya, A.I. Isakov, L. Yu. Kruchkova, A.A. Zolotarev Jr., and N.V. Platonova for collaboration.

Conflicts of Interest: The authors declare no conflict of interest.

References

1. Nelson, D.L.; Cox, N.N. *Lehninger Principles of Biochemistry*, 4th ed.; W. H. Freeman: New York, NY, USA, 2005; p. 1125.
2. Giron, D. Polymorphism in the pharmaceutical industry. *Therm. Anal. Calorim.* **2001**, *64*, 37–60. [CrossRef]
3. Murakami, H. From Racemates to Single Enantiomers–Chiral Synthetic Drugs over the last 20 year. *Top. Curr. Chem.* **2006**, *269*, 273–299.
4. Bredikhin, A.; Bredikhina, Z.; Zakharychev, D. Crystallization of chiral compounds: Thermodynamical, structural and practical aspects. *Mendeleev Commun.* **2012**, *22*, 171–180. [CrossRef]
5. Robins, J.; Jones, M.; Matisoo-Smith, E. *Amino Acid Racemization Dating in New Zealand: An Overview and Bibliography*; Auckland Univ.: Auckland, New Zealand, 2010.
6. Killops, S. Introduction to Organic Geochemistry, 2end edn (paperback). *Geofluids* **2005**, *5*, 236–237. [CrossRef]
7. Torres, T.; Ortiz, J.; Arribas, I.; Delgado, A.; Julia, R.; Martín-Rubí, J. Geochemistry of Persististrombus latus Gmelin from the Pleistocene Iberian Mediterranean realm. *Lethaia* **2010**, *43*, 149–163. [CrossRef]
8. Saha, B.K. Thermal Expansion in Organic Crystals. *J. Indian Inst. Sci.* **2017**, *97*, 177–191. [CrossRef]
9. Kotelnikova, E.; Isakov, A.; Lorenz, H. Thermal deformations of crystal structures formed in the systems of malic acid enantiomers and L-valine–L-isoleucine enantiomers. *CrystEngComm.* **2018**, *20*, 2562–2572. [CrossRef]
10. Jessen, S.M.; Kuppers, H. The precision of thermal expansion tensors of triclinic and monoclinic crystals. *J. Appl. Crystallogr.* **1991**, *24*, 239–242. [CrossRef]
11. Evans, J.S.O. Negative thermal expansion materials. *J. Chem. Soc. Dalton Trans.* **1999**, *19*, 3317–3326. [CrossRef]
12. Miller, W.; Smith, C.W.; Mackenzie, D.S.; Evans, K.E. Negative thermal expansion: A review. *J. Mater. Sci.* **2009**, *44*, 5441–5451. [CrossRef]
13. Nakata, K.; Takaki, Y.; Sakurai, K. Structure of the D form of DL-α-amino-n-butyric acid. *Acta Cryst.* **1980**, *36*, 504–506. [CrossRef]
14. Coles, S.J.; Gelbrich, T.; Griesser, U.J.; Hursthouse, M.B.; Pitak, M.; Threlfall, T. The Elusive High Temperature Solid-State Structure of D,L-Norleucine. *Cryst. Growth Des.* **2009**, *9*, 4610–4612. [CrossRef]
15. Görbitz, C.H.; Qi, L.; Mai, N.T.K.; Kristiansen, H. Redetermined crystal structure of α-DL-methionine at 340 K. *Acta Cryst.* **2014**, *70*, 337–340. [CrossRef]
16. Görbitz, C.H.; Karen, P.; Dušek, M.; Petříček, V. An exceptional series of phase transitions in hydrophobic amino acids with linear side chains. *IUCrJ.* **2016**, *3*, 341–353. [CrossRef] [PubMed]
17. Taratin, N.; Lorenz, H.; Binev, D.; Seidel-Morgenstern, A.; Kotelnikova, E. Solubility equilibria and crystallographic characterization of the L-threonine/L-allo-threonine system. Part 2: Crystallographic characterization of solid solutions in the threonine diastereomeric system. *Cryst. Growth Des.* **2015**, *15*, 137–144. [CrossRef]
18. Isakov, A.; Kotelnikova, E.; Bocharov, S.; Zolotarev, A.J.; Lorenz, H. Thermal deformations of the crystal structures of L-valine, L-isoleucine and discrete compound V$_2$I. In Proceedings of the 23rd International Workshop on Industrial Crystallization (BIWIC-2016), Magdeburg, Germany, 6–8 September 2016; pp. 7–12.
19. Kotelnikova, E.N.; Sadovnichii, R.V.; Kryuchkova, L.Y.; Lorenz, H. Limits of Solid Solutions and Thermal Deformations in the L-Alanine–L-Serine Amino Acid System. *Crystals* **2020**, *10*, 618. [CrossRef]
20. Derissen, J.L.; Endeman, H.J.; Peerdeman, A.F. The crystal and molecular structure of L-aspartic acid. *Acta Cryst. Sect. B Struct. Crystallogr. Cryst. Chem.* **1968**, *24*, 1349–1354. [CrossRef]
21. Rao, S.T. Refinement of DL-Aspartic Acid. *Acta Cryst.* **1973**, *29*, 1718–1720. [CrossRef]
22. Ruggiero, M.T.; Sibik, J.; Zeitler, J.A.; Korter, T.M. Examination of l-Glutamic Acid Polymorphs by Solid-State Density Functional Theory and Terahertz Spectroscopy. *J. Phys. Chem. A* **2016**, *120*, 7490–7495. [CrossRef]
23. Kitamura, M. Polymorphism in the Crystallization of L-Glutamic Acid. *J. Cryst. Growth* **1989**, *96*, 541–546. [CrossRef]
24. Sugita, Y. Polymorphism of L-Glutamic Acid Crystals and Inhibitory Substance for β- Transition in Beet Molasses. *Agric. Biol. Chem.* **1988**, *52*, 3081–3085.
25. CSD files (identifiers): LASPRT (L-asp), DLASPA02 (DL-asp) and LGLUAC01 (L-glu). Available online: https://www.ccdc.cam.ac.uk (accessed on 6 August 2021).
26. Filatov, S.; Krivovichev, S.; Bubnova, R. *General Crystal Chemistry*; Publishing House of St. Petersburg University: St. Petersburg, Russia, 2018; pp. 222–224. (In Russian)
27. Filatov, S.; Bubnova, R. The nature of special points on unit cell parameters temperature dependences for crystal substances. *Z. Kristallogr.* **2007**, *26*, 447–452. [CrossRef]

Article

Comparison of the Nucleation Parameters of Aqueous L-glycine Solutions in the Presence of L-arginine from Induction Time and Metastable-Zone-Width Data

Lie-Ding Shiau [1,2]

1 Department of Chemical and Materials Engineering, Chang Gung University, Taoyuan 333, Taiwan; shiau@mail.cgu.edu.tw; Tel.: +886-3-2118800 (ext. 5291)
2 Department of Urology, Linkou Chang Gung Memorial Hospital, Taoyuan 333, Taiwan

Abstract: Induction time and metastable-zone-width (MSZW) data for aqueous L-glycine solutions in the presence of L-arginine impurity were experimentally measured using a turbidity probe in this study. The nucleation parameters, including the interfacial free energy and pre-exponential nucleation factor, obtained from induction time data, were compared with those obtained from MSZW data. The influences of lag time on the nucleation parameters were examined for the induction time data. The effects of L-arginine impurity concentration on the nucleation parameters based on both the induction time and MSZW data were investigated in detail.

Keywords: crystallites; impurities; induction time; metastable zone width; nucleation parameters

---

Citation: Shiau, L.-D. Comparison of the Nucleation Parameters of Aqueous L-glycine Solutions in the Presence of L-arginine from Induction Time and Metastable-Zone-Width Data. *Crystals* **2021**, *11*, 1226. https://doi.org/10.3390/cryst11101226

Academic Editors: Heike Lorenz, Alison Emslie Lewis, Erik Temmel and Jens-Petter Andreassen

Received: 14 September 2021
Accepted: 11 October 2021
Published: 12 October 2021

Publisher's Note: MDPI stays neutral with regard to jurisdictional claims in published maps and institutional affiliations.

Copyright: © 2021 by the author. Licensee MDPI, Basel, Switzerland. This article is an open access article distributed under the terms and conditions of the Creative Commons Attribution (CC BY) license (https://creativecommons.org/licenses/by/4.0/).

1. Introduction

In crystal growth, the induction time is defined as the time interval between the establishment of the supersaturated state and the formation of detectable nuclei. The metastable-zone-width (MSZW) limit is defined as the time taken at a given cooling rate between the establishment of the supersaturated state and the formation of detectable nuclei. Nucleation is the initial process for the formation of crystals in liquid solutions. Thus, both the induction time and MSZW data are related to the nucleation rate of the crystallized substance in solutions. In classical nucleation theory (CNT) [1–3], the nucleation rate is expressed in the Arrhenius form, governed by two nucleation parameters, including the interfacial free energy and pre-exponential nucleation factor. The interfacial free energy is the energy required to create a new solid/liquid interface for the formation of crystals in liquid solutions, while the pre-exponential factor is related to the attachment rate of solute molecules to a cluster in the formation of crystals. The influences of impurities on the nucleation parameters have long been investigated using induction time or MSZW data with the addition of different impurities in solutions for a variety of compounds [4–14].

The nucleation parameters of a crystallized substance have been traditionally determined from induction time data by assuming $t_i^{-1} \propto J$, where J is nucleation rate [1]. Recently, various methods have been proposed to calculate the nucleation parameters from MSZW data [15–21]. Although the induction time and MSZW processes are two different temperature-controlling methods for determination of the nucleation parameters in a crystallization system, a model should be available to relate the induction time and MSZW data with the nucleation parameters. Furthermore, as a cooling process is applied first to reach the desired operating temperature and then a constant temperature is adopted in the induction time measurements, there always exists a lag time between the prepared supersaturated solution being at a higher temperature and it being cooled to the desired lower constant temperature. For simplicity, the lag time is usually neglected in determining the nucleation parameters from the induction time data.

The nucleation process can behave differently. For certain systems, induction time cannot even be considered due to sharp phase transition, while for some cases there is induction time governed by different material properties. For example, by evaporating a cellulose nanocrystal-based cholesteric drop, the drop edges are pinned to the substrate, which leads to nonequilibrium sliding of the individual cholesteric fragment with active ordering [22]; following the induction period of cholesteric collagen tactoids, phase separation goes through the nucleation process during which multiple chiral nuclei spontaneously emerge and grow throughout the continuous isotropic phase [23]. In the present work, a model was proposed based on CNT to relate the induction time and MSZW data with the nucleation parameters for the systems with an experimentally measurable nucleation point. The proposed model was then applied to determine the nucleation parameters for the aqueous L-glycine solutions in the presence of L-arginine impurity from the induction time and MSZW data. The effects of lag time on the nucleation parameters within the induction time data were investigated. L-glycine was adopted in this work as it is the simplest amino acid and is often used as a model compound in the study of solution nucleation [24–30]. L-arginine is another amino acid which was randomly chosen as impurity in the aqueous L-glycine solutions.

2. Theory

The nucleation rate according to CNT is expressed as [1–3]

$$J = A \exp\left[-\frac{16\pi v^2 \gamma^3}{3k_B{}^3 T^3 \ln^2 S}\right], \qquad (1)$$

where A is the nucleation pre-exponential factor, γ is the interfacial free energy, k_B is the Boltzmann constant, $v = M_w/\rho_c N_A$ is the molecular volume, T is the temperature, and S is the supersaturation.

A model is derived based on CNT to determine γ and A by relating the induction time and MSZW data with J as follows. If a solution saturated at T_0 is cooled to T_m at a constant cooling rate b within the time period $t = 0$ to t_m and then the temperature is kept at T_m within the time period t_m to $t_m + t_i$, the nucleation event for this combined process is assumed to be detected at $t = t_m + t_i$. If t_m is small compared to t_i, this combined process can be regard as the induction time process with consideration of the lag time t_m, which is the time required for the solution saturated at T_0 to cool to T_m at cooling rate b. Thus, $\Delta T_m = T_0 - T_m$ and the lag time is given by $t_m = \Delta T_m/b$. This combined process for $t_m = 0$ corresponds to the induction time process without consideration of the lag time. On the other hand, this combined process for $t_i = 0$ corresponds to the MSZW process.

Figure 1 depicts the MSZW process for a saturated solution of C_0 cooled at a constant cooling rate b, where T_0 is the initial saturated temperature at $t = 0$, T_m is the nucleation temperature at t_m, C_0 is the saturated concentration at T_0, C_m is the saturated concentration at T_m, $C_{eq}(T)$ is the solubility, and $S(T) = C_0/C_{eq}(T)$ is the supersaturation. As $C_{eq}(T)$ generally decreases with decreasing temperature, $S(T)$ increases and subsequently J increases with time. For the nucleation point at t_m, S_m is the supersaturation at T_m defined as $S_m = C_0/C_{eq}(T_m) = C_0/C_m$. The nucleation rate at T_m is given by

$$J_m = A \exp\left[-\frac{16\pi v^2 \gamma^3}{3k_B{}^3 T_m{}^3 \ln^2 S_m}\right]. \qquad (2)$$

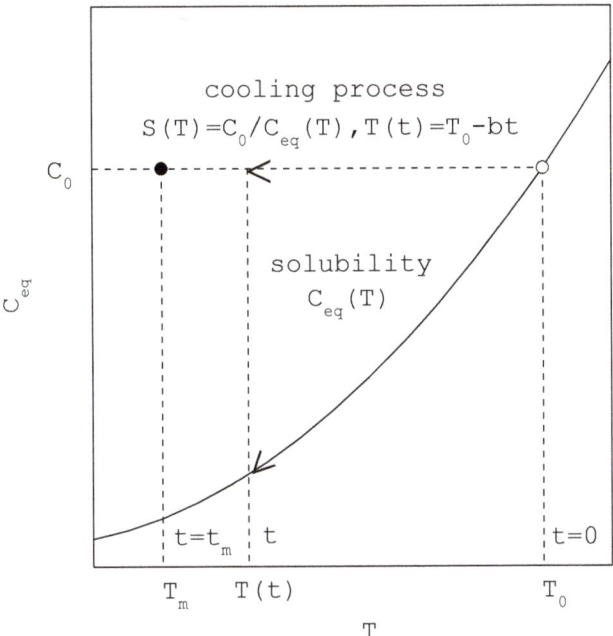

Figure 1. A schematic diagram showing the increasing of supersaturation during the cooling process reproduced from Shiau [31], where $C_{eq}(T)$ is the temperature-dependent solubility (○ represents the starting point and • represents the nucleation point).

Note that both S_m and ΔT_m are measures of the MSZW.

As the first appearance of nuclei can be regarded as a random process, the stochastic process of nucleation can be described by the Poisson's law [32–34]. For the combined process described above, as the temperature is cooled from T_0 to T_m within the time period $t = 0$ to t_m, $S(T)$ increases and J increases with time; and as the temperature is kept at T_m within the time period t_m to $t_m + t_i$, the supersaturation remains the same at S_m and J remains the same at J_m. Based on the given reasoning, the average number of expected nuclei N in a solution volume V within the time period $t = 0$ to $t_m + t_i$ is proposed in this study as

$$N = \left(\int_0^{t_m} JV dt\right) + J_m V t_i. \tag{3}$$

where the first term on the right-hand side represents the average number of expected nuclei generated within the time period $t = 0$ to t_m and the second term on the right-hand side represents the average number of expected nuclei generated within the time period t_m to $t_m + t_i$.

Based on the two-point trapezoidal rule for computing the value of a definite integral, one can derive [35]

$$\int_0^{t_m} JV dt = \frac{1}{2}(J_0 + J_m)V t_m = \frac{J_m V \Delta T_m}{2b}, \tag{4}$$

where J_0 and J_m represent the nucleation rate at $t = 0$ and $t = t_m$, respectively. Note that $J_0 = 0$ at $t = 0$ when $S(T_0) = 1$ and $t_m = \Delta T_m/b$.

According to the single nucleus mechanism (SNM) proposed by some researchers through experimental validation [32–34], a single primary nucleus is formed in a supersaturated solution, which grows out to a particular size and undergoes secondary nucleation by crystal-stirring-impeller or crystal-wall collision. Based on the assumptions that the growth time between the formation of nucleus and growth to the minimum size for secondary nucleation is negligible, and one secondary nucleation is enough to generate detectable crystal volume increase in a negligible amount of time, the nucleation event is detected after the secondary nucleation of the single primary nucleus. Thus, the nucleation event for the combined process occurs at $t = t_m + t_i$ when the first nucleus is formed. By substituting $N = 1$ in Equation (3), combining Equations (2)–(4) leads to

$$\ln\left(\frac{\Delta T_m}{2b} + t_i\right) = -\ln(AV) + \frac{16\pi v^2 \gamma^3}{3k_B{}^3 T_m{}^3 \ln^2 S_m}. \tag{5}$$

Thus, Equation (5) can be applied to determine the nucleation parameters from the induction time data, t_i, with consideration of the lag time, $\Delta T_m/b$. A plot of $\ln(\Delta T_m/2b + t_i)$ versus $\ln^2 S_m$ should give a straight line, the slope and intercept of which permit determination of γ and A, respectively.

Equation (5) for $\Delta T_m/b = 0$ reduces to

$$\ln t_i = -\ln(AV) + \frac{16\pi v^2 \gamma^3}{3k_B{}^3 T_m{}^3 \ln^2 S_m}, \tag{6}$$

Which corresponds to the conventional method adopted in determination of γ and A from the induction time data without consideration of the lag time. Equation (5) for $t_i = 0$ reduces to

$$\ln\left(\frac{\Delta T_m}{2b}\right) = -\ln(AV) + \frac{16\pi v^2 \gamma^3}{3k_B{}^3 T_m{}^3 \ln^2 S_m}, \tag{7}$$

Which can be applied to determine γ and A from the MSZW measurements, where a solution saturated at T_0 is cooled at a constant rate b from $t = 0$ to t_m and the nucleation event is detected at T_m.

If the temperature-dependent solubility is described in terms of the van't Hoff Equation (1), one obtains

$$\ln S_m = \ln\left(\frac{C_0}{C_m}\right) = \frac{-\Delta H_d}{R_G}\left(\frac{1}{T_0} - \frac{1}{T_m}\right) = \left(\frac{\Delta H_d}{R_G T_0}\right)\left(\frac{\Delta T_m}{T_m}\right), \tag{8}$$

where ΔH_d is the heat of dissolution and R_G is the gas constant. Substituting $\ln S_m$ in Equation (8) into Equation (7) yields

$$\left(\frac{T_0}{\Delta T_m}\right)^2 = \frac{3}{16\pi}\left(\frac{k_B T_0}{v^{2/3}\gamma}\right)^3 \left(\frac{\Delta H_d}{R_G T_0}\right)^2 \left[\ln\left(\frac{\Delta T_m}{b}\right) + \ln\left(\frac{AV}{2}\right)\right]. \tag{9}$$

A plot of $(T_0/\Delta T_m)^2$ versus $\ln(\Delta T_m/b)$ based on the MSZW data should give a straight line, the slope and intercept of which permit determination of γ and A, respectively. Equation (9) is consistent with the result developed by Shiau and Wu [21] in determination of γ and A from the MSZW data.

3. Experimental Methods

Deionized water, L-glycine (>99%, Alfa Aesar) and L-arginine (>98%, ACROS) were used to prepare the desired supersaturated solution for the specified impurity concentration. The experimental apparatus adopted by Shiau and Lu [18] was used in the study of nucleation, which consists of a 250 mL crystallizer equipped with a magnetic stirrer at a constant stirring rate of 350 rpm, immersed in programmable thermostatic water. A turbidity probe with a near-infrared source (Crystal Eyes manufactured by HEL limited, Hertford, UK) was used to detect the nucleation event.

The solubility of L-glycine in water from 303 K to 318 K was measured in this work. The solubility measurements indicated that the solubility of L-glycine in water was nearly not influenced by the presence of L-arginine ranging from $C_{im} = 0$–10 kg arginine/m^3 solution, which corresponds to 0–0.02 mol arginine/mol glycine. The measured solubility of L-glycine in water was consistent with the solubility data reported by Park et al. [36]. In terms of the van't Hoff equation for the measured solubility, one obtains $\Delta H_d = 10.2$ kJ/mol with $C_{eq}(303\ K) = 215$ kg/m^3 and $C_{eq}(318\ K) = 261$ kg/m^3 in this work.

For the induction time and MSZW experiments, a 200 mL aqueous L-glycine solution $(V = 2 \times 10^{-4}\ m^3)$ at the desired concentration was held at 5 K above the saturated temperature for 20 min to ensure a complete dissolution at the beginning of the experiments, which was also confirmed by the turbidity measurement. In the induction time experiments, the induction time and lag time data were measured by rapidly cooling the supersaturated solution at various supersaturations to 303 K. In the MSZW experiments, MSZW data were measured by cooling the solution saturated at 318 K with different constant cooling rates. Each run was carried out at least three times at each condition for the solubility, the induction time, and the MSZW measurements.

Although L-glycine can be crystallized in different polymorphs, including α-form, β-form and γ-form, α-form is usually obtained from pure aqueous L-glycine solutions [24–30]. In this work, the final dried crystals at the end of the experiments were analyzed using Raman spectroscopy (P/N LSI-DP2-785 Dimension-P2 System, 785 nm, manufactured by Lambda Solutions, INC., Seattle, WA, USA) to validate the polymorph of the L-glycine crystals. By comparing with the Raman spectra of α-form crystals reported by Murli et al. [37], it was found that α-form L-glycine crystals were formed from aqueous L-glycine solutions in this work for various supersaturations without and with the presence of L-arginine impurity. Figure 2 shows some Raman spectra of the L-glycine crystals obtained in this work at $S = 1.07$ and $S = 1.12$ for $C_{im} = 0$ and $C_{im} = 10$ kg/m^3, respectively.

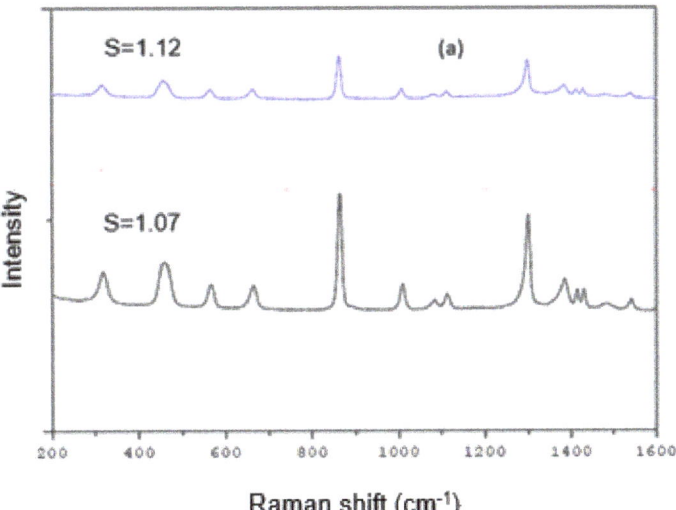

Figure 2. Cont.

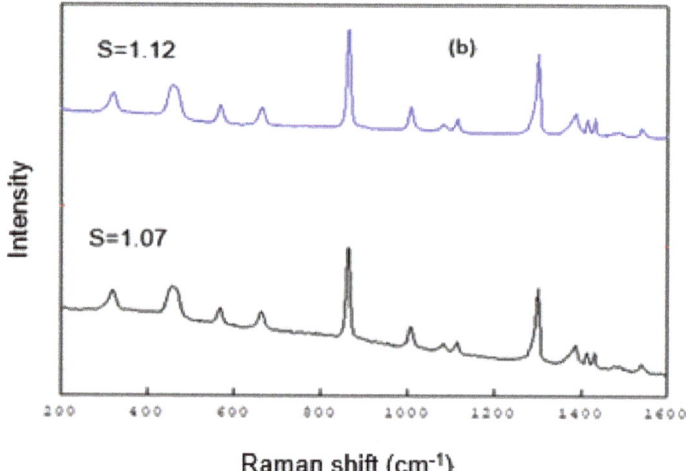

Figure 2. The Raman spectra of the produced L-glycine crystals at $S = 1.07$ and $S = 1.12$ for (**a**) $C_{im} = 0$ and (**b**) $C_{im} = 10$ kg/m³, respectively.

4. Results and Discussion

The induction time data of aqueous L-glycine solutions were measured for various supersaturations at 303 K in the presence of L-arginine for various impurity concentrations, C_{im}. The average induction times are listed in Table 1. The average lag times for the induction time data are listed in Table 2, which were measured based on $b \cong 0.038$ K/s adopted for cooling the heated supersaturated solution to the desired constant temperature. The lag time corresponds to the time required for the heated solution to be lowered to 303 K. Thus, as the temperature range ΔT_m increases, the lag time increases. The MSZW data of aqueous L-glycine solutions saturated at $T_0 = 318$ K were measured for various b in the presence of L-arginine for $C_{im} = 0$–10 kg/m³. The average MSZWs are listed in Table 3. Note that $M_w = 0.075$ kg/mol, $\rho_c = 1607$ kg/m³, and $v = 7.757 \times 10^{-29}$ m³ for L-glycine.

Table 1. The average induction times, t_i, in the induction time measurements for various impurity concentrations, C_{im}, and supersaturations, S, at 303 K. The standard deviations in the least significant digits are given in parentheses.

C_{im} (kg/m³)	t_i (×10² s)			
	$S = 1.07$	$S = 1.08$	$S = 1.10$	$S = 1.12$
0	27 (5.9)	14 (2.8)	8.3 (2.5)	4.4 (2.0)
2	62 (15)	33 (7.3)	16 (4.6)	8.2 (2.5)
5	107 (17)	48 (8.9)	23 (5.1)	12 (3.0)
10	154 (21)	62 (11)	31 (8.8)	16 (4.7)

Table 2. The average lag time, $\Delta T_m/b$, based on $b \cong 0.038$ K/s in the induction time measurements for various impurity concentrations, C_{im}, and supersaturations, S, at 303 K, where ΔT_m corresponds to the temperature range for a solution with concentration C_0 saturated at T_0 and cooled to 303 K. Note that $\Delta T_m = T_0 - 303$ K and $S = C_0/C_{eq}(303$ K$)$. The standard deviations in the least significant digits are given in parentheses.

C_{im} (kg/m³)	$\Delta T_m/b$ (s)			
	$S = 1.07$ ($\Delta T_m = 5.1$ K)	$S = 1.08$ ($\Delta T_m = 5.8$ K)	$S = 1.10$ ($\Delta T_m = 7.2$ K)	$S = 1.12$ ($\Delta T_m = 8.7$ K)
0	135 (14)	153 (16)	177 (16)	236 (19)
2	127 (12)	144 (13)	195 (17)	221 (17)
5	141 (13)	159 (15)	182 (16)	232 (17)
10	146 (15)	163 (14)	188 (18)	241 (22)

Table 3. The average MSZWs, ΔT_m, in the MSZW measurements for a solution saturated at $T_0 = 318$ K cooled at various impurity concentrations, C_{im}, and cooling rates. The standard deviations in the least significant digits are given in parentheses.

C_{im} (kg/m³)	ΔT_m (K)			
	$b = 0.00417$ K/s	$b = 0.00833$ K/s	$b = 0.01111$ K/s	$b = 0.01389$ K/s
0	6.9 (1.6)	8.5 (1.7)	9.1 (2.1)	9.9 (2.2)
2	8.4 (1.8)	10.3 (2.0)	11.7 (2.3)	12.2 (2.3)
5	9.7 (2.1)	12.2 (2.5)	13.8 (2.9)	14.4 (3.1)
10	11.5 (2.3)	13.9 (2.4)	16.1 (2.7)	18.8 (3.3)

Table 1 indicates that t_i increases significantly with increasing C_{im} for each S and decreases with increasing S for each C_{im}. Thus, L-arginine exerts a nucleation inhibition effect in aqueous L-glycine solutions, which increases with increasing C_{im}. Table 2 indicates that $\Delta T_m/b$, increases slightly with increasing S for each C_{im} and remains nearly independent of C_{im}. Note that ΔT_m corresponds to the temperature range for a solution saturated at T_0 cooled to 303 K, where T_0 increases with increasing S and remains nearly independent of C_{im}. For example, $\Delta T_m/b = 236$ s is quite significant compared with $t_i = 442$ s at $S = 1.12$ ($\Delta T_m = 8.7$ K) for $C_{im} = 0$. On the other hand, $\Delta T_m/b = 135$ s is negligible compared with $t_i = 2672$ s at $S = 1.07$ ($\Delta T_m = 5.1$ K) for $C_{im} = 0$.

Figure 3 shows plots of $\ln t_i$ against $\ln^2 S_m$ for each C_{im} according to Equation (6) based on the induction time data without consideration of the lag time. Figure 4 shows plots of $\ln(\Delta T_m/2b + t_i)$ against $\ln^2 S_m$ for each C_{im} according to Equation (5) based on the induction time data with consideration of the lag time. Calculated values of γ and A from the slope and intercept of the best-fit plots for each C_{im} are listed in Table 4. Note that the regression coefficient, R^2, with the lag time is generally greater than that without the lag time for each C_{im}, which indicates that Equation (5) with the lag time fits the induction time data better than Equation (6) without the lag time.

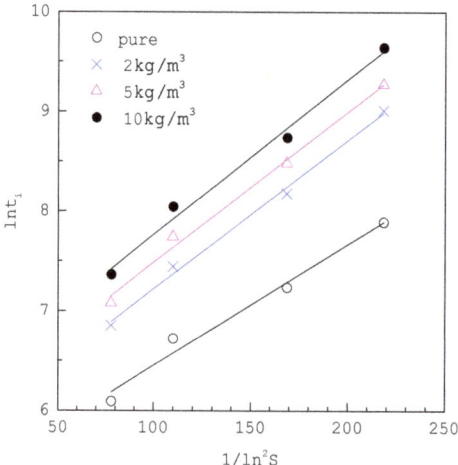

Figure 3. Plots of $\ln t_i$ against $\ln^2 S_m$ for various impurity concentrations, C_{im}, according to Equation (6) based on the induction time data without consideration of the lag time.

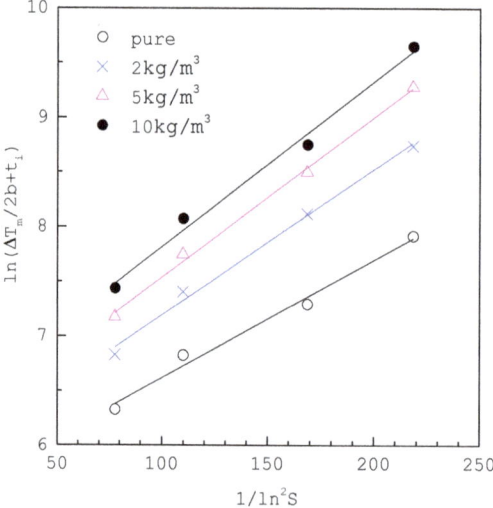

Figure 4. Plots of $\ln(\Delta T_m/2b + t_i)$ against $\ln^2 S_m$ for various impurity concentrations, C_{im}, according to Equation (5) based on the induction time data with consideration of the lag time.

Table 4. Calculated values of γ and A with the regression coefficients, R^2, based on the induction time data. The number before the slash represents the value without consideration of the lag time and the number after the slash represents the value with consideration of the lag time.

C_{im} (kg/m³)	γ (mJ/m²)	A (m⁻³s⁻¹)	R^2
0	2.07/1.99	26.5/19.7	0.981/0.987
2	2.17/2.13	16.9/14.3	0.991/0.994
5	2.22/2.20	12.7/11.6	0.993/0.997
10	2.24/2.22	10.0/9.1	0.989/0.990

As indicated in Table 4, one can note that the value of γ with the lag time, γ_{lag}, is lower by about 2% than that without the lag time, γ, while the value of A with the lag time,

A_{lag}, is lower by about 15% than that without the lag time, A. These findings are consistent with $\gamma_{lag} < \gamma$ and $A_{lag} < A$ derived in Supplementary Materials.

Table 3 indicates that ΔT_m increases with increasing C_{im} for each b and increases with increasing b for each C_{im}. Thus, as similar to the results from the induction time data, L-arginine exerts a nucleation inhibition effect in aqueous L-glycine solutions, which increases with increasing C_{im}. Figure 5 shows plots of $(T_0/\Delta T_m)^2$ against $\ln(\Delta T_m/b)$ for various C_{im} according to Equation (9) based on the MSZW data. Calculated values of γ and A from the slope and intercept of the best-fit plots for each C_{im} are listed in Table 5.

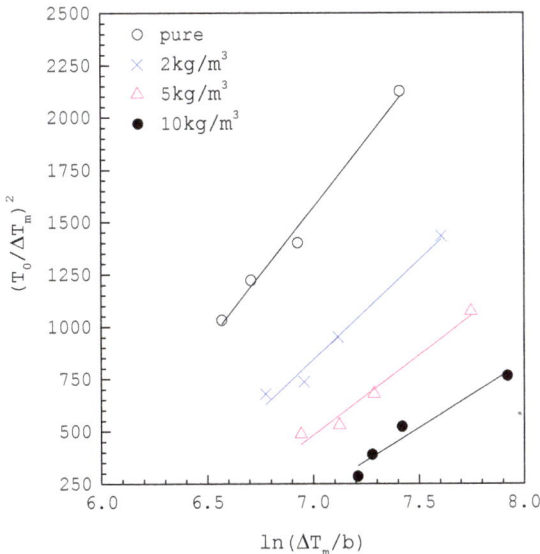

Figure 5. Plots of $(T_0/\Delta T_m)^2$ against $\ln(\Delta T_m/b)$ for various impurity concentrations, C_{im}, according to Equation (9) based on the MSZW data.

Table 5. Calculated values of γ and A with the regression coefficients, R^2, based on the MSZW data.

C_{im} (kg/m³)	γ (mJ/m²)	A (m⁻³s⁻¹)	R^2
0	2.13	30.9	0.990
2	2.36	22.1	0.981
5	2.54	17.2	0.975
10	2.71	12.6	0.953

The values of γ and A obtained from the MSZW data in Table 5 are consistent with those obtained from the induction time data in Table 4. They all indicate that, as C_{im} increases, γ increases slightly while A decreases quite significantly. For example, as C_{im} increases from 0 to 10 kg/m³, γ only increases slightly in the range of 10% to 30%, while A decreases significantly in the range of 50% to 60%. It is speculated that the presence of L-arginine in the aqueous L-glycine solution leads to some L-arginine molecules adsorbed on the nucleus surface of L-glycine, which suppresses nucleation and results in a higher γ compared to that without L-arginine adsorbed on the nucleus surface of L-glycine. On the other hand, the presence of L-arginine in the aqueous L-glycine solution suppresses nucleation and results in a lower A compared to that without L-arginine in the aqueous L-glycine solution. As the effects of L-arginine impurity on γ and A become more profound at a greater concentration of L-arginine impurity, a greater C_{im} results in a higher γ and a lower A. This trend is consistent with the finding reported by Heffernan et al. [8] for the nucleation of curcumin in propan-2-ol due to the presence of demethoxycurcumin and bisdemethoxycurcumin.

5. Conclusions

A model was proposed based on CNT to determine the nucleation parameters from both the induction and MSZW data. The unique feature is that the derivation of this model for both the induction and MSZW data is based on the same assumption that the nucleation point corresponds to the formation of a single primary nucleus in a supersaturated solution. This model results in two different equations. One is derived for the induction data while the other is derived for the MESZW data. The proposed model was applied to calculate the interfacial free energy and pre-exponential nucleation factor from both the induction time data and the MSZW data for the aqueous L-glycine solutions in the presence of L-arginine impurity. The results indicated that the values of interfacial free energy and pre-exponential nucleation factor obtained from the MSZW data are consistent with those obtained from the induction time data. The induction time data with consideration of the lag time lead to a lower interfacial free energy and a lower pre-exponential nucleation factor than those for the induction time data without consideration of the lag time. As the impurity concentration increases, the interfacial free energy increases slightly while the pre-exponential nucleation factor decreases quite significantly based on both the induction time and MSZW data.

Supplementary Materials: The following are available online at https://www.mdpi.com/article/10.3390/cryst11101226/s1, The derivation of $\gamma_{lag} < \gamma$ and $A_{lag} < A$.

Funding: This research was funded by Chang Gung Memorial Hospital (CMRPD2K0012) and Ministry of Science and Technology of Taiwan (MOST108-2221-E-182-034-MY2).

Data Availability Statement: Not applicable.

Acknowledgments: The author would like to thank Chang Gung Memorial Hospital (CMRPD2K0012) and Ministry of Science and Technology of Taiwan (MOST108-2221-E-182-034-MY2) for financial support of this research. The author also expresses his gratitude to Pin-Jhu Li and Dai-Rong Wu for their experimental work.

Conflicts of Interest: The authors declare no conflict of interest.

Notation

A = pre − exponential nucleation factor $(m^{-3}s^{-1})$
b = cooling rate (K/s)
C_0 = initial saturated concentration at T_0 (kg/m^3)
$C_{eq}(T)$ = saturated concentration at T (kg/m^3)
C_m = saturated concentration at T_m (kg/m^3)
C_{im} = concentration of impurity (kg/m^3)
J = nucleation rate $(m^{-3}s^{-1})$
J_0 = nucleation rate at $t = 0$ $(m^{-3}s^{-1})$
J_m = nucleation rate at t_m $(m^{-3}s^{-1})$
k_B = Boltzmann constant $(= 1.38 \times 10^{-23}$ J/K$)$
M_W = molar mass (kg/mol)
N = average number of expected nuclei $(-)$
N_A = Avogadro number $\left(= 6.02 \times 10^{23} \text{ mol}^{-1}\right)$
R_G = gas constant $\left(= 8.314 \text{ J mol}^{-1}\text{K}^{-1}\right)$
S = supersaturation $(-)$
S_m = supersaturation at t_m $(-)$
T = temperature (K)
T_0 = initial saturated temperature (K)
T_m = temperature at t_m (K)
t = time (s)

t_i = induction time (s)
t_m = time at the MSZW limit (s)
V = solution volume (m^3)

Greek Letters

ρ_c = crystal density (kg/m^3)
v = volume of the solute molecule (m^3)
γ = interfacial free energy (J/m^2)
ΔH_d = heat of dissolution (J/mole)
ΔT_m = MSZW (K)

References

1. Mullin, J.W. *Crystallization*, 3rd ed.; Butterworth-Heinemann: Oxford, UK, 1993.
2. Kashchiev, D. *Nucleation: Basic Theory with Applications*; Butterworth-Heinemann: Oxford, UK, 2000.
3. Kashchiev, D.; van Rosmalen, G.M. Review: Nucleation in solutions revisited. *Cryst. Res. Technol.* **2003**, *38*, 555–574. [CrossRef]
4. Dhanasekaran, P.; Srinivasan, P. Nucleation control, separation and bulk growth of metastable α-L-glutamic acid single crystals in the presence of L-tyrosine. *J. Cryst. Growth.* **2013**, *364*, 23–29. [CrossRef]
5. Peng, J.; Dong, Y.; Wang, L.; Li, L.L.; Li, W.; Feng, H. Effect of impurities on the solubility, metastable zone width, and nucleation kinetics of borax decahydrate. *Ind. Eng. Chem. Res.* **2014**, *53*, 12170–12178. [CrossRef]
6. Siepermann, C.A.P.; Huang, S.; Myerson, A.S. Nucleation inhibition of benzoic acid through solution complexation. *Cryst. Growth Des.* **2017**, *17*, 2646–2653. [CrossRef]
7. Yang, L.; Cao, J.; Luo, T. Effect of Mg^{2+}, Al^{3+}, and Fe^{3+} ions on crystallization of type α hemi-hydrated calcium sulfate under simulated conditions of hemi-hydrate process of phosphoric acid. *J. Cryst. Growth* **2018**, *486*, 30–37. [CrossRef]
8. Heffernan, C.; Ukrainczyk, M.; Zeglinski, J.; Hodnett, B.K.; Rasmuson, A.C. Influence of structurally related impurities on the crystal nucleation of curcumin. *Cryst. Growth Des.* **2018**, *18*, 4715–4723. [CrossRef]
9. Su, N.; Wang, Y.; Xiao, Y.; Lu, H.; Lou, Y.; Huang, J.; He, J.; Li, Y.; Hao, H. Mechanism of influence of organic impurity on crystallization of sodium sulfate. *Ind. Eng. Chem. Res.* **2018**, *57*, 1705–1713. [CrossRef]
10. Bodnar, K.; Hudson, S.P.; Rasmuson, A.C. Promotion of mefenamic acid nucleation by a surfactant additive, docusate sodium. *Cryst. Growth Des.* **2019**, *19*, 591–603. [CrossRef]
11. Keshavarz, L.; Steendam, R.R.E.; Blijlevens, M.A.R.; Pishnamazi, M.; Frawley, P.J. Influence of impurities on the solubility, nucleation, crystallization, and compressibility of paracetamol. *Cryst. Growth Des.* **2019**, *19*, 4193–4201. [CrossRef]
12. Huang, Y.; Lu, J.; Chen, H.; Du, W.; Wang, X. Effects of succinic acid and adipic acid on the metastable width of glutaric acid in acetic acid. *J. Cryst. Growth.* **2019**, *507*, 1–9. [CrossRef]
13. Chen, J.; Peng, J.; Wang, X.; Dong, Y.; Li, W. Effects of CO$_3^{2-}$ and OH$^-$ on the solubility, metastable zone width and nucleation kinetics of borax decahydrate. *R. Soc. Open Sci.* **2019**, *6*, 181862. [CrossRef]
14. Luo, M.; Liu, C.; Song, X.; Yu, J. Effects of Al$_2$(SO$_4$)$_3$ and K$_2$SiO$_3$ impurities on the crystallization of K$_2$SO$_4$ from aqueous solutions. *Cryst. Res. Technol.* **2021**, *56*, 2000052. [CrossRef]
15. Kubota, N. A new interpretation of metastable zone widths measured for unseeded solutions. *J. Cryst. Growth* **2008**, *310*, 629–634. [CrossRef]
16. Sangwal, K. Effects of impurities on the metastable zone width of solute-solvent systems. *J. Cryst. Growth* **2009**, *311*, 4050–4061. [CrossRef]
17. Shiau, L.D.; Lu, T.S. A model for determination of the interfacial energy from the measured metastable zone width by the polythermal method. *J. Cryst. Growth* **2014**, *402*, 267–272. [CrossRef]
18. Shiau, L.D.; Lu, T.S. A model for determination of the interfacial energy from the induction time or metastable zone width data based on turbidity measurements. *CrystEngComm* **2014**, *16*, 9743–9752. [CrossRef]
19. Yang, H.; Florence, A.J. Relating induction time and metastable zone width. *CrystEngComm* **2017**, *19*, 3966–3978. [CrossRef]
20. Li, Z.; Si, A.; Yan, Y.; Zhang, X.; Yang, H. Interaction of metastable zone width and induction time based on nucleation potential. *Ind. Eng. Chem. Res.* **2020**, *59*, 22597–22604.
21. Shiau, L.D.; Wu, D.R. Effect of L-valine impurity on the nucleation parameters of aqueous L-glutamic acid solutions from metastable zone width. *J. Cryst. Growth* **2020**, *546*, 125790. [CrossRef]
22. Chu, G.; Vilensky, R.; Vasilyev, G.; Martin, P.; Zhang, R.; Zussman, E. Structure evolution and drying dynamics in sliding cholesteric cellulose nanocrystals. *J. Phys. Chem. Lett.* **2018**, *9*, 1845–1851. [CrossRef]
23. Khadem, S.A.; Rey, A.D. Nucleation and growth of cholesteric collagen tactoids: A time-series statistical analysis based on integration of direct numerical simulation (DNS) and long short-term memory recurrent neural network (LSTM-RNN). *J. Colloid Interface Sci.* **2021**, *582*, 859–873. [CrossRef] [PubMed]
24. Towler, C.S.; Davey, R.J.; Lancaster, R.W.; Price, C.J. Impact of molecular speciation on crystal nucleation in polymorphic systems: the conundrum of γ glycine and molecular 'self-poisoning'. *J. Am. Chem. Soc.* **2004**, *126*, 13347–13353. [CrossRef] [PubMed]
25. Sun, X.; Garetz, B.A.; Myerson, A.S. Supersaturation and polarization dependence of polymorph control in the nonphotochemical laser-induced nucleation(NPLIN) of aqueous glycine solutions. *Cryst. Growth Des.* **2006**, *6*, 684–689. [CrossRef]

26. Bouchard, A.; Hofland, G.W.; Witkamp, G.J. Solubility of glycine polymorphs and recrystallization of β-glycine. *J. Chem. Eng. Data* **2007**, *52*, 1626–1629. [CrossRef]
27. Srinivasan, P.; Indirajith, R.; Gopalakrishnan, R. Growth and characterization of α and γ-glycine single crystals. *J. Cryst. Growth* **2011**, *318*, 762–767. [CrossRef]
28. Han, G.; Chow, P.S.; Tan, R.B.H. Direct comparison of α- and γ-glycine growth rates in acidic and basic solutions: New insights into glycine polymorphism. *Cryst. Growth Des.* **2012**, *12*, 2213–2220. [CrossRef]
29. Yani, Y.; Chow, P.S.; Tan, R.B.H. Glycine open dimers in solution: New insights into α-glycine nucleation and growth. *Cryst. Growth Des.* **2012**, *12*, 4771–4778. [CrossRef]
30. Devi, K.R.; Srinivasan, K. A novel approach to understand the nucleation kinetics of α and γ polymorphs of glycine from aqueous solution in the presence of a selective additive through charge compensation mechanism. *CrystEngComm* **2014**, *16*, 707–722. [CrossRef]
31. Shiau, L.D. Comparison of the interfacial energy and pre-exponential factor calculated from the induction time and metastable zone width data based on classical nucleation theory. *J. Cryst. Growth* **2016**, *450*, 50–55. [CrossRef]
32. Jiang, S.; ter Horst, J.H. Crystal nucleation rates from probability distributions of induction times. *Cryst. Growth Des.* **2011**, *11*, 256–261. [CrossRef]
33. Kadam, S.S.; Kramer, H.J.M.; ter Horst, J.H. Combination of a single primary nucleation event and secondary nucleation in crystallization processes. *Cryst. Growth Des.* **2011**, *11*, 1271–1277. [CrossRef]
34. Kadam, S.S.; Kulkarni, S.A.; Coloma Ribera, R.; Stankiewicz, A.I.; ter Horst, J.H. A new view on the metastable zone width during cooling crystallization. *Chem. Eng. Sci.* **2012**, *72*, 10–19. [CrossRef]
35. Green, D.W.; Perry, R.H. *Perry's Chemical Engineers' Handbook*, 8th ed.; McGraw Hill Book Company: New York, NY, USA, 2008.
36. Park, K.; Evans, J.M.B.; Myerson, A.S. Determination of Solubility of Polymorphs Using Differential Scanning Calorimetry. *Cryst. Growth Des.* **2003**, *3*, 991–995. [CrossRef]
37. Murli, C.; Thomas, S.; Venkateswaran, S.; Sharma, S.M. Raman spectroscopic investigation of α-glycine at different temperatures. *Phys. B* **2005**, *364*, 233–238. [CrossRef]

Article

Contact-Mediated Nucleation of Subcooled Droplets in Melt Emulsions: A Microfluidic Approach

Gina Kaysan [1], Alexander Rica [1], Gisela Guthausen [2,3] and Matthias Kind [1,*]

[1] Institute for Thermal Process Engineering, Karlsruhe Institute of Technology, 76131 Karlsruhe, Germany; gina.kaysan@kit.edu (G.K.); alexander.rica@gmail.com (A.R.)
[2] Institute for Mechanical Engineering and Mechanics, Karlsruhe Institute of Technology, 76131 Karlsruhe, Germany; gisela.guthausen@kit.edu
[3] Water Science and Technology, Engler-Bunte Institute, Karlsruhe Institute of Technology, 76131 Karlsruhe, Germany
* Correspondence: matthias.kind@kit.edu; Tel.: +49-721-608-42390

Abstract: The production of melt emulsions is mainly influenced by the crystallization step, as every single droplet needs to crystallize to obtain a stable product with a long shelf life. However, the crystallization of dispersed droplets requires high subcooling, resulting in a time, energy and cost intensive production processes. Contact-mediated nucleation (CMN) may be used to intensify the nucleation process, enabling crystallization at higher temperatures. It describes the successful inoculation of a subcooled liquid droplet by a crystalline particle. Surfactants are added to emulsions/suspensions for their stabilization against coalescence or aggregation. They cover the interface, lower the specific interfacial energy and form micelles in the continuous phase. It may be assumed that micelles and high concentrations of surfactant monomers in the continuous phase delay or even hinder CMN as the two reaction partners cannot get in touch. Experiments were carried out in a microfluidic chip, allowing for the controlled contact between a single subcooled liquid droplet and a single crystallized droplet. We were able to demonstrate the impact of the surfactant concentration on the CMN. Following an increase in the aqueous micelle concentrations, the time needed to inoculate the liquid droplet increased or CMN was prevented entirely.

Keywords: crystallization; microfluidic; contact-mediated nucleation; melt emulsion

1. Introduction

Emulsions are dispersions in which two mutually insoluble liquid substances are present. The dispersed droplet phase for oil-in-water emulsions is represented by oil and the continuous phase by water. According to Bancroft's rule [1], the type of emulsion is determined by the solubility of surface-active substances (e.g., emulsifiers). The phase in which the emulsifier is more soluble forms the continuous phase.

A thermodynamic description of emulsions is provided by Sharma et al. [2]. The stability of emulsions is influenced by different physical effects, such as Ostwald ripening, flocculation, aggregation, sedimentation (creaming), phase inversion and coalescence. The time scales in which these effects influence the stability of emulsions may vary profoundly [3].

One potential method of stabilizing emulsions for a longer time is the addition of emulsifiers. Emulsifiers are excipients with a characteristic molecular structure. According to the chemical structure of the surfactant, they accumulate at the interface of the two phases and decrease the specific interfacial energy [4].

Emulsifiers are amphiphilic molecules with a hydrophilic (head group) and a hydrophobic part (tail). The head is polar, while the tail usually consists of long, nonpolar hydrocarbon chains. The emulsifiers at the oil-water interface arrange themselves according to their affinity to the solvent, i.e., the head groups are on the aqueous side, whereas the tails reach into the oil phase [5]. The hydrophilic part of an emulsifier can be charged and classified as ionic (anionic, cationic), zwitterionic or nonionic [4].

The phase distribution of the emulsifier in thermodynamic equilibrium occurs at the equivalence of temperature, pressure and chemical potentials of the surfactant in both phases, at the interface and in the micelles [6]. A quantitative description of thermodynamic equilibrium between the oil and the water phase is possible using the Nernst partition coefficient [7]. Accordingly, the distribution coefficient at phase equilibrium can be calculated using the concentrations of the component of these two phases.

1.1. Contact-Mediated Nucleation

Contact-mediated nucleation (CMN) describes a mechanism of crystallization, triggered by means of contact between a liquid, subcooled droplet and an already crystallized droplet (= particle).

McClements et al. [8] found that the fraction of solidified droplets increased during the experimental time when an equivalent initial distribution of solidified particles and liquid droplets was provided for a quiescent subcooled n-hexadecane-water emulsion with emulsifier Tween®20 (diffusional motion only). This occurred despite the fact that subcooling should have prevented spontaneous nucleation. Measurements by nuclear magnetic resonance (NMR) showed that no crystallization occurred after 175 h within an emulsion with only liquid n-hexadecane droplets. McClements et al. [9] found that the higher the available surfactant concentration in the emulsion, the higher the final solids fraction of the droplets. Additionally, Dickinson et al. [10] determined an increase in the solid content relative to an increase in the emulsifier concentration in the continuous phase for the same material system. They estimated that one in 10^7 collisions between crystallized and liquid droplets resulted in CMN. Hindle et al. [11] found a steady increase in the fraction of solidified droplets of a subcooled n-hexadecane-water emulsion with emulsifier Tween®20. The increase occurred after crystallized n-hexadecane droplets were added. Complete crystallization of the dispersed phase occurred during the 15 days after the addition of crystallized droplets. They concluded that, according to their results, micelles could not mediate nucleation and any increases in the solid fraction were due to the contact between a liquid droplet and a solid particle. They also state that micelles may affect CMN.

Regarding experiments performed in the field of coalescence, Dudek et al. [12] determined longer coalescence times with increasing surfactant concentrations. This is contrary to the results described above, because increasing micellular numbers should improve the coalescence process as more dispersed phase can be transported from one droplet to another. Additionally, with increasing aqueous surfactant concentrations, higher energy barriers must be overcome to achieve direct contact between the two collision partners [13]. The latter is discussed in further detail later in this paper.

In order to enhance the contact between crystallized particles and liquid droplets, achieving a relative motion between each particle and droplet is advantageous. According to Vanapalli et al. [14], the relative motion between droplets and particles can be classified into orthokinetic (externally imposed velocity fields) or perikinetic (Brownian motion). In this work, the relative motion is governed by the orthokinetic mechanism due to the microfluidic setup and droplet sizes used.

In analogy to the coalescence theory, three external flow factors can influence CMN: contact time, contact force and collision frequency [15]. The contact time and contact force depend on further flow phenomena, such as the flattening of the film radius, film drainage and film rupture [16]. The surfactant may play an important role for CMN because it influences the specific interfacial energy and forms micelles in the continuous phase which may inhibit the contact of colliding partners.

1.2. Microfluidics

Microfluidic systems have become an important tool in emulsion research [17]. The application of microfluidics offers many advantages, for example, laminar flow conditions, high surface-to-volume ratios, small fluid volumes, the possibility of droplet manipulation, easy access for microscopical analysis, and excellent control over mass and heat transport. Microfluidics in crystallization research appears as a supporting platform for fundamental research into crystallization processes and crystal formation [18,19]. Hence, new possibilities have been made available to investigate processes within relevant time scales [20,21], for example, the investigation of CMN. Single droplet experiments are also used to investigate coalescence and to correlate specific interfacial energy to droplet stability [22–25].

1.3. Theoretical Description of Contact-Mediated Nucleation—An Approach

Similarities can be found between the modelling of coalescence processes and CMN [15]. So far, there are no established, specific theoretical descriptions of CMN in the literature. The process of coalescence takes places between liquid droplets or between gaseous bubbles. No phase change takes place upon coalescence. There is an asymmetry in the aggregate state in CMN, since one of the two contact partners is already crystallized and the other is a liquid, subcooled droplet. This results in the following general assumption of contact-mediated nucleation: the interface of the liquid droplet can take several states, from mobile to rigid, depending on the interfacial surfactant concentration, whereas the interface of the crystallized particle is only rigid.

According to [26], the interstitial film needs to fall below a critical thickness for coalescence to take place. Below that specific thickness, the Van der Waals attraction between two droplets approaching each other is stronger than any possible repulsive forces. Different mechanisms are detailed in the following passage, which could counteract the film thinning in the material system of an n-hexadecane–water emulsion used, stabilized with Tween®20.

The Derjaguin-Landau-Verwey-Overbeek (DLVO) theory [27,28] is a theoretical description of the stability of colloidal systems, such as emulsions, that account for attractive and repulsive forces. Regarding stability factors [29] larger than 1, repulsion dominates the interaction of two suspended particles. Electrostatic repulsion between colloids requires equally charged surfaces. Therefore, an increased charge of equal polarity between colloids could counteract film thinning. Dimitrova et al. [30] measured the forces between ferrofluidic suspension droplets in an aqueous solution stabilized with Tween®20. When the concentration of the emulsifier increased, the electrostatic repulsion also increased. The authors only found agreement with the DLVO theory for low concentrations of the emulsifier. The repulsive force measured at higher surfactant concentrations showed higher values at shorter distances than the predicted values calculated via the DLVO theory. In order to explain this discrepancy, the authors suspected a steric component of the repulsive force. This was thought to be caused by the presence of Tween®20 micelles between the droplets.

The influence of micelles in the continuous phase on the contact between colloids could be explained by the oscillating structural and depletion forces (OSF) [31]. These forces occur in the interstitial film when the film thickness decreases in the presence of small, dissolved entities in the film liquid, for example, micelles.

Taking OSF into account, Basheva et al. [13] derived a nonlinear relationship between the diameter of spherical micelles, the volume fraction of these micelles, the distance of the two colliding droplets and the interaction energy between the two droplets. This approach is used to estimate emulsion stability as a function of the number of micelles present in the solution. We calculated the interaction energy U_{total} analogously with the literature presented regarding the occurrence of OSF (Figure 1). The corresponding material system parameters are listed in Table 1. U_{total} was estimated by the sum of the oscillatory component (U_{osc}) and the Van der Waals interaction (U_{vdW}). A full description of the equations can be found in [13].

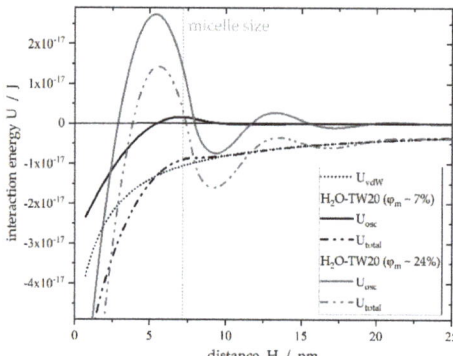

Figure 1. Simulated interaction energy of two spherical droplets with the distance H according to Trokhymchuk et al. [32] and Basheva et al. [13]. It is assumed that the film between the droplets contains hard, spherical micelles. Parameters used for calculation are summarized in Table 1. The volume fraction of the micelles in the continuous phase φ_m for a given aqueous surfactant concentration $\tilde{c}_{TW20}^{H2O}$ was approximated with a linear expression according to data obtained by Basheva et al. [13]: $\varphi_m = 1.7 \times 10^{-3} \, m^3 mol^{-1} \cdot \tilde{c}_{TW20}^{H2O}$.

Table 1. Parameters used for calculation of the total interaction energy U.

Parameter	Value
Micelle diameter of Tween®20 [13]	7.2×10^{-9} m
Radius n-hexadecane droplet in the microfluidic device	3×10^{-4} m
Temperature	290.35 K
Length of emulsifier's brush layer (determined by Avogardo, Version 1.2.0 [33])	1.2×10^{-9} m
Hamaker constant for an emulsion system [13]	4×10^{-21} J

Figure 1 indicates the possible influence of micelles in the aqueous phase because they are able to hinder or delay CMN by increasing the repulsive forces. As indicated, a particular volume density of micelles is necessary to achieve stability factors larger than 1.

We used a microfluidic setup for our studies to investigate the efficiency of CMN depending on the specific interfacial energy and micellar concentration in the continuous phase. We suggest increasing induction times and decreasing wetting by increasing aqueous micellular concentrations. Moreover, we assume that a link can be made between the coalescence theory and CMN to characterize the latter regarding its crystallization efficiency.

2. Materials and Methods

2.1. Materials Used

Microfluidic experiments were performed with n-hexadecane ($C_{16}H_{34}$, Hexadecane ReagentPlus®, Sigma-Aldrich, Schnelldorf, Germany, purity: 99%) as the dispersed phase and ultrapure water (OmniTap®, stakpure GmbH, Niederahr, Germany; electrical conductivity 0.057 μS cm^{-1}) as the continuous phase. The surfactant used in this study was Polyoxyethylen(20)-sorbitan-monolaurat (Tween®20, Merck KGaA, Darmstadt, Germany) in different concentrations.

2.2. Microfluidic Measurement Setup

A microfluidic system based on the type of continuous-flow emulsion-based droplet microfluidics was applied [34]. The multiple phase flow appeared as the so-called Taylor flow [35]. A characteristic sequence of liquid droplets was formed, separated by slugs of the continuous phase. These slugs were sections of the aqueous phase [36]. The droplets had an average droplet volume of around 25 nL and mean equivalent diameters of approximately 525 μm. The droplet volume was calculated according to [37].

The microfluidic chip acted as the central element of the setup (Figure 2). It consisted of a transparent polycarbonate plate (thickness 2 mm) into which several microchannels have been milled. The channels were 300 µm wide and 200 µm deep. Additionally, a channel for a temperature sensor was milled into the side of the chip. The channels received their characteristic rectangular cross-sectional area by bonding a thin 250 µm polycarbonate foil on top. In order to guarantee that wetting of n-hexadecane did not occur on the channel walls, the latter were hydrophilized according to [38]. The microfluidic chip was fixed on a water-tempered aluminum cooling block with two independently controlled temperature zones. The temperature of the microfluidic chip was measured with a temperature sensor (Pt 100, ES Electronic Sensor GmbH, Heilbronn, Germany) close to the location of the contact experiment. Crystallization processes were tracked with a high-speed camera (sCMOS pco.edge 5.5®, Excelitas PCO GmbH, Kelheim, Germany) connected to a stereo microscope (SZ61, OLYMPUS EUROPA® Se & Co. KG, Hamburg, Germany) with an integrated polarization filter. Two silicon wafers were installed directly beyond the microfluidic chip to support the polarization filter in highlighting crystalline structures. Volume flow rates of the continuous and dispersed phases were adjusted by a low-pressure injection pump system (Nemesys, CETONI GmbH, Korbußen, Germany). This syringe pump system was connected to the computer via a BASE120 base module.

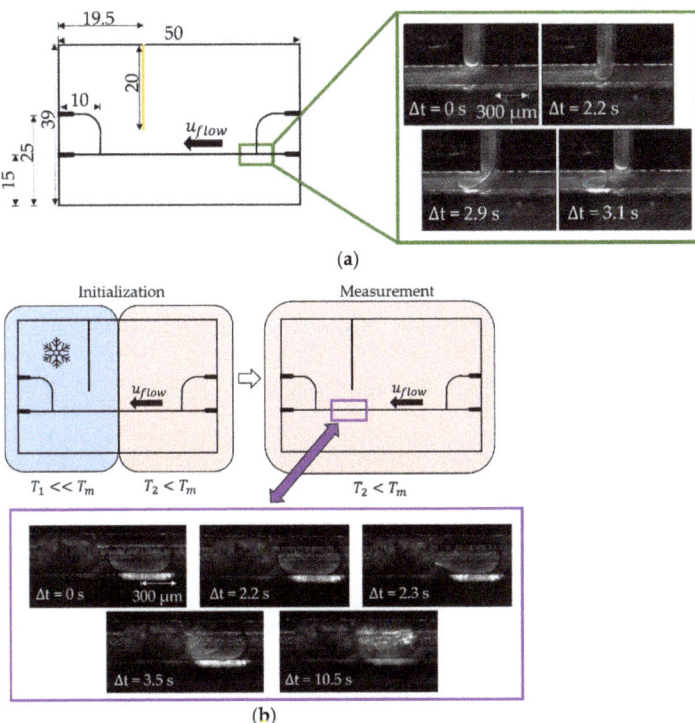

Figure 2. (a) Schematic view of the microfluidic chip with rectangular channels and sectional temperature regulation with scale (mm). The temperature sensor was inserted sideways into the chip through a fitting channel (yellow). The n-hexadecane droplets were formed at the T-junction (green frame) at a temperature above the melting point of n-hexadecane T_m. The produced droplets flowed along the channel in the direction indicated (u_{flow}). (b) Half of the chip was cooled (blue area) to a subcooling of around $\Delta T_1 = T_m - T_1 = 7.6$ K for spontaneous nucleation (initialization). The other half was kept below the melting point at $\Delta T_2 = 1.1$ K (red area); the droplets on this side retained liquid. The whole microfluidic chip was kept at ΔT_2 (red area) for the contact-mediated nucleation (CMN) measurement. The purple frame exemplarily displays a time-resolved CMN.

The microfluidic T-junction allowed for the formation of reproducible droplet sizes (Figure 2). At the beginning of the droplet formation process, the dispersed phase began to fill the channel cross-section almost completely, and when a critical proportion between droplet size and channel cross-sectional area was reached, droplets of the dispersed phase were formed.

The aim of the experiments was to investigate the CMN as a function of surfactant concentration. The temperature profile of the microfluidic chip during one collision followed a predefined protocol, as shown in Figure 3.

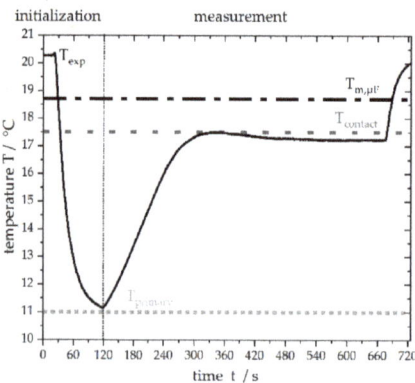

Figure 3. Temperature profile of the microfluidic chip T_{exp} over time measured by the incorporated temperature sensor. After initiating spontaneous, primary nucleation at a subcooling of $\Delta T_1 = T_{m,\mu F} - T_{primary} = 7.6$ K, the temperature was set to a subcooling of $\Delta T_2 = T_{m,\mu F} - T_{contact} = 1.1$ K for the observation of CMN. The melting point of n-hexadecane $T_{m,\mu F}$ was defined as 18.6 °C.

In order to initialize the experiment, droplets were formed at the T-junction and, as soon as both plates were covered with droplets, one side of the chip was cooled to around $T_{primary} = 11$ °C ($\Delta T_1 = 7.6$ K) to enforce spontaneous droplet crystallization. To avoid spontaneous crystallization during the experiments, both plates were thermostated at $T_{contact} = 17.5$ °C ($\Delta T_2 = 1.1$ K), which is below the melting point of n-hexadecane, and, as a result, the frozen particles did not thaw. Following this, the continuous phase was initiated, and the liquid droplet moved towards the solid particle. Volume flow rates ranging from 15 to 400 µL h^{-1} were applied. The solid particle was fixed on the channel walls as a result of crystallization. Due to the rectangular cross-sectional area of the channel and the round particle, the aqueous phase was still able to flow around the solid particle. This experimental design allowed for the controlled contact of two collision partners. A detailed experimental protocol is shown in Figure 4.

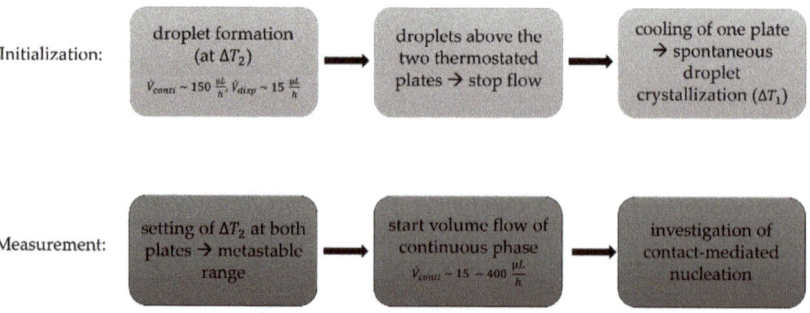

Figure 4. Experimental protocol for microfluidic experiments designed to investigate CMN.

A high-speed camera allowed time-resolved detection of the contact progress (for example, relative velocity of the droplet and the particle Δu and wetting evolution) and of the nucleation events. All collisions were tracked with the camera at a frame rate of 100 frames per second. The relative velocity Δu was determined by the distance travelled by the subcooled liquid droplet within a specific timeframe, while the solid particle had a fixed position (compare Figure 2).

2.3. Melting Point Measurements

In order to quantify the possible impact of the microfluidic system on the melting point of n-hexadecane, the presence of water or emulsifier was determined in the microchannel to $T_{m,\mu F}$ = 18.6 ± 0.2 °C. A broad range of melting points can be found in the literature, varying between 16.7 and 20.0 °C [39–44]. Our results are in good agreement with the available literature and, therefore, no impact of the setup or the presence of the water phase and surfactant was identified. In the following, $T_{m,\mu F}$ was used to calculate subcooling.

2.4. Specific Interfacial Energy Measurements

Droplet formation within the microfluidic device was possible with and without the usage of an additional surfactant. Therefore, it was not possible to estimate the specific interfacial energy and, consequently, the droplet surface coverage achieved by the emulsifier directly during the microfluidic experiment. However, surface coverage may play an important role for CMN as interfacial energy is known to greatly influence coalescence [45].

Measurements of the specific interfacial energy were obtained via the pendant drop method (OCA 25, DataPhysics Instruments GmbH, Filderstadt, Germany) to quantify the time needed by the surfactants to cover the liquid-liquid interface completely. A syringe with an outer diameter of 0.91 mm was used for the generation of the droplet, and the temperature was set to 20 ± 0.2 °C. Measurements without surfactant resulted in specific interfacial energies of 48.6 ± 0.5 mN·m^{-1}, which is in good agreement with the specific interfacial energies described in the literature ([46]: 43.16 mN·m^{-1}, [47]: 47.0 mN·m^{-1}). Time-resolved specific interfacial energies were considered to characterize the adsorption process of the surfactant to the liquid-liquid interface and to outline any dependencies of the surfactant concentration in the aqueous or oil phase (Tables 2 and 3, Figure 4).

Table 2. Specific interfacial energies approximately 4 h after interface formation for different surfactant concentrations dissolved in the continuous phase ($\tilde{c}^{H2O}_{TW20}(t=0)$).

$\tilde{c}^{H2O}_{TW20}(t=0)$/mol m^{-3}	γ_{LL}/mN m^{-1}
8.2	3.5 ± 0.3
16.6	2.9 ± 0.2

Table 3. Specific interfacial energies about 4 h after interface formation for different surfactant concentrations added at the beginning to the dispersed phase ($\tilde{c}^{hex}_{TW20}(t=0)$).

$\tilde{c}^{hex}_{TW20}(t=0)$/mol m^{-3}	γ_{LL}/mN m^{-1}
6.4	1.9 ± 0.4
12.9	0.9 ± 0.3

As Figure 5 and Tables 2 and 3 indicate, the phase in which the surfactant is dissolved at the beginning of the experiment ($t = 0$) plays a major role. The convolution of the surfactant distribution between the aqueous and oil phases will be discussed in Section 3. The measurements were obtained in a time frame of approximately 4 h (data not shown), without reaching a constant specific interfacial energy. The microfluidic experiments were performed within around 0.2 h. For $t < 0.2$ h, the final distribution of the surfactant between the continuous phase, dispersed phase and liquid-liquid interface was not reached according to the specific interfacial energy measurements. It should also be mentioned that the specific interfacial energy tension reduced from around 48 mN·m^{-1} to values between approximately 1 and 4 N m^{-1} within the first seconds after droplet formation. This suggests that most surfactants adsorbed to the interface shortly after droplet formation and any further changes were only due to the ad- and desorption of a smaller number of molecules.

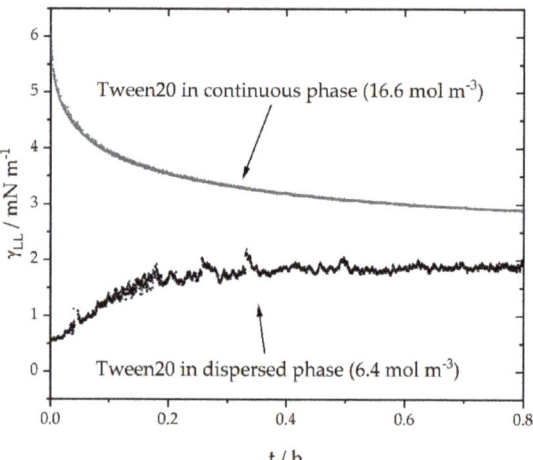

Figure 5. Two exemplary time-resolved specific interfacial energy (γ_{LL}) measurements with surfactant supported in either the aqueous, continuous ($\bar{c}_{TW20}^{H2O}(t=0) = 16.6$ mol·m^{-3}) or dispersed oil phase ($\bar{c}_{TW20}^{hex}(t=0) = 6.4$ mol·m^{-3}).

The increasing specific interfacial energy for systems in which the emulsifier was previously dissolved in the dispersed phase, shows the convective transport of the emulsifier from the interface to the surrounding continuous phase. As soon as thermodynamic equilibrium would be reached, the specific interfacial energies should be the same, without any differences regarding initial surfactant concentration gradients between the aqueous and oil phase, providing there is enough surfactant to completely cover the interface.

When transferring these findings to an idealistic model, we assumed different, instationary surfactant distributions between the dispersed and continuous phase, depending on the initial surfactant concentration (Figure 6). We assumed that the depicted surfactant distribution represented the situation during our microfluidic experiments.

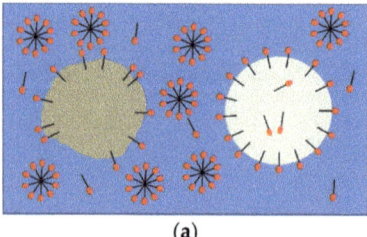 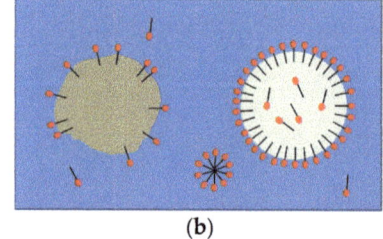

(a)   (b)

Figure 6. (a) Schematic system description when the surfactant was initially dissolved in the continuous phase (water). The subcooled droplet is presented in a light color, the solid particle is shown as a brown sphere. (b) Schematic system description when the surfactant was initially dissolved in the dispersed phase (n-hexadecane). Both systems had not yet reached the thermodynamic equilibrium.

Within the experimental time range, a higher emulsifier concentration at the interface can be assumed for systems where the surfactant is dissolved in the oil phase at $t = 0$ due to lower specific interfacial energies (Figures 5 and 6b). In addition, the concentration of the surfactant affects the specific interfacial energy and, thus, the emulsifier concentration at the interface across relevant time scales. Decreasing specific interfacial energies with increasing surfactant concentration (even above the critical micellular concentration: CMC) are also described in the literature [12,48]. During the microfluidic experiments, we assumed that, when the surfactant was initially supported in the water phase, more micelles were present in the aqueous phase compared to the number of micelles when the surfactant was initially dissolved in the oil phase.

2.5. Wetting

Different types of wetting between the liquid droplet and the solid particle were observed (Figure 7). The efficiency of CMN can be obtained by taking the wetting angle φ of the two collision partners into account.

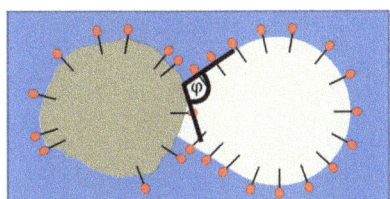

Figure 7. Definition of the wetting angle φ between a liquid, subcooled droplet and a solid particle. Both contact partners are stabilized with surfactant.

In this work, four different forms of contact are identified by their corresponding wetting angle and their efficiency in initiating nucleation. A brief overview is provided in Table 4.

Table 4. Classification of CMN based on the wetting angle φ. The differentiation is made based on whether nucleation occurs or not. Additionally, a schematic abstraction and the experimental observation are shown. In the abstraction, solid structures are light brown (left particle), and liquid droplets (right) are visualized in white. For the experimental illustration, solid structures are either light white, in cases where crystallization had just taken place, or dark gray. The liquid droplet volume is represented by the transparent parts. We assumed that solid–liquid interfaces are partially composed of surfactant. In the case of $\varphi = 0°$, hug, the red circle highlights the part of the liquid droplet that has already crossed the solid particle and appears directly behind the latter. The first two rows illustrate that contact occurred when the surfactant was supported in the continuous phase; the last two rows show wetting angles for the initial surfactant support in the dispersed phase.

Wetting Angle	Nucleation?	Wetting?	Abstraction	Experimental
$\varphi = 180°$, blank	no	no		
$\varphi = 0°$, hug	yes	not initially		

Table 4. *Cont.*

Wetting Angle	Nucleation?	Wetting?	Abstraction	Experimental
$0° < \varphi < 180°$	yes	yes		
$\varphi = 0°$	yes	yes		

Crystallization did not occur when the wetting angle was 180°, and we assume that no wetting occurred either. When the surfactant was dissolved in the aqueous phase, crystallization was visible with increasing relative velocities and a decreasing aqueous surfactant concentration, but not immediately after the first contact. The liquid droplet even surrounded the solid particle at certain points (Table 4, $\varphi = 0°$, hug, red circle).

3. Results

3.1. Surfactant Distribution between the Water and Oil Phase and the Liquid-Liquid Interface

The CMN may be influenced by the surface coverage of the droplet and particle with emulsifier. The influence of the specific interfacial energy on coalescence has already been discussed in the literature (e.g., [45]). In addition, the appearance of micelles may play an important role for the successful inoculation of the subcooled droplet. Due to the above-mentioned reasons, it is of great importance to estimate the surfactant distribution between the aqueous continuous and the dispersed oil phase.

In order to determine whether a diffusion process of surfactant from the water to the oil phase occurs, that is driven by concentration gradients due to different chemical potentials, NMR measurements were performed using a 400 MHz spectrometer (Avance Neo, Bruker BioSpin GmbH). Samples were prepared with surfactant concentrations up to $\tilde{c}_{TW20}^{H2O} = 180$ mol·m^{-3}. N-hexadecane was added at two different mass ratios to the continuous phase: [50:50] and [80:20] (water:n-hexadecane). After the addition of n-hexadecane, the samples were mixed with a stirring fish at 700 rpm for 2 min. Droplets between 10 and 500 µm were produced. The samples were left untouched for one week. During this time, a phase separation occurred, and three different phases became visible. Large n-hexadecane droplets were found at the top. The intermediate phase consisted of smaller n-hexadecane droplets and the bottom phase was aqueous. The aqueous phase was carefully separated with a syringe to determine the surfactant distribution after one week. This sample was analyzed by ^{1}H NMR spectroscopy (Figure 8).

The measured samples are in good agreement with the calibration curve up to a concentration of $\tilde{c}_{TW20}^{H2O}(t=0) = 90$ mol·m^{-3}. Therefore, up to the specified concentration, no measurable number of surfactants dissolved in the n-hexadecane phase. This supports our hypothesis that surfactant diffusion due to concentration gradients and, thus, chemical potential differences between the continuous and dispersed phase did not play a major role within our experimental time frame ($\tilde{c}_{TW20}^{H2O}(t=0) < 50$ mol·m^{-3}). Moreover, we did not notice an influence of the volume ratios of water and oil on the distribution of the surfactant. The concentration of surfactant at the interface was too small to be measurable with NMR. Pendant drop measurements must be considered to obtain information about the interfacial surfactant concentration (Figure 5, Tables 2 and 3).

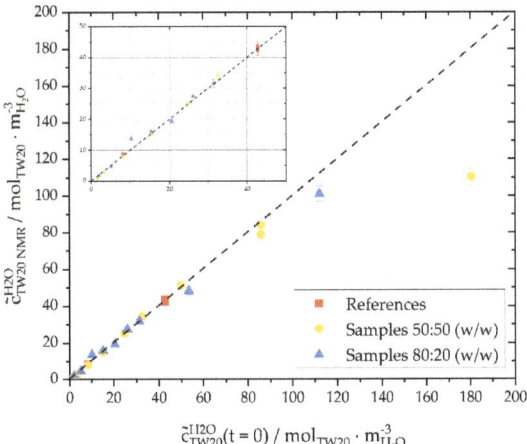

Figure 8. The concentration of Tween®20 in the water phase determined via NMR spectra $\tilde{c}_{TW20,NMR}^{H2O}$ as a function of the concentration of the weighed-in components $\tilde{c}_{TW20}^{H2O}(t=0)$ of samples with a 50:50 or 80:20 mass ratio of the continuous phase to the dispersed phase (n-hexadecane). In addition, reference samples of Tween®20 in ultrapure water with defined concentrations are displayed (red). The samples were measured at 20 °C. Inset: Zoom-in to concentrations up to $\tilde{c}_{TW20}^{H2O}(t=0) = 50$ mol·m^{-3}. The relative measurement error of the spectrometer was 2%.

The measured surfactant concentrations, for initial concentrations higher than $\tilde{c}_{TW20}^{H2O}(t=0) = 90$ mol·m^{-3}, differ from the reference calibration line. This may indicate that the molecular solubility limit of Tween®20 in the continuous phase had been exceeded. The information on the solubility of Tween®20 in water in the literature spans a certain range although a concrete solubility limit could not be found. Data sheets [49,50] mention the solubility limit at emulsifier concentrations of $c_{TW20}^{H2O} = 2 \times 10^{-3}$ g·L^{-1} and 100 g·L^{-1}. These correspond to the following values: $\tilde{c}_{TW20}^{H2O} = 2 \times 10^{-9}$ mol·m^{-3} and $1 \cdot 10^{-4}$ mol·m^{-3}, respectively, with $\rho_{H2O}(20\,C) = 998.2$ kg·m^{-3}. These values are much lower than the emulsifier concentrations investigated and used in the present series of experiments. Pollard et al. [51] found Tween®20 to be 'extremely soluble' in water and other solvents. In their experiments, Tween®20 was completely solvable in water up to 671 g·L^{-1} at 20 °C ($\tilde{c}_{TW20}^{H2O} = 546$ mol·m^{-3}), and the authors did not investigate further additions of emulsifier. The maximum solubility was not reached at this concentration. No phase separation of Tween®20 in water, that was measurable by eye, could be identified in any of our experiments. Nonetheless, a nonvisible phase separation could have taken place. Another possible explanation may be the diffusion of surfactant from the aqueous to the oil phase. With an increasing aqueous surfactant concentration, the difference of the chemical potential between Tween®20 in water and n-hexadecane increased, leading to a faster rate of surfactant diffusion into the oil phase.

Long-term experiments were performed over a period of 43 days to support the findings of the slow equilibrium adjustment of the surfactant between the continuous and the dispersed phase (Figure 9) and to estimate the distribution coefficient K_{TW20} of Tween®20 between water and n-hexadecane. K_{TW20} is defined as

$$K_{TW20} = \frac{\tilde{c}_{TW20}^{hex}}{\tilde{c}_{TW20}^{H2O}}, \qquad (1)$$

where $\tilde{c}_{TW20}^{hex}$ represents the molar concentration of Tween®20 in the n-hexadecane phase and $\tilde{c}_{TW20}^{H2O}$ represents the molar concentration in the water phase.

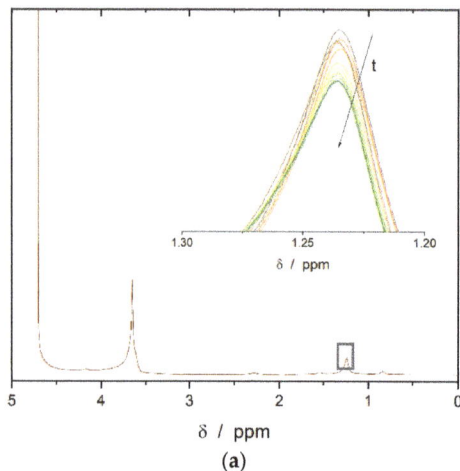
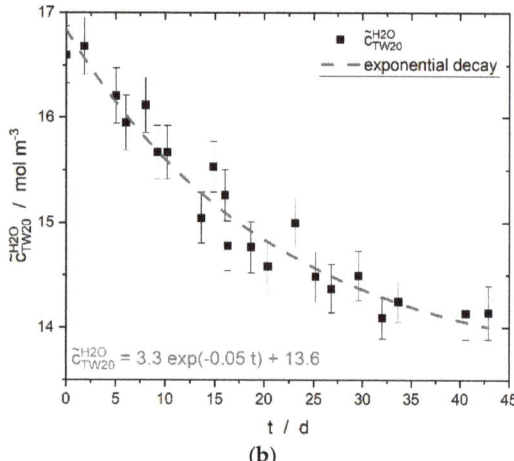

Figure 9. (a) ^{1}H spectra of the aqueous phase, where the surfactant Tween®20 was dissolved at the beginning of the experiment. The water peak appears at about 4.8 ppm; all other visible peaks are related to Tween®20. Inset: time-resolved evolution of the CH$_2$ peak of Tween®20. (b) Time-resolved measurements of Tween®20 concentration in the continuous phase over a period of 43 days, determined by NMR spectroscopy. The volume ratio of water and n-hexadecane was 50:50 and the interface had an area of 13.2 mm^2. Additionally, the fitted exponential decay curve is visible as a dashed line.

A 300 MHz spectrometer (nanobay, Bruker BioSpin GmbH) was used to determine the distribution coefficient. The measured sample consisted of two separated phases, namely of pure n-hexadecane, and water with an initial total surfactant concentration of 16.6 mol·m^{-3}. The phase volumes were both 300 µL. The sample was positioned in the sensitive measuring range so that only the water phase was measured. The CH$_2$ peak was used for evaluation to avoid overlapping and the influence of the water peak on the peak of the ethoxylate group of the emulsifier at 3.7 ppm. Modelling of the experimental data with an exponential decay of the first order (Figure 9b) led to an equilibrium concentration of $\tilde{c}_{TW20}^{H2O}(t \to \infty) = 13.6$ mol·m^{-3}. The value of $\tilde{c}_{TW20}^{hex}(t \to \infty)$ for the extrapolated equilibrium state was calculated as $\tilde{c}_{TW20}^{hex}(t \to \infty) = 3.0$ mol·m^{-3} by mass balance calculations of the two-phase system observed. The distribution coefficient (Equation (1)) was found to be $K_{TW20} = 0.22$ for an initial aqueous surfactant concentration of $\tilde{c}_{TW20}^{H2O}(t = 0) = 16.6$ mol·m^{-3}, similar volumes of water and n-hexadecane, and an interfacial area of approximately 13.2 mm^2.

3.2. Effect of Tween®20 Distribution on Contact-Mediated Nucleation

As has been mentioned previously, the contact time and contact force required for nucleation may be influenced by the surfactant concentrations present in the water and oil phases and at the separating interface. This may be due to potentially prolonged film drainage times or the rearrangement of the surfactant at the interface when the two collision partners approach one another. Regarding industrial processes, wetting effects should be minimized to exclude partial coalescence and achieve comparable droplet/particle size distributions before and after the crystallization step. Partial coalescence describes when two particles are connected by a small bridge, but they do not form a single, spherical particle.

Within the experimental microfluidic time frame of around 10 min, the specific interfacial energy depended on the distribution of the surfactant between the continuous and dispersed phase (Figure 5). Different concentrations of Tween®20 in the dispersed and continuous phase were used for the collision experiments to outline the effect of the specific interfacial energy and the influence of micelles on the CMN. The number of micelles or single molecules per unit volume in the continuous phase and the specific interfacial energy may influence the contact force needed for crystallization. Here, we used the relative velocity between the subcooled droplet and the crystalline particle as an indirect indicator of contact force because the latter cannot be measured directly. This velocity difference will be transformed into a contact force per contact area and thus into, for example, a contact pressure. The contact time is not limited in the microfluidic chip because the continuous phase constantly pushes the liquid droplet towards the solid particle. The induction time needed for crystallization is presented in Section 3.4.

The distribution of the surfactant cannot be determined within the microfluidic setup itself. An empirical approach was used to estimate the distribution. A diffusion-controlled model was applied for the adsorption of Tween®20 at the water–n-hexadecane interface. According to [52], this assumption is valid because the droplets had an average size larger than 10 µm. Tween®20 was assumed to be a highly surface-active molecule. Our calculations showed that a complete coverage of the interface was reached after at least 6 min for $\tilde{c}_{TW20}^{H2O} = 8 \times 10^{-3}$ mol·m^{-3} and 1.2×10^{-5} s for $\tilde{c}_{TW20}^{H2O} = 42.8$ mol·m^{-3}. The maximum surface loading was reached within less than 1 s ($t_{max} \ll 1$ s) for all initial surfactant concentrations higher than the CMC. For the calculations according to [53], a maximal surface loading Γ_{max} of 1.79×10^{-6} mol·m^{-2} (calculated according to [54], assuming a surfactant monolayer at the liquid-liquid interface and a surface pressure equal to zero; 100% of interface covered with surfactant) and a diffusion coefficient of the single surfactant molecules in water of $D_{TW20}^{H2O} = 2.6 \times 10^{-10}$ m^2·s^{-1} (own measurements via NMR diffusion measurements [400 MHz spectrometer, Avance Neo, Bruker BioSpin GmbH]) was used. Assuming a fully loaded interface before starting the collision experiments, the continuous surfactant concentration still represents more than 99% of the initial bulk concentration of TW20 ($\tilde{c}_{TW20}^{H2O}(t=0) \sim \tilde{c}_{TW20}^{H2O}(t)$, $\tilde{c}_{TW20}^{H2O} > CMC$). Ad- and desorption processes from the interface into the n-hexadecane droplet phase can also be neglected within the experimental time range, as is shown in Figure 8.

When Tween®20 was initially dissolved in n-hexadecane, measurements of the specific interfacial energy suggested fast adsorption to the interface followed by a desorption to the continuous phase. The diffusion coefficient of Tween®20 in n-hexadecane was measured as $D_{TW20}^{hex} \sim 2.0 \times 10^{-10}$ m^2·s^{-1} for surfactant concentrations between 0.2 and 360 mol·m^{-3}, which is comparable to D_{TW20}^{H2O}. We therefore assume that only single molecules are present in the n-hexadecane droplet and no inverse micelles are formed. Calculations of the maximum time needed for complete interfacial coverage, according to [53], revealed that a complete coverage can be assumed in periods significantly shorter than 10 min. We, furthermore, assume that no or only a few micelles are formed within the continuous phase until contact crystallization occurs when Tween®20 was initially dissolved in the oil phase. This hypothesis will be verified later, because micelles and single molecules in the continuous phase hinder contact crystallization tremendously. Nucleation occurred for all experiments where the surfactant was initially dissolved in the dispersed phase.

The wetting angle φ (Figure 10a,b) and crystallization probability P_c (Figure 10c) depended on the surfactant's concentration and the distribution of the surfactant throughout the system. The crystallization probability P_c represents the ratio between the crystallized droplets and the total number of droplets.

There was no initial nucleation visible by eye at first contact for any experiments in which the surfactant was added to the continuous phase alone at the beginning of the experiment. Higher relative velocities and, thus, higher shear rates and higher contact forces were necessary (e.g., Figure 10a, 7 · CMC) to trigger nucleation. The liquid droplet often surrounded the particle before nucleation occurred. We assume that the new liquid–liquid droplet surface can be refilled faster with an increasing surfactant concentration in the continuous phase. This, consequently, prevents the direct contact of crystalline structures with subcooled liquids and, therefore, hinders crystallization. Less free surfactant is available at a lower surfactant concentration and the new interface cannot be completely covered quickly enough. Consequently, crystallization occurs.

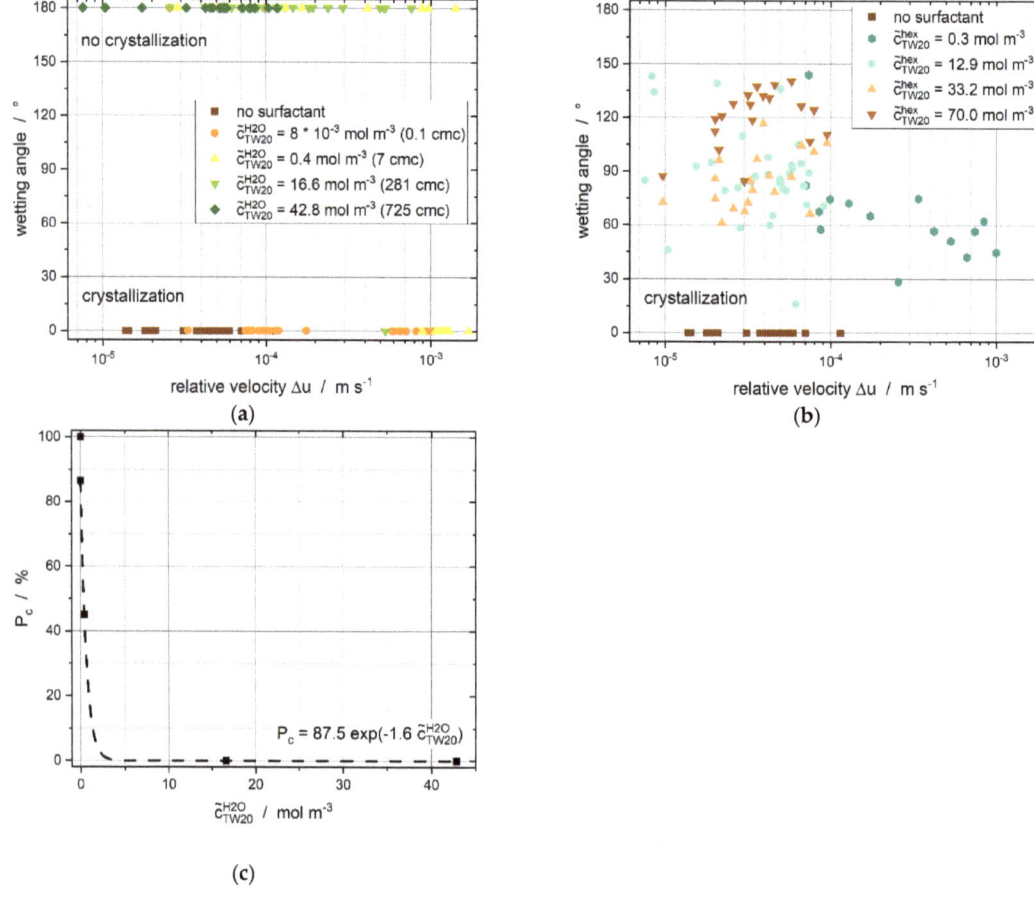

Figure 10. (a) Wetting angle as a function of the relative velocity (difference of velocities between droplets and particles) for different surfactant concentrations. Tween®20 was added to the continuous phase. (b) Wetting angle as a function of relative velocities for different surfactant concentrations as Tween®20 was dissolved in the dispersed phase. All experiments led to crystallization, but differences in the contact form were found according to the surfactant concentration. (c) Crystallization probability P_c as Tween®20 was added to the continuous phase for relative velocities ranging from 10^{-5} up to 2×10^{-3} m s^{-1}.

Only the two lower surfactant concentrations ($\tilde{c}_{TW20}^{H2O}$ = 8 × 10^{-3} and 0.4 mol·m^{-3}) led to nucleation within the experimental observation period of around 60 s. The crystallization probability $P(c)$ (Figure 10c) decays exponentially as a function of the initial aqueous surfactant concentration. We, thus, assume that micelles in the continuous phase hinder nucleation, as the highest crystallization probability was reached at $\tilde{c}_{TW20}^{H2O}$ = 8 × 10^{-3} mol·m^{-3}, which is below the CMC of Tween®20 in water (CMC_{TW20} = 0.059 mol·m^{-3} [55]). Experiments without the surfactant were performed to prove this hypothesis. Crystallization occurred in all the experiments without the emulsifier. This confirms our assumption that micelles and increasing monomer concentrations in the continuous phase can weaken or even prevent CMN.

When the surfactant was initially added to the dispersed phase alone, all contacts resulted in nucleation, independent of the surfactant concentration and relative velocity. A differentiation can be made according to the presented wetting angles (Figure 10b).

The surfactant's equilibrium distribution between the two phases was investigated to determine the dominant factor in CMN. The phase composition is provided by K_{TW20}, which was acquired from the long-term spectroscopic measurements (Figure 9). If the influence of the micelles is dominant, no crystallization should take place, whereas if the emulsifier in the dispersed phase and at the interface has a stronger influence, crystallization should take place in all collisions (Figure 11).

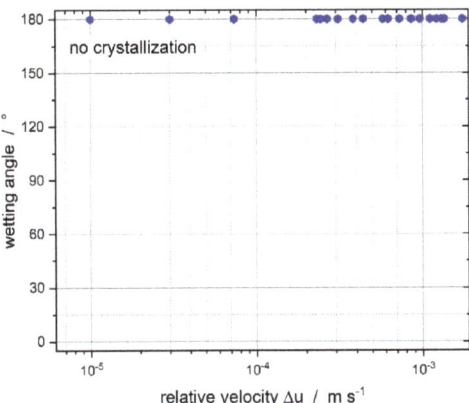

Figure 11. Wetting angle as a function of the relative droplet velocity for the equilibrium surfactant distribution between the continuous and dispersed phase, determined from long-term spectroscopy measurements (compare Figure 9).

Successful nucleation was not observed for relative velocities up to 1.7×10^{-3} m s^{-1}. This is a clear indication of the prevention of CMN by the presence of micelles in the continuous phase. Bera et al. [56] did not identify the occurrence of coalescence when the surfactant concentration was above the CMC, which is in good agreement with our results as we did not detect any CMN when the surfactant concentration was above $7 \cdot$ CMC in the continuous phase.

3.3. Formation of Liquid Bridges before Contact-Mediated Nucleation

A formation of liquid bridges between the liquid droplet and the solid particle was detected during some experiments, when the surfactant was initially dissolved in the oil phase. Regarding the desirable separation of the two reaction partners after the collision in industrial processes, small or no liquid bridges are required. Otherwise, the shelf life of the product would be reduced, or the product properties may change due to partial coalescence. Separation after collision could not be achieved for the experiment setup used due to the limitations presented by the experimental execution.

The size of the liquid bridge, which was formed between droplet and particle, changed as a function of the surfactant concentration in the dispersed phase (Figure 12). A very clear decrease in the mean size of the liquid bridge is visible. A significant difference between the population means and the population variances (Levene's test) was observed at a level of 0.05 by means of an analysis of variance of the experimental data. Nonetheless, the Tukey's post hoc test showed no significant difference between the two data sets of $\tilde{c}_{TW20}^{hex} = 12.9$ and 33.2 mol·m^{-3} at a level of 0.05.

A decrease in the diameter of the liquid bridge with an increasing emulsifier concentration is also described by Nowak et al. [57] for two coalescing droplets. Since the addition of surfactant causes decreasing specific interfacial energies, smaller driving forces are needed for coalescence and, hence, for CMN in the experiments presented here. The specific interfacial energy has a gradient on the droplet surface during the CMN due to the movement of surfactants at the interface. Consequently, Marangoni flow [58] develops. This allows for a homogeneous surfactant distribution at the interface. Thus, there are two effects promoting small liquid bridges at higher surfactant concentrations. Firstly, since there is more emulsifier in the dispersed phase, diffusion-limited transport to the interface is faster. Secondly, the gradient on the surface is greater because more emulsifier is present at the interface and, thus, the Marangoni flow is stronger. Higher surfactant concentrations also allow for a faster refilling of the interface when there is a concentration gradient between the continuous and dispersed phase, resulting in the desorption of surfactant molecules from the interface to the continuous phase. Chesters [16] hypothesizes that coalescence is favored for low viscosities of the dispersed phase. This may also promote the formation of larger liquid bridges at smaller surfactant concentrations and, thus, lower dispersed phase viscosities. Regarding industrial processes, smaller liquid bridges or even no bridges are favorable to avoid partial coalescence, coalescence or agglomeration so as to maintain the product quality.

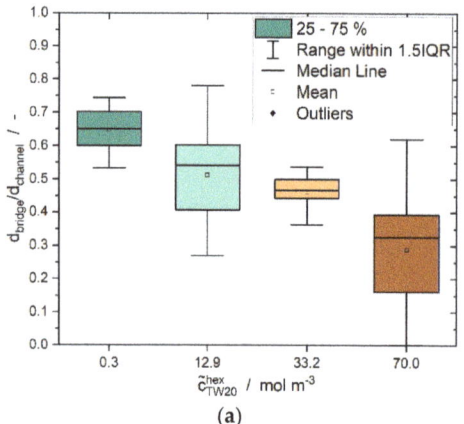
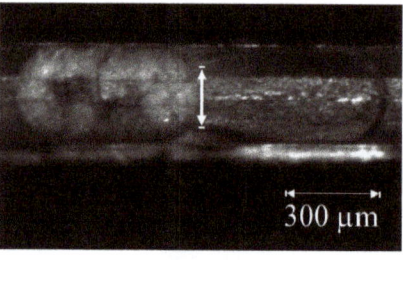

Figure 12. (**a**) Dimensionless diameter of the liquid bridge between the solid particle and the liquid droplet shortly after formation for all relative velocities tested. The emulsifier was dissolved in the dispersed phase at different concentrations at the beginning of the experiment. (**b**) Determination of the diameter of the liquid bridge formed during CMN between a solid and liquid n-hexadecane droplet in a microfluidic channel. The images used for the determination of the diameter of the bridge were taken directly after contact of the droplets, with the maximum error between the moment of contact and the picture shot being $t = 0.01$ s.

3.4. Induction Time

Stirred vessels are used with wide shear rate distributions for industrial melt emulsion production and storage. Melt emulsification is a top-down approach that can produce suspensions with μm-sized particles, while overcoming the disadvantages of the energy- and time-consuming wet-milling process [59,60]. The contact time is inversely proportional to the shear rate [16], therefore, it is important to know the required induction time t_{ind} to trigger crystallization. Once t_{ind} is obtained, it can be used to optimize process flows. For the collision experiments, t_{ind} was determined as a function of the surfactant concentration in either the continuous or dispersed phase (Figure 13). Furthermore, t_{ind} is defined as the time between the first visible contact and the detection of the first crystal.

When the experimental aqueous concentration of the surfactant was found to be above the CMC micelles were detected in the continuous phase with a volume fraction >0.1% (Figure 13, i: water) and the induction time was up to ten times higher than without or with very few micelles and aqueous single molecules (Figure 13, i: n-hexadecane). Without any surfactant, the induction time ranged from 0.1 to 0.4 s (data not shown), which highlights the crystallization-impeding effect of the aqueous emulsifier micelles or single molecules. A wide range of induction times measured is apparent. Aqueous surfactant concentrations of 16.6 mol·m^{-3} (φ_m ~2.8%) and 42.8 mol·m^{-3} (φ_m ~7.1%) are not shown because crystallization did not occur within 60 s.

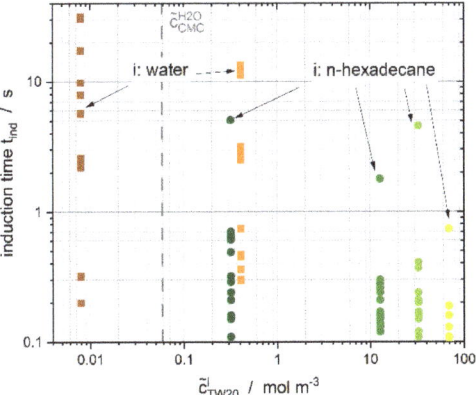

Figure 13. Induction time t_{ind}, defined as the delay between nucleation and the first visible contact between the droplet and particle, as a function of surfactant concentration in either dispersed or continuous phase for relative velocities ranging from 6×10^{-6} up to 4×10^{-3} m s^{-1}.

According to [61–63], we calculated the theoretical coalescence times for two n-hexadecane droplets with a radius r in water without any surfactant. The theoretical coalescence times ranged from 0.3 s to 1.1 s at room temperature $\left(\eta = 30.3 \text{ mPa s}, \rho_{hex} = 773 \text{ kg m}^{-3}, \rho_{water} = 998 \text{ kg m}^{-3}, \gamma_{LL} = 47 \text{ mN m}^{-1}, r = 184 \text{ μm}\right)$. Adding surfactant to the system increased the theoretical coalescence times by up to 43 s ($\gamma_{LL} \sim 4$ mN m^{-1}). Taboada et al. [25] measured coalescence times ranging from 10 s up to more than 30 min for different emulsifiers in single droplet experiments with water as the continuous phase. Regarding nonionic surfactants, Leister et al. [64] obtained coalescence times between 5 and around 100 s. The measurements from Taboada et al. [25] and Leister et al. [64] are within the same range as our calculated theoretical coalescence times. We expect induction times to be within the mentioned range, providing that the surfactant is dissolved in water, which is in good agreement with our experimental data (Figure 13). Moreover, the theoretical coalescence time without any surfactant and the determined experimental induction time are in the same range. Dudek et al. [12] described increasing mean coalescence times and the increasing distributional width of the coalescence times with increasing surfactant concentrations. We, therefore, assume that mechanisms that prevent coalescence also hinder crystallization, for example, an increasing surfactant concentration in the continuous phase.

When Tween®20 was dissolved in the dispersed phase during our experiments, induction times were comparable to the theoretical coalescence times without any surfactant. This highlights the influence of surfactant within the continuous phase. With increasing surfactant concentration in the oil phase (and, thus, decreasing specific interfacial energies), slightly shorter induction times were measured, and the span of the induction times decreased.

4. Discussion

The results presented show that higher relative velocities and, thus, higher contact forces are needed to ensure crystallization if the surfactant concentration in the continuous aqueous phase increases. Regarding aqueous surfactant concentrations of Tween®20 higher than $\tilde{c}_{TW20}^{H2O} = 23$ mol·m^{-3} (aqueous micellular volume fraction φ_m~3.8%), an oscillatory, repulsive force by micelles is considered likely, which is shown by calculations according to [13], where OSF are considered (Figure 1). Connecting this repulsive force with the Van der Waals forces, an aqueous surfactant concentration higher than around 90 mol·m^{-3} (φ_m ~15%) would show an energy barrier that must be overcome by contact force. Crystallization should take place for all aqueous surfactant concentrations presented here because the stability factor is smaller than 1 and attractive forces should dominate. This hypothesis cannot be verified by the experimental results. From all the continuous surfactant concentrations tested above the CMC, only 7 · CMC shows CMN at relative velocities higher than 8×10^{-4} m s^{-1}. Oscillatory forces are not expected at this surfactant concentration in the aqueous phase, and crystallization should take place even for low contact forces (=low relative velocities).

Consequently, in addition to the volume fraction of micelles, their stability also plays a major role. Christov et al. [65] concluded from atomic force microscopy measurements that Tween®20 micelles are significantly more unstable when exposed to hydrodynamic shear than, for example, micelles from Brij 35. The micelles dissolve into single molecules even in cases where only a small force is applied, therefore, a considerably larger number of single molecules are displaced from the gap between the droplet and the particle, which results in a greater time requirement for nucleation.

The single aqueous surfactant molecules can also occupy newly formed interfaces. The droplet spread around the particle and increased its interface for relative velocities higher than 7.5×10^{-5} m s^{-1} and $\tilde{c}_{TW20}^{H2O} > 8 \times 10^{-3}$ mol·m^{-3}. This effect was only observed when Tween®20 was dissolved in the continuous, aqueous phase and became more dominant with increasing relative velocities. Depending on the freely available, single molecule concentration, this newly formed interface is unlikely to be occupied fast enough and, thus, nucleation took place. Higher relative velocities then triggered crystallization. One possible reason is the increasing contact force relative to increasing relative velocities. Moreover, emulsifier could be detached from the droplet/particle surfaces due to the increased shear forces, resulting in an unoccupied interface.

Furthermore, pH and conductivity measurements (data not shown) revealed an electrical loading for the surfactants' head groups. Although nonionic emulsifiers were used by Dudek et al. [12] amongst others, a negative surface charge occurred in their experiments due to the deprotonated hydroxide groups of the emulsifiers' head groups. They attribute their observation of longer mean coalescence times to a stronger rejection of the head groups at an increased emulsifier concentration. An increase in the negative charge of oil-in-water emulsions with an increasing Tween®20 concentration has also been reported by Hsu et al. [66]. The electrostatic repulsion could explain the smaller bridge diameters (Figure 12), increased surfactant concentration in the oil phase and longer induction times (Figure 13) with an increasing continuous surfactant concentration because a higher number of surfactant molecules leads to increasing repulsion. Accordingly, the specific interfacial energy plays less of a role compared to the micelles. This hypothesis is further supported by the fact that the induction times for the experiments with surfactant dissolved in the dispersed phase are very similar to those determined without any surfactant.

It is well-known that, for biological systems, nonpolar substances attract each other in water due to hydrophobic interactions, although the mechanism behind this is not yet fully understood (e.g., [67]). In the literature, attraction was evidenced within distances ranging from 100 to 6500 Å [68,69] and even up to around 3.5 mm for rough (super)hydrophobic surfaces [70]. Nonetheless, any hydrophobic interactions between n-hexadecane particles and droplets are attenuated by the micelles to a greater degree in the continuous phase than by emulsifiers at the interface. The micelles seem to shield the two reaction partners from each other. When no micelles or only very few single molecules are present in the continuous phase, the hydrophobic interactions can be detected by the liquid bridge formation.

Krawczyk et al. [71] state that the rupture of the liquid film between two colliding partners is greatly affected by the phase in which the surfactant is dissolved. This is due to different surfactant transportation mechanisms to the interface. Surfactants dissolved in the continuous phase can delay film rupture, because the surfactant must travel from the film perimeter into the film center, where the lowest interfacial surfactant concentration occurs. This additional flux can influence the film rupture. Emulsifiers that are dissolved in the oil phase are required to travel shorter distances and are, therefore, more efficient in equalizing local specific interfacial energy gradients (as long as the film radius << the droplet radius).

The authors describe similar behavior of systems with surfactant in the dispersed phase and systems without any surfactant. This can also be seen in our measured induction times, which are similar for systems without emulsifier and with surfactants dissolved in the dispersed oil phase. Comparable behavior may be explained by Bancroft's rule [1], which describes a stable emulsion as a system in which the emulsifier is preferentially dissolved in the continuous phase.

Another possible reason for limited nucleation could be the absence of a monolayer at the liquid–liquid interface and the presence of a multilayer or even micelles that attach to the interface (e.g., [72]). This additional shield could prevent or decrease the inoculation efficiency of the CMN.

5. Conclusions

Considering all our experimental results, we found that the variation of the locus of the initial dissolution and the concentration of the emulsifier had a significant influence on the contact form observed and on the efficiency of the CMN. Importantly, the aqueous surfactant concentration and the relative velocity between the droplet and the particle significantly impacted the CMN in the microfluidic system.

We were able to show that increasing surfactant concentration in the dispersed oil phase can trigger nucleation. In applying these results to industrial melt emulsion production, the processes after collision and crystallization must be considered as pivotal for influencing the particle size distribution and, thus, product properties. The higher the contact area is, the higher the probability that partial coalescence occurs, and, in the case of temperature fluctuations during, for example, transportation, coalescence may occur. Therefore, based on the observations of this study, dissolution of the water-soluble emulsifier in the hydrophobic dispersed phase prior to the experiment can be useful to trigger CMN. Partial coalescence could also be decreased with high micellular concentrations in the continuous phase, but, at the same time, CMN also becomes hindered.

Further experiments will test the newly stated hypothesis that the coalescence theory can be transferred to CMN. The emulsifier, for example, will be varied, and ionic emulsifiers will be used to investigate a possible ionic repulsion of the head groups. Furthermore, a differential pressure sensor will be connected to the microfluidic setup to calculate the contact force needed for crystallization and to determine the dependency of the contact force on the relative velocity and surfactant concentration. Moreover, experiments that take place in a new microfluidic device containing a larger liquid reservoir, where droplets and particle can freely collide without the geometric restrictions of a single channel, are planned. Similar experimental setups are described in the literature for coalescence-time experiments [56,73].

Author Contributions: Conceptualization, G.K., M.K.; methodology, G.K.; software, G.K., A.R., G.G.; validation, G.K., A.R., G.G., M.K.; formal analysis, G.K., A.R.; investigation, G.K., A.R.; data curation, G.K., G.G., M.K.; writing—original draft preparation, G.K.; writing—review and editing, G.G., M.K.; visualization, G.K., A.R.; supervision, M.K. All authors have read and agreed to the published version of the manuscript.

Funding: This research received no external funding.

Institutional Review Board Statement: Not applicable.

Informed Consent Statement: Not applicable.

Data Availability Statement: The data and methods used in the research are presented in sufficient detail in the document for other researchers to replicate the work.

Acknowledgments: We acknowledge support from the KIT-Publication Fund of the Karlsruhe Institute of Technology. We thank the Deutsche Forschungsgesellschaft for the substantial financial contribution in the form of NMR instrumentation and access to the instrumental facility Pro^{2}NMR. Special thanks to Nico Leister and Jasmin Reiner from the Institute of Process Engineering in Life Sciences, Chair I: Food Process Engineering, who permitted the surface tension measurements and engaged in helpful discussions. Additionally, this work would not have been possible without the support of the workshop employees at the Institute of Thermal Process Engineering, especially Max Renaud. Anisa Schütze, who is a student, also performed valuable experiments for this article.

Conflicts of Interest: The authors declare no conflict of interest.

References

1. Bancroft, W.D. The Theory of Emulsification, V. *J. Phys. Chem.* **1913**, *17*, 501–519. [CrossRef]
2. Sharma, M.K.; Shah, D.O. Introduction to Macro- and Microemulsions. In *Proceedings of the ACS Symposium Series*; American Chemical Society (ACS): Washington, DC, USA, 1985; Volume 28, pp. 1–18.
3. Tadros, T.F. Emulsion Formation, Stability, and Rheology. In *Emulsion Formation and Stability*; Tadros, T.F., Ed.; Wiley-VCH Verlag GmbH & Co. KGaA: Weinheim, Germany, 2013; pp. 1–75.
4. Rosen, M.J.; Kunjappu, J.T. *Surfactants and Interfacial Phenomena*; John Wiley & Sons: Hoboken, NJ, USA, 2012; pp. 1–38.
5. Lauth, G.J.; Kowalczyk, J. *Einführung in die Physik und Chemie der Grenzflächen und Kolloide*; Springer Spektrum: Heidelberg/Berlin, Germany, 2016; pp. 53–89.
6. Callen, H.B. *Thermodynamics And An Introduction To Thermostatics*; John Wiley & Sons: New York, NY, USA, 1985; pp. 35–58.
7. Nernst, W. Ueber die Berechnung chemischer Gleichgewichte aus thermischen Messungen. *Nachr. Ges. Wiss. Göttingen Math. Phys. Kl.* **1906**, *1906*, 1–40.
8. McClements, J.D.; Dickinson, E.; Povey, M. Crystallization in hydrocarbon-in-water emulsions containing a mixture of solid and liquid droplets. *Chem. Phys. Lett.* **1990**, *172*, 449–452. [CrossRef]
9. McClements, D.J.; Dungan, S.R. Effect of Colloidal Interactions on the Rate of Interdroplet Heterogeneous Nucleation in Oil-in-Water Emulsions. *J. Colloid Interface Sci.* **1997**, *186*, 17–28. [CrossRef] [PubMed]
10. Dickinson, E.; Kruizenga, F.-J.; Povey, M.J.; van der Molen, M. Crystallization in oil-in-water emulsions containing liquid and solid droplets. *Colloids Surf. A Physicochem. Eng. Asp.* **1993**, *81*, 273–279. [CrossRef]
11. Hindle, S.A.; Povey, M.J.W.; Smith, K. Kinetics of Crystallization in n-Hexadecane and Cocoa Butter Oil-in-Water Emulsions Accounting for Droplet Collision-Mediated Nucleation. *J. Colloid Interface Sci.* **2000**, *232*, 370–380. [CrossRef]
12. Dudek, M.; Fernandes, D.; Helno Herø, E.; Øye, G. Microfluidic method for determining drop-drop coalescence and contact times in flow. *Colloids Surf. A Physicochem. Eng. Asp.* **2019**, *586*, 124265. [CrossRef]
13. Basheva, E.S.; Kralchevsky, P.A.; Danov, K.D.; Ananthapadmanabhan, K.P.; Lips, A. The colloid structural forces as a tool for particle characterization and control of dispersion stability. *Phys. Chem. Chem. Phys.* **2007**, *9*, 5183–5198. [CrossRef] [PubMed]
14. Vanapalli, S.A.; Coupland, J.N. Orthokinetic Stability of Food Emulsions. In *Food Emulsions*; Friberg, S., Ed.; CRC Press: Boca Raton, FL, USA, 2003; pp. 327–352.
15. Kaysan, G.; Schork, N.; Herberger, S.; Guthausen, G.; Kind, M. Contact-mediated nucleation in melt emulsions investigated by Rheo-NMR. *Magn. Reson. Chem.* **2021**. [CrossRef]
16. Chesters, A. The modelling of coalescence processes in fluid-liquid dispersions: A review of current understanding. *Chem. Eng. Res. Des.* **1991**, *69*, 259–270.
17. Bremond, N.; Bibette, J. Exploring emulsion science with microfluidics. *Soft Matter* **2012**, *8*, 10549–10559. [CrossRef]
18. Puigmartí-Luis, J. Microfluidic platforms: A mainstream technology for the preparation of crystals. *Chem. Soc. Rev.* **2013**, *43*, 2253–2271. [CrossRef] [PubMed]
19. Whitesides, G.M. The origins and the future of microfluidics. *Nature* **2006**, *442*, 368–373. [CrossRef] [PubMed]
20. Muijlwijk, K.; Berton-Carabin, C.; Schroën, K. Cross-flow microfluidic emulsification from a food perspective. *Trends Food Sci. Technol.* **2016**, *49*, 51–63. [CrossRef]
21. Schroën, K.; Bliznyuk, O.; Muijlwijk, K.; Sahin, S.; Berton-Carabin, C.C. Microfluidic emulsification devices: From micrometer insights to large-scale food emulsion production. *Curr. Opin. Food Sci.* **2015**, *3*, 33–40. [CrossRef]
22. Won, J.Y.; Krägel, J.; Makievski, A.V.; Javadi, A.; Gochev, G.; Loglio, G.; Pandolfini, P.; Leser, M.E.; Gehin-Delval, C.; Miller, R. Drop and bubble micro manipulator (DBMM)—A unique tool for mimicking processes in foams and emulsions. *Colloids Surf. A Physicochem. Eng. Asp.* **2013**, *441*, 807–814. [CrossRef]
23. Dickinson, E.; Murray, B.S.; Stainsby, G. Coalescence stability of emulsion-sized droplets at a planar oil–water interface and the relationship to protein film surface rheology. *J. Chem. Soc. Faraday Trans. 1 Phys. Chem. Condens. Phases* **1988**, *84*, 871–883. [CrossRef]
24. Neumann, S.M.; van der Schaaf, U.S.; Karbstein, H.P. Investigations on the relationship between interfacial and single droplet experiments to describe instability mechanisms in double emulsions. *Colloids Surf. A Physicochem. Eng. Asp.* **2018**, *553*, 464–471. [CrossRef]

25. Taboada, M.; Leister, N.; Karbstein, H.P.; Gaukel, V. Influence of the Emulsifier System on Breakup and Coalescence of Oil Droplets during Atomization of Oil-In-Water Emulsions. *ChemEngineering* **2020**, *4*, 47. [CrossRef]
26. Shinnar, R.; Church, J.M. Statistical Theories of Turbulence in Predicting Particle Size in Agitated Dispersions. *Ind. Eng. Chem.* **1960**, *52*, 253–256. [CrossRef]
27. Verwey, E.J.W. Theory of the stability of lyophobic colloids. *J. Phys. Colloid Chem.* **1947**, *51*, 631–636. [CrossRef]
28. Derjaguin, B.; Landau, L. Theory of the stability of strongly charged lyophobic sols and of the adhesion of strongly charged particles in solutions of electrolytes. *Prog. Surf. Sci.* **1993**, *43*, 30–59. [CrossRef]
29. Fuchs, N. Über die Stabilität und Aufladung der Aerosole. *Eur. Phys. J. A* **1934**, *89*, 736–743. [CrossRef]
30. Dimitrova, T.D.; Leal-Calderon, F. Forces between Emulsion Droplets Stabilized with Tween 20 and Proteins. *Langmuir* **1999**, *15*, 8813–8821. [CrossRef]
31. Israelachvili, J.N. *Intermolecular and Surface Forces*, 3rd ed.; Academic Press: Burlington, MA, USA, 2011; pp. 191–499.
32. Trokhymchuk, A.; Henderson, D.; Nikolov, A.; Wasan, D.T. A Simple Calculation of Structural and Depletion Forces for Fluids/Suspensions Confined in a Film. *Langmuir* **2001**, *17*, 4940–4947. [CrossRef]
33. Hanwell, M.D.; Curtis, D.E.; Lonie, D.C.; Vandermeersch, T.; Zurek, E.; Hutchison, G.R. Avogadro: An advanced semantic chemical editor, visualization, and analysis platform. *J. Cheminform.* **2012**, *4*, 17. [CrossRef] [PubMed]
34. Chou, W.-L.; Lee, P.-Y.; Yang, C.-L.; Huang, W.-Y.; Lin, Y.-S. Recent Advances in Applications of Droplet Microfluidics. *Micromachines* **2015**, *6*, 1249–1271. [CrossRef]
35. Angeli, P.; Gavriilidis, A. Hydrodynamics of Taylor flow in small channels: A Review. *Proc. Inst. Mech. Eng. Part C: J. Mech. Eng. Sci.* **2008**, *222*, 737–751. [CrossRef]
36. Baroud, C.N.; Gallaire, F.; Dangla, R. Dynamics of microfluidic droplets. *Lab Chip* **2010**, *10*, 2032–2045. [CrossRef]
37. Musterd, M.; van Steijn, V.; Kleijn, C.R.; Kreutzer, M.T. Calculating the volume of elongated bubbles and droplets in microchannels from a top view image. *RSC Adv.* **2015**, *5*, 16042–16049. [CrossRef]
38. Selzer, D.; Spiegel, B.; Kind, M. A Generic Polycarbonate Based Microfluidic Tool to Study Crystal Nucleation in Microdroplets. *J. Cryst. Process. Technol.* **2018**, *8*, 1–17. [CrossRef]
39. McClements, D.; Dungan, S.; German, J.; Simoneau, C.; Kinsella, J. Droplet Size and Emulsifier Type Affect Crystallization and Melting of Hydrocarbon-in-Water Emulsions. *J. Food Sci.* **1993**, *58*, 1148–1151. [CrossRef]
40. Vélez, C.; Khayet, M.; Ortiz de Zárate, J.M. Temperature-dependent thermal properties of solid/liquid phase change even-numbered n-alkanes: N-Hexadecane, n-octadecane and n-eicosane. *Appl. Energy* **2015**, *143*, 383–394. [CrossRef]
41. Zou, G.L.; Tan, Z.C.; Lan, X.Z.; Sun, L.X.; Zhang, T. Preparation and characterization of microencapsulated hexadecane used for thermal energy storage. *Chin. Chem. Lett.* **2004**, *15*, 729–732.
42. González, J.A.; Zawadzki, M.; Domanska, U. Thermodynamics of mixtures containing polycyclic aromatic hydrocarbons. *J. Mol. Liq.* **2008**, *143*, 134–140. [CrossRef]
43. Zhang, P.; Ma, Z.; Wang, R. An overview of phase change material slurries: MPCS and CHS. *Renew. Sustain. Energy Rev.* **2010**, *14*, 598–614. [CrossRef]
44. Spiegel, B.; Käfer, A.; Kind, M. Crystallization Behavior and Nucleation Kinetics of Organic Melt Droplets in a Microfluidic Device. *Cryst. Growth Des.* **2018**, *18*, 3307–3316. [CrossRef]
45. Schroën, K.; de Ruiter, J.; Berton-Carabin, C. The Importance of Interfacial Tension in Emulsification: Connecting Scaling Relations Used in Large Scale Preparation with Microfluidic Measurement Methods. *ChemEngineering* **2020**, *4*, 63. [CrossRef]
46. Hashimoto, M.; Garstecki, P.; Stone, H.A.; Whitesides, G.M. Interfacial instabilities in a microfluidic Hele-Shaw cell. *Soft Matter* **2008**, *4*, 1403–1413. [CrossRef] [PubMed]
47. van der Graaf, S.; Schroën, C.G.P.H.; van der Sman, R.G.M.; Boom, R.M. Influence of dynamic interfacial tension on droplet formation during membrane emulsification. *J. Colloid Interface Sci.* **2004**, *277*, 456–463. [CrossRef]
48. Miller, R.; Aksenenko, E.; Fainerman, V. Dynamic interfacial tension of surfactant solutions. *Adv. Colloid Interface Sci.* **2017**, *247*, 115–129. [CrossRef]
49. Tween®20; Safety Data Sheet for Tween®20 (Polysorbat) 817072 [Online]; MERCK: Darmstadt, Germany, 5 December 2020; Available online: https://www.merckmillipore.com/DE/de/product/msds/MDA_CHEM-817072 (accessed on 31 May 2021).
50. Tween®20; Product Information on Polyoxyethylenesorbitan Monolaurate (Tween20) [Online]; Sigma Aldrich: St. Louis, MO, USA, 29 July 2021; Available online: https://www.sigmaaldrich.com/content/dam/sigma-aldrich/docs/Sigma/Product_Information_Sheet/1/p6585pis.pdf (accessed on 24 November 2021).
51. Pollard, J.M.; Shi, A.J.; Göklen, K.E. Solubility and Partitioning Behavior of Surfactants and Additives Used in Bioprocesses. *J. Chem. Eng. Data* **2006**, *51*, 230–236. [CrossRef]
52. Jin, F.; Balasubramaniam, R.; Stebe, K.J. Surfactant Adsorption to Spherical Particles: The Intrinsic Length Scale Governing the Shift from Diffusion to Kinetic-Controlled Mass Transfer. *J. Adhes.* **2004**, *80*, 773–796. [CrossRef]
53. Staszak, M. A Linear Diffusion Model of Adsorption Kinetics at Fluid/Fluid Interfaces. *J. Surfactants Deterg.* **2016**, *19*, 297–314. [CrossRef] [PubMed]
54. Chanamai, R.; McClements, D. Isothermal titration calorimetry measurement of enthalpy changes in monodisperse oil-in-water emulsions undergoing depletion flocculation. *Colloids Surf. A Physicochem. Eng. Asp.* **2001**, *181*, 261–269. [CrossRef]

55. Helenius, A.; McCaslin, D.R.; Fries, E.; Tanford, C. Properties of detergents. In *Biomembranes Part G: Bioenergetics: Biogenesis of Mitochondria, Organization, and Transport*; Kaplan, N.P., Colowick, N.P., Fleischer, S., Sies, H., Eds.; Elsevier: Amsterdam, The Netherlands, 1979; pp. 734–749.
56. Bera, B.; Khazal, R.; Schroën, K. Coalescence dynamics in oil-in-water emulsions at elevated temperatures. *Sci. Rep.* **2021**, *11*, 1–10. [CrossRef]
57. Nowak, E.; Kovalchuk, N.; Che, Z.; Simmons, M. Effect of surfactant concentration and viscosity of outer phase during the coalescence of a surfactant-laden drop with a surfactant-free drop. *Colloids Surf. A Physicochem. Eng. Asp.* **2016**, *505*, 124–131. [CrossRef]
58. Scriven, L.E.; Sterling, C.V. The Marangoni Effects. *Nature* **1960**, *187*, 186–188. [CrossRef]
59. Schuchmann, H.P.; Danner, T. Emulgieren: Mehr als nur Zerkleinern. *Chemie Ingenieur Technik* **2004**, *76*, 364–375. [CrossRef]
60. Köhler, K.; Hensel, A.; Kraut, M.; Schuchmann, H.P. Melt emulsification—Is there a chance to produce particles without additives? *Particuology* **2011**, *9*, 506–509. [CrossRef]
61. Chen, J.-D.; Hahn, P.S.; Slattery, J.C. Coalescence time for a small drop or bubble at a fluid-fluid interface. *AIChE J.* **1984**, *30*, 622–630. [CrossRef]
62. Mackay, G.D.M.; Mason, S.G. The gravity approach and coalescence of fluid drops at liquid interfaces. *Can. J. Chem. Eng.* **1963**, *41*, 203–212. [CrossRef]
63. Hodgson, T.; Woods, D. The effect of surfactants on the coalescence of a drop at an interface. II. *J. Colloid Interface Sci.* **1969**, *30*, 429–446. [CrossRef]
64. Leister, N.; Karbstein, H.P. Influence of Hydrophilic Surfactants on the W1–W2 Coalescence in Double Emulsion Systems Investigated by Single Droplet Experiments. *Colloids Interfaces* **2021**, *5*, 21. [CrossRef]
65. Christov, N.C.; Danov, K.D.; Zeng, Y.; Kralchevsky, P.A.; von Klitzing, R. Oscillatory Structural Forces Due to Nonionic Surfactant Micelles: Data by Colloidal-Probe AFM vs. Theory. *Langmuir* **2010**, *26*, 915–923. [CrossRef] [PubMed]
66. Hsu, J.-P.; Nacu, A. Behavior of soybean oil-in-water emulsion stabilized by nonionic surfactant. *J. Colloid Interface Sci.* **2003**, *259*, 374–381. [CrossRef]
67. Hammer, M.U.; Anderson, T.H.; Chaimovich, A.; Shell, M.S.; Israelachvili, J. The search for the hydrophobic force law. *Faraday Discuss.* **2010**, *146*, 299–308. [CrossRef] [PubMed]
68. Meyer, E.E.; Rosenberg, K.J.; Israelachvili, J. Recent progress in understanding hydrophobic interactions. *Proc. Natl. Acad. Sci. USA* **2006**, *103*, 15739–15746. [CrossRef]
69. Zhang, X.; Zhu, Y.; Granick, S. Softened hydrophobic Attraction between Macroscopic Surfaces in Relative Motion. *J. Am. Chem. Soc.* **2001**, *123*, 6736–6737. [CrossRef]
70. Singh, S.; Houston, J.; van Swol, F.; Brinker, C.J. Superhydrophobicity: Drying transition of confined water. *Nature* **2006**, *442*, 526. [CrossRef]
71. Krawczyk, M.A.; Wasan, D.T.; Shetty, C. Chemical demulsification of petroleum emulsions using oil-soluble demulsifiers. *Ind. Eng. Chem. Res.* **1991**, *30*, 367–375. [CrossRef]
72. Opawale, F.O.; Burgess, D.J. Influence of Interfacial Properties of Lipophilic Surfactants on Water-in-Oil Emulsion Stability. *J. Colloid Interface Sci.* **1998**, *197*, 142–150. [CrossRef] [PubMed]
73. Muijlwijk, K.; Colijn, I.; Harsono, H.; Krebs, T.; Berton-Carabin, C.; Schroën, K. Coalescence of protein-stabilised emulsions studied with microfluidics. *Food Hydrocoll.* **2017**, *70*, 96–104. [CrossRef]

Article

Reactive Crystallization Kinetics of K$_2$SO$_4$ from Picromerite-Based MgSO$_4$ and KCl

Abad Albis [1,2,*], Yecid P. Jiménez [2,3], Teófilo A. Graber [2] and Heike Lorenz [4]

[1] Carrera de Química, Facultad de Ciencias Puras, Universidad Autónoma Tomás Frías, Potosí 53820, Bolivia
[2] Departamento de Ingeniería Química y Procesos de Minerales, Facultad de Ingeniería, Universidad de Antofagasta, Av. Angamos 601, Antofagasta 1270300, Chile; yecid.jimenez@uantof.cl (Y.P.J.); teofilo.graber@uantof.cl (T.A.G.)
[3] Centro de Economía Circular en Procesos Industriales (CECPI), Facultad de Ingeniería, Universidad de Antofagasta, Av. Angamos 601, Antofagasta 1270300, Chile
[4] Max Planck Institute for Dynamics of Complex Technical Systems, 39106 Magdeburg, Germany; lorenz@mpi-magdeburg.mpg.de
* Correspondence: abadalcis11z@gmail.com

Abstract: In this work, the kinetic parameters, the degrees of initial supersaturation (S_0) and the profiles of supersaturation (S) were determined for the reactive crystallization of K$_2$SO$_4$ from picromerite (K$_2$SO$_4$·MgSO$_4$·6H$_2$O) and KCl. Different reaction temperatures between 5 and 45 °C were considered, and several process analytical techniques were applied. Along with the solution temperature, the crystal chord length distribution (CLD) was continuously followed by an FBRM probe, images of nucleation and growth events as well as the crystal morphology were captured, and the absorbance of the solution was measured via ATR-FTIR spectroscopy. In addition, the ion concentrations were analyzed. It was found that S_0 is inversely proportional to the reactive crystallization temperature in the K$^+$, Mg^{2+}/Cl$^-$, SO$_4^{2-}$//H$_2$O system at 25 °C, where S_0 promotes nucleation and crystal growth of K$_2$SO$_4$ leading to a bimodal CLD. The CLD was converted to square-weighted chord lengths for each S_0 to determine the secondary nucleation rate (B), crystal growth rate (G), and suspension density (M_T). By correlation, from primary nucleation rate (B_b) and G with S_0, the empirical parameters b = 3.61 and g = 4.61 were obtained as the order of primary nucleation and growth, respectively. B versus G and M_T were correlated to the reaction temperature providing the rate constants of B and respective activation energy, E = 69.83 kJ·mol^{-1}. Finally, a general Equation was derived that describes B with parameters K_R = 13,810.8, i = 0.75 and j = 0.71. The K$_2$SO$_4$ crystals produced were of high purity, containing maximal 0.51 wt% Mg impurity, and were received with ~73% yield at 5 °C.

Keywords: potassium sulfate; picromerite; nucleation; crystal growth; supersaturation

Citation: Albis, A.; Jiménez, Y.P.; Graber, T.A.; Lorenz, H. Reactive Crystallization Kinetics of K$_2$SO$_4$ from Picromerite-Based MgSO$_4$ and KCl. *Crystals* **2021**, *11*, 1558. https://doi.org/10.3390/cryst11121558

Academic Editor: Shujun Zhang

Received: 31 October 2021
Accepted: 8 December 2021
Published: 14 December 2021

Publisher's Note: MDPI stays neutral with regard to jurisdictional claims in published maps and institutional affiliations.

Copyright: © 2021 by the authors. Licensee MDPI, Basel, Switzerland. This article is an open access article distributed under the terms and conditions of the Creative Commons Attribution (CC BY) license (https://creativecommons.org/licenses/by/4.0/).

1. Introduction

Potassium is one of the three types of macronutrients [1]. Its natural sources are mineral salt deposits containing KCl in sylvinite or K$_2$SO$_4$ in kainite or picromerite and natural brines from salt flats such as Atacama-Chile [2,3] and Uyuni-Bolivia [4]. There are several known and traditional methods to produce K$_2$SO$_4$ as, for example, the reaction of KCl with sulphate containing compounds such as H$_2$SO$_4$ [5], Phosphogypsum in an NH$_4$OH and isopropanol medium [6], Na$_2$SO$_4$ [7,8], and (NH$_4$)$_2$SO$_4$ [9], or from seawater [10] and leonite or langbeinite minerals to produce intermediate picromerite by hydration and final transformation to K$_2$SO$_4$ [11].

Producing K$_2$SO$_4$ from MgSO$_4$ and KCl by reactive crystallization in an aqueous solution involves the formation of picromerite and K$_2$SO$_4$, using the Jänecke projection that describes the reactions via the phase diagram of the reciprocal quaternary system K$^+$, Mg^{2+}/Cl$^-$, SO$_4^{2-}$//H$_2$O at 25 °C [12]. The latter also allows estimating the proportions of reactants and products. Jannet et al. [12], Voigt [13], Fezei et al. [14], and

Goncharik et al. [15] depict the reactive crystallization process based on the underlying phase equilibria in two stages: first, the formation of picromerite and second, the generation of K_2SO_4. The work mentioned above on the conversion processes and the K_2SO_4 yield improvement in a batch crystallizer do not explicitly consider the reactive crystallization kinetics of K_2SO_4.

Crystallization is the first and most crucial step to produce, purify, and isolate high purity products in batch and continuous processes, and batch crystallization is one of the unit operations widely used in the chemical, pharmaceutical, and food industries [16,17]. However, batch crystallization has disadvantages associated with its control and, in particular batch-to-batch variability, expressed in terms of product purity, crystal morphology, mean particle size, crystal size distribution (CSD), bulk density, filterability, and flow properties of the dry solids that depend on the profile of supersaturation achieved during the crystallization, drying, packaging, and transport process [18–20].

Therefore, to design the operating conditions with optimal control of a crystallizer, the crystallization kinetics must first experimentally be determined based on the population density [17]. This requires the collection of CSD data that directly relate the nucleation and crystal growth events occurring simultaneously. Furthermore, new crystals can result from two different nucleation mechanisms, classified as primary and secondary nucleation or a combination of both [18].

Nemdili et al. [21], Bari et al. [22], and Luo et al. [23] measured the metastable zone width (MSZW) of a K_2SO_4-water solution by means of a turbidimeter and an ultrasonic probe and used an ultrasonic and Focused Beam Reflectance Measurement (FBRM) probe, respectively, to determine mainly the primary nucleation parameters. Bari and Pandit [24] simulated the order and the primary and secondary nucleation constants of K_2SO_4 using the gPROMS program and data from MSZW. Luo et al. [23] concluded that the MSZW of the potassium sulfate solution increases with the increasing concentration of aluminium and silicon ions as impurities because they suppress nucleation and block the active sites of K_2SO_4 crystals. They reported a nucleation activation energy of 33.99 kJ·mol^{-1}.

In relation to the K_2SO_4 growth kinetics, Nemdili et al. [21] determined the desupersaturation curve of a K_2SO_4-H_2O solution when cooling from 40 to 27 °C and using a seed size of 250–355 µm. They derived the order of growth and concluded that growth occurs via the spiral growth mechanism. Gougazeh et al. [25] investigated the growth kinetics of a K_2SO_4 single crystal under a microscope analyzing the obtained images. They concluded that the order of growth is not affected by temperature, and the single crystal growth is governed by the spiral growth mechanism. They specified the activation energy of growth as 39.4 kJ·mol^{-1} and claimed that the presence of 5 ppm Cr^{3+} blocks the active growth sites of K_2SO_4. Kubota et al. [26] confirmed this mechanism leading to suppressed growth. Mullin and Gaska [27] determined the growth kinetics of a K_2SO_4 single crystal in two selected directions, width (001) and length (100), finding that the linear growth rate is greater for axis 001 and minor for axis 100, due to flow effects of the supersaturated aqueous solution circulating through the cell. Furthermore, they reported that crystal growth of K_2SO_4 under the conditions used is diffusion-controlled.

Secondary nucleation occurs in the presence of "father" crystals and their interaction with themselves, the crystallizer walls, and the agitator. It results from seeding, presence of dendrites, breakage, and fluid shear forces according to mechanical, hydrodynamic, and supersaturation forces. In industry, secondary nucleation is the mechanism that governs the crystallization of soluble substances from a singular solute solution (e.g., salt-water) [28].

Chianese et al. [29] studied the effect of secondary nucleation on CSD for the K_2SO_4-H_2O system in a seeded batch crystallizer. They found that the generation of fragments depends on suspension density and supersaturation. They incorporated the effects of secondary nucleation due to attrition in their model and established that the model does not predict the experimental behaviour at all. Mohamed et al. [30] simultaneously determined the kinetic parameters of secondary nucleation and growth of K_2SO_4 in its aqueous solution with seeding based on the population density data. They performed the population balance and solved the population equilibrium Equation using the "s" plane analysis method. They correlated the magma density (M_T), growth rate (G), solution circulation rate (Q), and absolute temperature (T). They reported the relative parameters of secondary nucleation rate (B) according to the model $B = K_R G^i M_T^j Q^k \exp(-E/RT)$ with values of relative kinetic order, $i = 0.90$, exponent of magma density, $j = 0.57$, exponent of solution circulation rate, $k = 1.49$, and apparent secondary nucleation activation energy, $E = -3.36$ kJ·mol^{-1}.

The effect of secondary nucleation on CSD is generally determined in binary systems because supersaturation and seeding can be adequately controlled. However, both primary and secondary nucleation mechanisms occur in the reactive crystallization process via the generation of supersaturation by reaction [31]. There, the reaction rate, the narrow variation in agitation, and the limited transfer of mass to the crystal interface make the study of secondary nucleation and growth complex. Despite these limitations, it is important to study secondary nucleation for soluble salts in reciprocal salt pair systems because many reactions produce intermediate compounds such as double salt hydrates as picromerite to which another reagent can be added to obtain the desired anhydrous product. Furthermore, these products can be redissolved to further improve the CSD by controlled cooling. There are a few related reports in the literature on reactive crystallization for sparingly soluble salts, such as the works of Taguchi et al. [31], Lu et al. [32] and Mignon et al. [33], but to the knowledge of the authors not on soluble salts in a reciprocal salt pair system.

Thus, the purpose of this work is to study the reactive crystallization kinetics between aqueous solutions (aq) of $KCl_{(aq)}$ and $MgSO_{4(aq)}$ (the latter from picromerite) to produce K_2SO_4. Reaction temperatures of 5, 15, 25, 35, and 45 °C were applied, and initial local supersaturations and the supersaturation profiles as a function of time were determined. Population density data in terms of crystal count, length, area, and volume were collected using an FBRM probe to simultaneously estimate primary (B_b) and secondary (B) nucleation kinetics, growth (G), and suspension density (M_T) over time. Since the CLD measured by FBRM is related but not similar to the CSD [34], the CLD data was converted to CSD data based on the works of Heath et al. [35]; Togkalidou et al. [17]; Trifkovic et al. [36]; Óciardhá et al. [37], and, Ajinkya et al. [34]. During reactive crystallization, the absorbance and temperature of the solution were followed in situ and in real time using ATR-FTIR and Pt-100, respectively. In addition, the change in ion concentrations was measured, and images of nucleation and growth events, as well as crystal morphology were captured.

This work contributes to determining the kinetic parameters of nucleation and growth for the reciprocal salt system under study and the degree of supersaturation based on the Pitzer ionic interaction parameters for the reactive crystallization process of $K_2SO_{4(s)}$. In addition, this research constitutes a potential tool for producing K_2SO_4 from picromerite.

1.1. Reaction, Thermodynamic and Crystallization Kinetics Framework

1.1.1. Dissolution and Reaction of Picromerite

The dissolution of picromerite is an incongruent process; it decomposes in the presence of water according to the following Equation:

$$K_2SO_4 \cdot MgSO_4 \cdot 6H_2O_{(s)}/H_2O \leftrightarrow nK_2SO_{4(s)} + (1-n)K_2SO_{4(aq)} + MgSO_{4(aq)} + 6H_2O_{(l)}/H_2O \quad (1)$$

In a second step, a $KCl_{(aq)}$ solution reacting with $MgSO_{4(aq)}$ from picromerite leads to the formation of $K_2SO_{4(s)}$ product crystals according to Equation (2) which forms the basis of the crystallization kinetics study in this work.

$$(1-n)K_2SO_{4(aq)} + MgSO_{4(aq)} + 2KCl_{(aq)} \leftrightarrow K_2SO_{4(s)} + (1-n)K_2SO_{4(aq)} + MgCl_{2(aq)} \quad (2)$$

1.1.2. Thermodynamic Supersaturation

The determination of the thermodynamic supersaturation is based on estimating the activity coefficients of the ionic pair K^+-SO_4^{2-} as a function of temperature in a mixture of electrolytes K^+-$Mg^{2+}//Cl^-$-SO_4^{2-}, based on the Pitzer model. The latter considers the ionic strength of the electrolyte mixture since supersaturation is the driving force that promotes nucleation and growth of crystals in crystallization processes.

Pitzer's model [38–41] requires the binary ionic interaction parameter values of $\beta_{ij}^{(0)}$, $\beta_{ij}^{(1)}$, $\beta_{ij}^{(2)}$, and $C_{ij}^{\varnothing}$ for reactant electrolytes and products as a function of temperature, where i and j represent the ionic pair cation and anion, respectively. The ionic interaction parameters of $KCl_{(aq)}$, $MgCl_{2(aq)}$, and $MgSO_{4(aq)}$ as a function of temperature are given in Table 1. They were obtained by Equations (S1)–(S3), whose empirical parameters are found in Tables S1–S3, established by Holmes et al. [42], De Lima and Pitzer. [43], and Phutela and Pitzer [44], respectively (see Supplementary Information (SI)).

Table 1. Pitzer ionic interaction parameters of electrolytes KCl, $MgCl_2$ and $MgSO_4$ at different temperatures.

Temperature [°C]	Ionic Pair	$\beta_{ij}^{(0)} \cdot 10^2$	$\beta_{ij}^{(1)} \cdot 10^2$	$\beta_{ij}^{(2)}$	$C_{ij}^{\varnothing} \cdot 10^4$
5	K^+-Cl^-	3.38	18.31		8
	Mg^{2+}-Cl^-	36.29	157.02		87
	Mg^{2+}-SO_4^{2-}	19.49	311.56	−19.88	23
15	K^+-Cl^-	4.16	20.35		−1
	Mg^{2+}-Cl^-	35.69	160.81		76
	Mg^{2+}-SO_4^{2-}	20.69	324.94	−28.60	20
25	K^+-Cl^-	4.80	21.88		-8
	Mg^{2+}-Cl^-	35.11	165.12		65
	Mg^{2+}-SO_4^{2-}	21.51	336.63	−32.77	17
35	K^+-Cl^-	5.33	23.18		−14
	Mg^{2+}-Cl^-	34.54	165.12		55
	Mg^{2+}-SO_4^{2-}	22.09	347.10	−32.97	15
45	K^+-Cl^-	5.75	24.37		−18
	Mg^{2+}-Cl^-	33.98	175.30		45
	Mg^{2+}-SO_4^{2-}	22.55	356.78	−29.71	13

Once the activity coefficients of the ion pair K^+-SO_4^{2-} are determined in the system K^+-$Mg^{2+}//Cl^-$-SO_4^{2-}, they are related to the molal concentration and the thermodynamic equilibrium constant K_{sp} of K_2SO_4, which was determined using the Pitzer model for isotherms of 5, 15, 25, 35, and 45 °C based on solubility data [45]. The K_{sp} data for K_2SO_4 estimated in this work are similar to those reported by Jiménez et al. [46] for isotherms of 15, 25, 35, and 45 °C. The K_{sp} results of K_2SO_4 found in Table S4 (SI) are replaced in Equation (3) to obtain the dimensionless degree of supersaturation "S" used in the reactive crystallization process of K_2SO_4. Then, S is given by:

$$S = \gamma_\pm \left(\frac{m_+^{v_+} m_-^{v_-}}{K_{sp}} \right)^{1/v} \quad (3)$$

where, $\gamma_\pm$ is the mean activity coefficient corresponding to K_2SO_4, $m_+^{v_+}$ and $m_-^{v_-}$ are the molal concentration of the K^+ and SO_4^{2-} ions raised to their respective stoichiometric coefficients and v is the total coefficient of K^+ and SO_4^{2-} $(v_+ + v_-)$.

1.1.3. Crystallization Kinetics

The CLD measured by the FBRM probe differs from the CSD measured by sieving, laser diffraction, and PVM. However, using the mean or mode average of the square-weighted chord length was found to be comparable to conventional sizing techniques for the range of about 50–400 µm [35]. Therefore, on that basis, it is possible to determine the population density and then exploit it for crystallization kinetic studies [17,35]. The mathematical Equations that allow obtaining CSD-based moments from CLD are presented below [17,35].

The jth moments from CLD data are calculated with Equation (4) according to Trifkovic et al. [36]:

$$\mu_j(t) \equiv \int_0^\infty L^j n(L,t) dl \approx \sum_{i=1}^{FBRM} L_{ave,i}^j N_i(L_{ave,i}, k) = N_k \quad (4)$$

where L_i is the chord length, N_i the number of particles in channel i, and k the discrete time counter. $L_{ave,i}$ is the arithmetic mean between the upper (L_i) and lower (L_{i-1}) channel size and is expressed as:

$$L_{ave,i} = \frac{L_i + L_{i-1}}{2} \quad (5)$$

Trifkovic et al. [36] used the total number of crystals in the entire FBRM size range (1 to 1000 microns) to calculate the nucleation rate. The number of particles as a function of time is given by Equation (6).

$$N(t) = \int_0^\infty n(t) dL \approx \sum_{i=1}^{FBRM} N_{i,k} = N_k \quad (6)$$

Furthermore, it is possible to determine the nucleation rate as a difference of the total number of particles at each moment per mass of solvent, according to Equation (7).

$$B_{exp}(t) = \frac{1}{M(t)} \frac{dN(t)}{dt} \equiv \frac{1}{M_k} \frac{\Delta N_k}{\Delta t} = B_{exp,k} \quad (7)$$

where $B_{exp,k}$ and M_k are the nucleation rate and the total mass of solvent in the kth time interval. From this Equation, the ratio of the particle count to the mass of the solvent can be considered. However, by adopting Equation (4) to determine the 0th moment, this would become a summation of CLD of original FBRM data, which does not consider the conversion from CLD to CSD. According to the report of Heath et al. [35], the population density is considered via the conversion of CLD to CSD as counts of square-weighted chord lengths by the following Equation:

$$N_{i,n} = N_{i,0} C_{i,A}^n \quad (8)$$

where $N_{i,n}$ is the number of n-weighted chord length counts in the ith channel, $N_{i,0}$ the count of unweighted chord lengths in the ith channel, and $C_{i,A}^n$ is the geometric mean length of the ith channel given by:

$$C_{i,A} = (C_{i,u} \cdot C_{i,l})^{1/2} \quad (9)$$

with $C_{i,u}$ and $C_{i,l}$ the length of the upper and lower limit of the ith channel, respectively. Then, by combining Equations (4) and (8), Equation (10) is constituted to determine the moments based on the population density as a function of time per unit mass of solvent. M is the mass of the solvent (H$_2$O).

$$\mu_j(t) \equiv \int_0^\infty L^j n(L,t) dl \approx \sum_{i=1}^{FBRM} L_{ave,i}^j N_{i,n}(L_{ave,i}, k) * \frac{1}{M} \quad (10)$$

Following the previous Equation, moments of the order 0, 1, 2, and 3 can be determined with the units (#/kgH$_2$O), (μm/kgH$_2$O), (μm^2/kgH$_2$O), and (μm^3/kgH$_2$O), respectively as a function of time, also for different degrees of initial local supersaturation reached at temperatures of 5, 15, 25, 35, and 45 °C.

Based on the population balance equation (PBE), different ordinary differential equations (ODE's) are obtained that allow for the determination of the nucleation and growth rate and suspension density with respect to time considering the conversion from CLD to counts of square-weighted chord lengths, which are presented below.

Primary nucleation rate.

In the reactive crystallization of K_2SO_4, primary and secondary nucleation (B_b and B, respectively) occur, whereas B_b results from the high supersaturation generated by the reaction. The nucleation rate (B) of K_2SO_4 is determined based on the 0th moment (μ_0) after conversion of the CLD to square-weighted counts [35] (Figures S1 and S2 (see SI)). Then, the B is estimated using the following Equation [47]:

$$\frac{d\mu_o}{dt} = B \qquad (11)$$

In the present work, B_b is related with the initial local supersaturation (S_0) since it has been obtained from the thermodynamic approach when considering the activities of the species K^+ and SO_4^{2-} in the multicomponent system $K^+, Mg^{2+}/Cl^-, SO_4^{2-}//H_2O$. To determine the kinetic parameters such as order and rate constant of primary nucleation rate, the following Equation [48] has been used:

$$B_b = k_b \cdot S_o^b \qquad (12)$$

where b is the primary nucleation order and k_b is the primary nucleation rate constant.

Secondary nucleation rate.

Secondary nucleation usually depends on suspension density (M_T) and supersaturation (S). Furthermore, growth (G) is also a function of S. Then, the relationship of B with G and M_T describes the secondary nucleation rate of K_2SO_4 in the reactive crystallization process with time and is given as follows [49]:

$$B = K_R \cdot G^i \cdot M_T^j \qquad (13)$$

Their i is the nucleation relative to the order of growth, j the suspension density exponent, and K_R the relative rate coefficient.

Crystal growth rate.

The crystal growth rate (G) of K_2SO_4 was simultaneously determined from the population density data, where G is linked to the 0th and first moment (Figure S3 (SI)), according to the following Equation [47]:

$$\frac{1}{\mu_o}\frac{d\mu_1}{dt} = G \qquad (14)$$

With the data of G for different degrees of initial supersaturation S_0, the empirical fit parameters are determined according to [48]:

$$G = k_g \cdot S_0^g \qquad (15)$$

Crystal suspension density.

The suspension density as a function of time, based on the data of the total volume of crystals μ_3 (μm^3) (Figure S5 (SI)), was determined as follows [35]:

$$\frac{dM_T}{dt} = 3 \cdot k_v \cdot \rho_c \cdot G \cdot \mu_2 \qquad (16)$$

where ρ_c is the K_2SO_4 crystals density (2.66×10^{-12} g/μm^3), k_v is the volume shape factor (0.69 for K_2SO_4), and μ_2 is the second moment (Figure S4 (SI)). However, the empirical Equation (17) was specified gravimetrically to corroborate the behaviour of the suspension density and through the mass balance (Equation (18)) using data from the third moment as follows:

$$M_{Ti} = \frac{\mu_{3i}}{\mu_{3f}} M_{Tf} \qquad (17)$$

where μ_{3i} and μ_{3f} are the initial and final total crystal volume, and M_{Ti} and M_{Tf} are the initial and final suspension density in the reactive crystallization. The mass balance Equation is given by [50]:

$$C(t) = C_0 - M_{Ti} \qquad (18)$$

with C_0 being the initial concentration of the solute resulting from mixing both reactive solutions, $C(t)$ the concentration of the solute over time, and M_{Ti} as given above.

Secondary nucleation activation energy.

In a chemical reaction, efficient collisions occur between atoms, molecules, or ions. The ions of the reciprocal system $K^+, Mg^{2+}Cl^-, SO_4^{2-}//H_2O$ require the minimum kinetic energy to form the active complex specified by the activation energy E_a and expressed by the Arrhenius Equation:

$$K = A \cdot \exp\left(-\frac{E_a}{RT}\right) \tag{19}$$

where K is the reaction rate constant, A is the frequency of molecular collisions, R is the universal gas constant, and T is the absolute temperature. Taking the logarithm, it gives:

$$\ln K = \ln A + \left(-\frac{E_a}{R}\right)\frac{1}{T} \tag{20}$$

The graphical relationship of $\ln K$ versus $1/T$ is linear, with the slope equals $(-E_a/R)$ and the intercept $\ln A$. Thus, the activation energy E_a can be obtained from the slope of this plot. For reactive crystallization, the dependence of the secondary nucleation rate constant K_R on the absolute temperature allows estimating the secondary nucleation activation energy (E) in [kJ/mol] [30].

2. Materials and Methods

2.1. Reagents

Magnesium sulfate (MgSO$_4$) was purchased from Acros Organic (Geel, Belgium; purity > 97 wt%), potassium sulfate (K$_2$SO$_4$) from Carl Roth GmbH & Co. (Karlsruhe, Germany; purity $\geq$ 98 wt%), and potassium chloride (KCl) from Merck (Darmstadt, Germany; purity 99.5–100 wt%). All three reagents were used as received. Solid picromerite was obtained by slow evaporation, filtration, and drying. Deionized ultrapure water used as a solvent in all experiments was provided by a Merck Millipore, Milli-Q® Advantage A10 system.

2.2. Experimental Setup and Process Analytical Technology (PAT)

The experimental setup for reactive crystallization studies is shown in Figure 1. The applied PAT, additional optical process monitoring using the Technobis Crystalline PV system, and further implemented offline analytics will be discussed below.

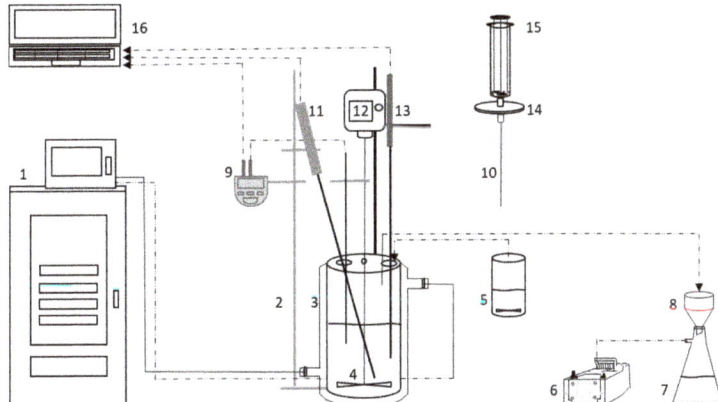

Figure 1. Scheme of the experimental setup. 1 Temperature control system; 2 Universal stand; 3 250 mL-Jacketed glass crystallizer; 4 Glass impeller; 5 Reservoir for saturated solution of KCl; 6 Vacuum pump; 7 Buchner flask; 8 Buchner funnel; 9 Pt-100 sensor; 10 Syringe needle; 11 FBRM probe; 12 Overhead stirrer; 13 ATR-FTIR probe; 14 Disposable syringe filter; 15 Syringe; 16 Data acquisition system.

2.2.1. Focused Beam Reflectance Measurement (FBRM)

For monitoring the change in the CLD of K_2SO_4 crystals during reactive crystallization, the Particle Track S400A FBRM® probe with focused beam reflectance measurement technology from Mettler Toledo, Gießen, Germany was used. It is based on the backscattering of a rotating laser beam tracking particles flowing in front of the probe window. Since particles can be scanned in parts smaller or larger than their average size, the detected value is called the "chord length". In this study, the terms chord length distribution and a "number of crystals" is used to determine the nucleation rate B, growth rate G, and suspension density M_T of K_2SO_4, but in CLD. The CLD of the crystals was measured in real time and fell in 100 channels. The CLD by channels varies on a logarithmic scale from 1 to 1000 µm, allowing a higher resolution of the CLD of particles <100 µm with the ability to characterize the presence of uni and bimodal particle size distributions. The counts of CLD data were captured every 10 s during 60 min of reactive crystallization. The operational variables of the FBRM to collect data consisted in adjusting the FBRM with an angle of inclination of 82° that also acts as a deflector, with the focal position (window) located above the tip of the impeller at a height less than 8 mm. To ensure the presence of crystal samples with the impeller speed used, the pulp concentration should not exceed 20 wt/wt%, and the CLD collection was established in fine mode. In addition, mechanical stirring was kept constant at 317 rpm allowing for a good suspension of crystals. Higher stirring speed generates vortices, small and large bubbles near the edge of the propellant and on the surface of the solution, respectively, which is a problem for moment data reads. All previous operations were based on Heath et al. [35].

2.2.2. Attenuated Total Reflectance-Fourier Transform Infrared (ATR-FTIR) Spectroscopy

The change in absorbance of the solution during reactive crystallization was monitored in real time using the Mettler Toledo ReactIR™ 15 ATR-FTIR probe (Mettler Toledo, Gießen, Germany). The probe immersed in the solution measured the absorbance every 60 s during the 60 min of the reactive crystallization process.

2.2.3. Temperature Measurement

A Pt-100 sensor (Ahlborn Mess- und Regelungstechnik GmbH, Holzkirchen, Germany) was used to follow the temperature of the solution before and during reactive crystallization. Temperature data were collected and recorded with a measurement precision ± 0.01 °C. The FBRM-S400A, ATR-FTIR, and Pt-100 probes are shown in Figure 1 as the main components collecting process data on the reactive crystallization of K_2SO_4.

2.2.4. Technobis Crystalline PV

For image monitoring of nucleation and growth of K_2SO_4 crystals, the Crystalline PV multi-reactor system with particle viewer module (Technobis Crystallization Systems, Alkmaar, The Netherlands) was used. Images were taken every 20 s during the 60 min of the reactive crystallization process of K_2SO_4. Digital photographic images of K_2SO_4 particles as a function of time are displayed directly on the computer screen.

2.2.5. X-ray Powder Diffraction Phase (XRPD) Analysis

Powder samples of the reactive crystallization products were subjected to XRPD analysis using a PANalytical X'Pert-Pro diffractometer (PANalytical GmbH, Kassel, Germany) with an X'Celerator detector and Cu Kα radiation. Samples were measured in a 2-Theta range of 10–100° with a step size of 0.0167° and a counting time of 30 s per step.

2.2.6. Chemical Analysis by Ion Chromatography (IC)

The concentrations of K^+, Mg^{2+}, Cl^- and SO_4^{2-} were determined by ion chromatography using the Dionex ICS 1100 IC system (Thermo Fischer Scientific, Dreieich, Germany). For calibration, standard solutions of known composition were prepared for the anionic and cationic pairs Cl^--SO_4^{2-} and K^+-Mg^{2+}, respectively. These standards allowed determining the concentrations of the cations and anions, with priority for Mg^{2+} as an impurity in the K_2SO_4 product crystals to evaluate the product quality.

2.3. Experimental Procedure

2.3.1. Preparation of the Saturated Solution of Magnesium Sulfate from Picromerite

There are two methods to prepare $MgSO_{4(aq)}$ from picromerite: (1) by dissolving the $MgSO_4$ from the picromerite crystal lattice by addition of water and (2) by isothermal synthesis. Both methods are based on the phase diagram for the ternary system K_2SO_4-$MgSO_4$-H_2O at 25 °C (see Figure S6 (SI)). In this work, method (2) has been adopted to prepare $MgSO_{4(aq)}$ from pure reagents K_2SO_4, $MgSO_4$ and H_2O without the need to prepare the picromerite. Since the dissolution of picromerite is an incongruent process, it decomposes in the presence of water, according to Equation (1). As a result, a solid phase of $K_2SO_{4(s)}$ is in equilibrium with a saturated liquid phase of K^+, Mg^{2+} and SO_4^{2-} ions in water. The solid potassium sulfate is filtered under a vacuum, and the resulting solution is used in the K_2SO_4 reactive crystallization process to determine the kinetic parameters. For the pulp synthesis according to point C in Figure S6 (see SI), pure reagents $K_2SO_{4(s)}$, $MgSO_{4(s)}$ and $H_2O_{(l)}$ were mixed to obtain a synthetic pulp constituted of 11.10, 12.54, and 76.34 wt/wt% of K_2SO_4, $MgSO_4$, and H_2O, respectively.

The pulp synthesis and the generation of the saturated $MgSO_{4(aq)}$ solution from picromerite was performed in a jacketed vessel. All the reagents were added on a precision analytical balance ($\pm 1 \times 10^{-4}$ g): 12.9660 g of $K_2SO_{4(s)}$, 8.9700 g of $MgSO_{4(s)}$, and 54.5820 g of $H_2O_{(l)}$. The vessel with the reagents was hermetically closed, taken into the thermostatted reactor at 25 °C and left stirring for 24 h to reach solid–liquid equilibrium and guarantee the pulp composition. The solid was separated from the pulp by vacuum filtration, and the particle-free solution was transferred to another closed vessel maintaining the temperature in a thermostatic bath. The product $K_2SO_{4(s)}$ was allowed to dry at room temperature under a fume hood for 24 h. This synthesis procedure was repeated for all the reactive crystallization runs of K_2SO_4.

2.3.2. Preparation of Saturated Solution of KCl

The $KCl_{(aq)}$ solution reacting with $MgSO_{4(aq)}$ from picromerite gives rise to the formation of crystals of $K_2SO_{4(s)}$ according to Equation (2). A saturated solution of $KCl_{(aq)}$ was prepared at 25 °C with an excess of 1% of $KCl_{(s)}$ to guarantee solid–liquid equilibrium. The dissolution of $KCl_{(s)}$ in water was carried out separately with magnetic stirring.

2.3.3. Operating Conditions of the Reactive Crystallization Process

The experimentation plan with the operating conditions used is given in Table S5 (SI). Based on Equation (1), the saturated solution of $MgSO_{4(aq)}$ from picromerite was prepared at 25 °C. For all the reactive crystallization runs, the concentration of 457 (g of solute/kg H_2O) was kept constant where the solute concentration was composed only of (1 − n) $K_2SO_{4(aq)}$ and $MgSO_{4(aq)}$ at 25 °C, with n = 0.38. On the other hand, the saturated solution of $KCl_{(aq)}$ was prepared according to Equation (2), the concentration was 343.98 (g solute/kg H_2O), which also remained constant for all reactive crystallization runs.

2.3.4. Mixing of Reagents and Generation of Initial Supersaturation (S_0)

For performing reactive crystallization between the saturated synthetic solution of $MgSO_{4(aq)}$ from picromerite with $KCl_{(aq)}$, it is essential to determine the degree of initial supersaturation S_0 and the supersaturation profile with time S to derive the nucleation rates, growth rates, and the suspension density during the reactive crystallization of potassium sulfate at different temperatures.

The initial supersaturation S_0 was generated by mixing the $MgSO_{4(aq)}$ solutions from picromerite at 25 °C, and the saturated solution of $KCl_{(aq)}$, also at 25 °C. Mixing of both solutions was carried out at each of the temperatures under study (5, 15, 25, 35, and 45 °C) in the experimental setup shown in Figure 1. In each run, the change of CLD, absorbance and temperature was recorded using the FBRM, ATR-FTIR, and Pt-100 probes, respectively. For offline monitoring of the change of ion concentrations K^+, Mg^{2+}, Cl^-, and SO_4^{2-} over time, aliquots of the solution were taken every 2.5 min until 5 min, then every 5 min until 30 min, and finally every 10 min until 60 min. The weight of the aliquots was 0.3291 g of particle-free solution. At the end of the reactive crystallization process programmed for 60 min, the crystallized solid was separated from the pulp by vacuum filtration and dried for 24 h at room temperature. Afterwards, samples of the products obtained were subjected to XRPD analysis to check the phase identity and to chemical analysis by IC.

For the optical reactive crystallization process monitoring of nucleation, growth, and crystal morphology via the Crystalline PV particle vision system, the same conditions described above were applied, with a difference in the reagent amounts. In this case, 4.00 g of $MgSO_{4(aq)}$ from picromerite with 2.19 g of a saturated solution of $KCl_{(aq)}$ were studied in a closed vessel using a hook type stirrer at 400 rpm.

3. Results and Discussion

3.1. Absorbance and Temperature in the Reactive Crystallization Process

The results of monitoring the change in absorbance of the solution by ATR-FTIR spectroscopy are shown in Figure S7 (SI). The absorbance decreases as a function of time as the reaction progresses and moves to smaller absolute values at lower temperatures, caused by the lower saturation concentrations of the solute. The absorbance peaks of SO_4^{2-} and H_2O are found in the wavenumber range of 1080–1100 cm^{-1} and 1645 cm^{-1}, respectively. The corresponding temperature profiles are shown in Figure S8 (SI). As expected, after the addition of $KCl_{(aq)}$, stored at 25 °C, to the $MgSO_4$ solution at 5, 15, 25, 35, and 45 °C, the temperature in the resulting solution increased for lower reaction temperatures of 5 and 15 °C to 10 and 17 °C, and, decreased for higher reaction temperatures of 35 and 45 °C to 32.5 and 41 °C, respectively, for a short time only. For the reaction isotherm at 25 °C, the temperature remained constant. However, in all cases, in less than 3 min, the initial reaction temperature is restored except for a slight difference at 45 °C.

3.2. Supersaturation Profiles, Reactive Crystallization Images and CLD

The supersaturation as a function of time, S, was determined based on the concentration of the ions K^+, Mg^{2+}, Cl^-, and SO_4^{2-}. Assuming that the Mg^{2+} and Cl^- concentrations in the solution remain constant but K^+ and SO_4^{2-} concentrations vary due to the formation of $K_2SO_{4(s)}$ crystals, the concentrations of K^+ and SO_4^{2-} were obtained by subtracting those integrated into $K_2SO_{4(s)}$ crystals from those entering the reaction. The molal concentrations of m_{K^+}, $m_{SO_4^{2-}}$, $m_{Mg^{2+}}$ and m_{Cl^-} obtained were used to determine the activity coefficients and S, according to Equation (3).

Figure 2 shows the trajectory of S starting from the initial supersaturation S_0, images of solution/crystals captured by Crystalline PV, and the respective CLD as a function of time during the reactive crystallization of K_2SO_4 at 5 °C. The S_0 value in the reaction mixture was 5.15, showing a strong supersaturation as a precondition to initiate nucleation and growth. After an induction time of ~5 min where the reaction solution reached thermal equilibrium and became homogenized, the supersaturation decreased as a result of crystallization, reaching a minimum value of S = 2.30 at 20 min. The Crystalline PV system detected at 1.44 and 5 min of reaction (Figure 2a,b) embryo clouds and tiny pseudohexagonal crystals due to nucleation and growth. At the same time, after ~5 min, a low CLD in terms of crystals counted by the FBRM per s <2.5 crystals/s with a chord length of 20–30 microns and a rather unimodal CLD is determined. Once the driving force has been exhausted, after 20 min, the supersaturation remains constant and reactive crystallization is finished, which is corroborated by chemical analysis of the K^+, Mg^{2+}, Cl^- and SO_4^{2-} shown in Figure S9 (SI). At ~10 min, Figure 2c shows growing K_2SO_4 crystals, separated from each other but overlapping in some parts with defined vertices, maintaining the pseudohexagonal and orthorhombic morphology. The crystal count reached 25 crystals/s with a chord length between 30–40 microns. A few particles between 2 and 5 microns having a crystal count of 8 crystals/s also appear (Figure 2f), thus leading to a bimodal CLD.

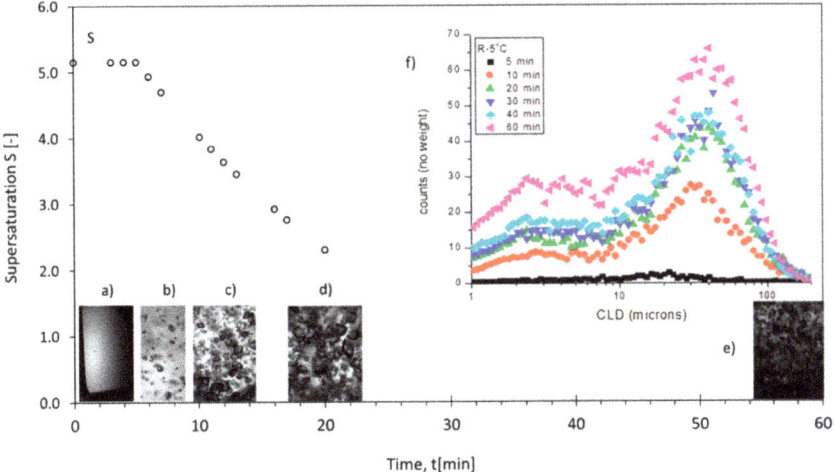

Figure 2. Supersaturation S, crystal/solution images and CLD for a reaction isotherm of 5 °C as a function of time at (**a**) 1.44 min, (**b**) 5 min, (**c**) 10 min, (**d**) 20 min, (**e**) 60 min and (**f**) CLD.

With the time at 20 min (Figure 2d), the crystals grow with a considerable presence of fines whose crystal count reached 42.5 crystals/s in the size range of 30–50 microns and 15 crystals/s in the range of 2–3 microns, keeping the bimodal CLD. Finally, at 6 min (Figure 2e), the pulp contains a considerable amount of fines that obstruct the image viewer of the Crystalline PV equipment; the crystal count reaches 70 crystals/s in the size range of 40–50 microns and 30 crystals/s in the 2–3 microns size range shown in Figure 2f. As seen, the counts of CLD increase with time, specifying product crystals with a bimodal CLD, which is attributed to breakage, attrition by friction or collision of crystals, or crystals with the propeller blades. Thus, crystal count increased but not growth.

At a reactive crystallization temperature of 25 °C, an initial supersaturation of $S_0 = 3.84$ is generated (Figure 3), which remains constant for ~10 min and then decreases slowly, reaching the final S of 2.53 at 30 min. This implies that the supersaturation was consumed in 20 min and the crystallization process was stopped. Figure 3a,b show the presence of very tiny crystals with a crystal count <2.5 crystals/s in the chord length range of 10–40 microns (Figure 3e). After 30 min (Figure 3c), it is impossible to identify the crystals' morphology due to their tiny sizes, with a crystal count of 20 crystals/s in a chord length range of 20–30 microns and 15 crystals/s with the size of 9–10 microns. After 60 min (Figure 3d), many tiny crystals are observed with a crystal count of 30 crystals/s between 30–40 microns size and 18.5 crystals/s between 3–6 microns size. Thus, again a bimodal CLD was developed as a result of the generation of fine particles from secondary nucleation mechanisms.

Comparing the reactive crystallization processes of K_2SO_4 at the two temperatures of 5 and 25 °C, as expected, the initial degree of supersaturation is greater at lower temperature and vice versa. As a result, for higher initial supersaturation, de-supersaturation and induction times were shorter, and crystal growth was enhanced, leading to a defined morphology, and a higher crystal count was reached with bimodal CLD. As the crystal images provide qualitative proof of the crystallization progress, in this report, the moment data collected by the FBRM probe were considered to determine the crystallization kinetics. The supersaturation profiles for all reaction temperatures used are compiled in Figure S10 (SI). They confirm the reaction crystallization trends discussed above.

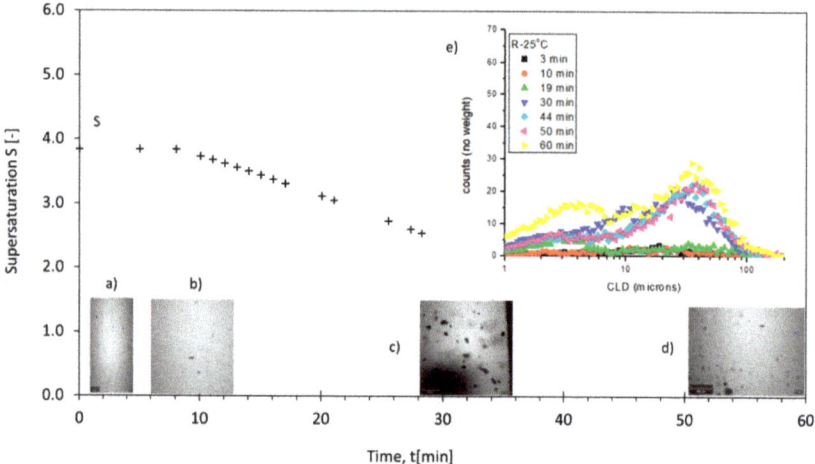

Figure 3. Supersaturation S, crystal/solution images and CLD for a reaction isotherm of 25 °C as a function of time at (**a**) 1.20 min, (**b**) 10 min, (**c**) 30 min, (**d**) 60 min, and (**e**) CLD.

The initial supersaturations S_0 generated by reactive crystallization in the reciprocal system are large with values of S_0 = 5.15, 4.13, 3.84, 3.54, and 3.24 at 5, 15, 25, 35, and 45 °C, respectively. In contrast, the S_0- values used in solution crystallization by cooling methods for the binary K_2SO_4-H_2O system are usually lower. For example, Bari and Pandit [24] worked in a range of S_0 = 1.07–1.11 when cooling a saturated solution of K_2SO_4-H_2O from 60 °C to a temperature between 53–49 °C. Gougazeh et al. [25] worked with $S \cong 0.00913$ at 50 °C and Lyczko et al. [51] with S_0 = 1.31 at 30 °C. Mohamed et al. [30] used an initial supersaturation of S_0 = 1.07 for a controlled cooling in a temperature range of 63.5 to 24.6 °C. Furthermore, Garside and Tavare [52] applied S_0 = 1.09, Mullin and Gaska [27] worked with $S \cong 1.07$ at 20 °C, and Garside et al. [53] with S = 1.15 at 30 °C. On the other hand, in reactive crystallization processes of sparingly soluble salts, the S_0 generated is much larger than for the reciprocal salt system studied in this work. Therefore, crystal nucleation and growth events occur within a few seconds of the induction time. Lu et al. [32] report an initial supersaturation of S_0 = 1596, with an induction time of 26 s at 15 °C to crystallize $Mg(OH)_2$; Taguchi et al. [31] an S_0 = 70 with an induction time of 60 s to crystallize $BaSO_4$, Steyer C. [54] an S_0 = 500 to crystallize $BaSO_4$, and Mignon et al. [33] values of S_0 = 771 and 960 to crystallize $SrSO_4$ and $CaCO_3$, respectively. This results from the fact that sparingly soluble salts exhibit minimal values of the thermodynamic equilibrium constant (K_{SP}). For example, the K_{SP} of $BaSO_4$ is 2.88×10^{-10} (mol/kg)^2 at 25 °C [31], compared with a K_{SP} of 0.0162 (mol/kg $H_2O)^3$ for K_2SO_4 according to the solubility data [45].

Figure 4 presents the CLDs of K_2SO_4 crystals after 5 min and the final product after 60 min at the different reaction temperatures investigated. Usually, for an industrial mass crystallization product, a narrow and unimodal CLD, as far as possible free of fine crystals, is targeted. However, in the experiments, a unimodal CLD is only observed at the beginning after 5 min of reaction (Figure 4a), where crystal counts are very low (1 to 4 crystals/s) with crystal sizes between 10–100 µm and fines <10 µm, which is found for all the isotherms studied. Likewise, Figure 4b shows the CLDs of the final K_2SO_4 product with a bimodal distribution. The region of fines of size <10 µm is specified by crystal counts <2 crystals/s at 45 °C, 30 crystals/s at 5 °C, and 10–15 crystals/s for isotherms in-between. In the coarse region, the size of the crystals is 15–200 µm with crystal counts of 5 crystals/s at 45 °C, 70 crystals/s at 5 °C, and 25–35 crystals/s for isotherms between 15 and 35 °C. The bimodal distribution is attributed to the effect of secondary nucleation due to the suspension density or breakage and attrition of the crystals. The difference between Figure 4a,b is related to uncontrolled depletion of supersaturation during isothermal reactive crystallization.

Therefore, supersaturation is independent and predominant in reactive crystallization as the crystals simultaneously grow while others are born due to the motive force. As a result, the final CLD of the K_2SO_4 product is not narrow and thus not favourable for subsequent downstream processes such as filtration, drying, storage and transport, and the resistance to relative humidity as well due to the presence of fines <10 μm that can generate lumps, dust release, moisture absorption, etc.

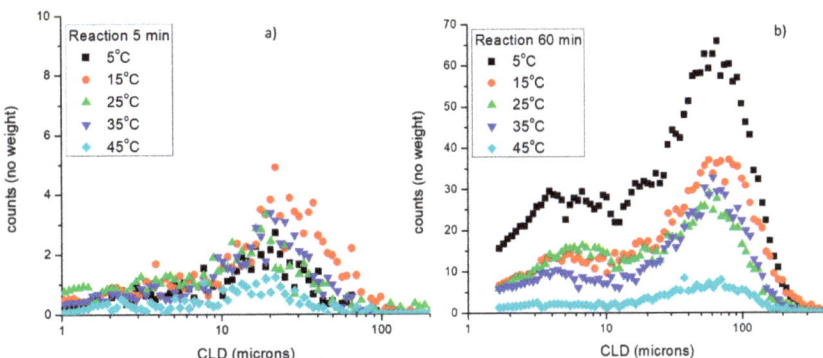

Figure 4. CLDs obtained (**a**) at 5 min and (**b**) at 60 min, i.e., the final product, for different reaction isotherms.

In this work, the maximum K_2SO_4 crystal size obtained according to the semiquantitative analysis of CLD was between 70 and 80 μm at 5 °C, which is small compared with crystallization from pure K_2SO_4-H_2O solutions. For example, Bari and Pandit [24] reported an average size of 286 μm via the isothermal method for low supersaturation and Bari et al. [22] an average size of 250 μm by the cooling method. Jones et al. [55] received an average size of 500 μm without fines using seeding and cooling, while via salting out with acetone, they obtained crystal sizes <350 μm with a greater presence of fines and agglomerates. It is concluded that reactive crystallization of K_2SO_4 from the quaternary K^+, Mg^{2+}, Cl^-, and SO_4^{2-} system provides smaller crystals as obtained in K_2SO_4-H_2O systems, which is due to the reaction-inherent high degree of initial local supersaturation.

3.3. Primary Nucleation Rate B_b

To determine the nucleation rate, the CLD data have been converted to CLD counts square-weighted as a function of time for each reaction isotherm. The results transformed to moment data of the order 0, 1, 2, and 3 by Equation (10) are shown in Figures S2–S5 (SI).

In the reactive crystallization of K_2SO_4, primary nucleation (B_b) and secondary nucleation (B) occur, while B_b results from the high supersaturation generated. The nucleation rate (B) of K_2SO_4 was determined by Equation (11). B_b is related with S_0 since it has been obtained from the thermodynamic approach when considering the activities of the species K^+ and SO_4^{2-} in the multicomponent system K^+, Mg^{2+}/Cl^-, $SO_4^{2-}//H_2O$. To determine the kinetic order and rate constant parameters of B_b, the nucleation rates obtained for each S_0 were averaged and then fitted to Equation (12) presented in Figure 5. The fitting parameters derived are $b = 3.61$ and $k_b = 83.68$ [#/min·kg H_2O] with a coefficient of determination $R^2 = 0.89$, verifying that the behaviour of B_b is proportional to S_0.

The primary nucleation order obtained for the reactive crystallization of K_2SO_4 is lower than found for the single solute system K_2SO_4-H_2O, as reported by Bari et al. [22], with $b = 6.5$ for the isothermal method (t_{ind}). Nemdili et al. [21] reported $b = 4.10$ and 4.68 when measuring MSZW and t_{ind}, respectively. Additionally, Bari and Pandit [24] reported $b = 6.5$ and 5.73 when determining MSZW via the conventional method and sonocrystallization, respectively. There, b is a physical parameter that describes the dependence of the MSZW on the cooling rate regardless of the method used. In contrast, high values for the primary nucleation order are observed in reactive crystallization kinetics of sparingly soluble salts, such as $SrSO_4$ and $CaCO_3$ with $b = 36.0$ and 12.0, respectively, reported by

Mignon et al. [33]. In this report, the empirical value of b = 3.61 for the reactive crystallization of K_2SO_4 has no relation to the MSZW but classifies into the range established by the value of the primary nucleation order for inorganic compounds obtained by cooling, which, as a general rule, are found between 0.98 and 8.3 [23].

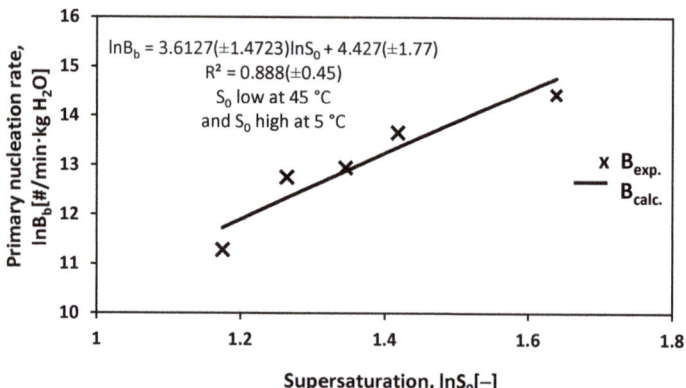

Figure 5. Correlation of the primary nucleation rate with S_0 for reactive crystallization of K_2SO_4.

3.4. Secondary Nucleation Rate B

The profiles of the nucleation rates B during reactive crystallization of K_2SO_4 at different temperatures are shown in Figure 6. At 5 °C (with S_0 = 5.15), B decreases from a maximum of 2.68 × 10^6 (#/min·kgH$_2$O) at 3 min to a minimum of 8.35 × 10^5 (#/min·kgH$_2$O) at 20 min, which implies the consumption of supersaturation due to spontaneous nucleation. The decrease in B is attributed to the presence of crystals within the solution since the growth of crystals consumes the supersaturation, the concentration resulting from the suspension density is higher, which is also observed for the 15 °C isotherms. However, at 25 °C (with S_0 = 3.84) B starts with 7.7 × 10^4 (#/min·kgH$_2$O) at 10 min reaching a maximum of 8 × 10^5 (#/min·kgH$_2$O) at ~30 min of crystallization. Clearly B increases, with the increase being slight and attributed to slow crystal growth consuming less supersaturation. The concentration resulting from the suspension density is lower, a behaviour also seen for the 35 and 45 °C isotherms. Therefore, it is concluded that the greater S_0, the greater is B, and it is depleted in less time than with less S_0. This statement agrees with the report by Bari and Pandit [24], also referring to reactive crystallization.

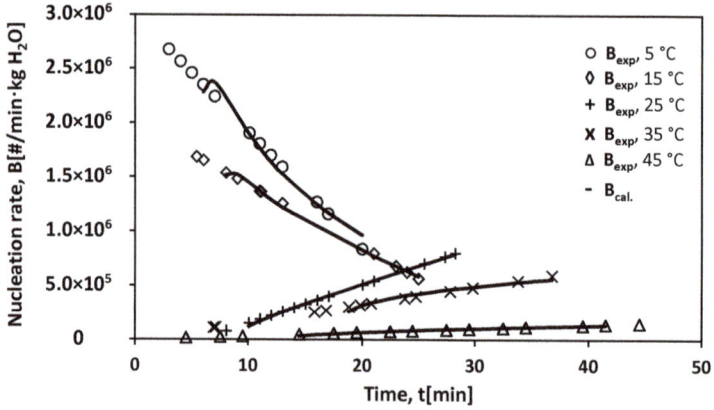

Figure 6. Experimental and calculated nucleation rates B_{exp} [#/min·kg H_2O] (symbols) and B_{cal} (line) as a function of time for different S_0 obtained at 5, 15, 25, 35 and 45 °C.

Once the nuclei appear, they become stable crystals and grow over time. However, nucleation persists assisted by crystals constituted as suspension density, supersaturation decreases to a low level, and it is there that secondary nucleation occurs (B). Therefore, secondary nucleation is often dependent on suspension density (M_T) and supersaturation (S). Furthermore, growth (G) is also a function of S. Then, the relationship of B with G and M_T describes the secondary nucleation rate of K_2SO_4 in the reactive crystallization process with time and is given by Equation (13). The parameters K_R, i, and j are estimated by adjusting the experimental data of G and M_T with time using multiple linear regression to obtain the secondary nucleation rate calculated (B_{cal}) for each S_0.

The results of B_{exp} and B_{cal} are shown in Figure 6. A good correlation of secondary nucleation with G and M_T is observed for most of the isotherms. The B_{exp} and B_{cal} values have practically the same tendencies, superimposed, at 5, 15, 25 and 45 °C (S_0 = 5.15, 4.15, 3.84 and 3.24), with a coefficient of determination of R^2 = 0.976, 0.997, 0.998, and 0.979, respectively, and a slightly lower correlation for the 35 °C isotherm (S_0 = 3.54), with R^2 = 0.943.

The empirical parameters K_R, i, and j obtained for each reaction isotherm are compiled in Table S6 (SI). From the results, it can be concluded that B depends on the density of the suspension, whose values of j are between 0.20 and 0.80. However, i has values of 1.18, 0.63 and 0.1 for the 5, 15, and 25–45 °C isotherms, which is attributed to the fact that B depends on G implied by the supersaturation, whose value of i is greater than j for isotherms of 5 and 15 °C. For the 25–45 °C isotherms, i is less than j, suggesting little dependence on B for G. On the other hand, the K_R values show irregular behaviour, as seen in the trends of B in Figure 6.

To finally evaluate secondary nucleation via Equation (13), the G and M_T values for different S_0 (Figure 6) were fitted to the following Equation:

$$B = 13810.83 \cdot G^{0.75} \cdot M_T^{0.71} \tag{21}$$

This general description of secondary nucleation in the reactive crystallization process of K_2SO_4 leads to a relatively low coefficient of determination R^2 = 0.11. Therefore it is difficult to attribute the dependence of B to the suspension density or crystal growth rate, whose values of i and j are 0.75 and 0.71, respectively, and thus, both are close to unity [20]. For the crystallization processes in the single solute K_2SO_4-H_2O system via cooling and seeding, Mohamed et al. [30] reported i = 0.9 and j = 0.57 and concluded that B depends on S and M_T.

The presence of $K_2SO_{4(s)}$ in the reactive crystallization solution contributes to nucleation so that the exponents i and j take on characteristic values. However, the primary nucleation order (b) depends solely on the degree of supersaturation and the method of detecting the appearance of nuclei, so the nucleation order is greater than the secondary one. Bari et al. [22] reported the dependence of secondary nucleation order (b2) on supersaturation and M_T. They found b2 = 2.25 and b = 6.5 for the K_2SO_4-H_2O system, obtained by the isothermal method and, as evident, b2 < b. Taguchi et al. [31] observed in the reactive crystallization process of $BaSO_4$, when mixing two equimolar solutions of $BaCl_2$ and Na_2SO_4 at 25 °C, that secondary nucleation is influenced by the stirring speed (N_s), M_T and S, whose exponents are 0.98, 0.84 and 1.72, respectively, with a multiple correlation coefficient of 0.61. The K_R parameter implies the dependence of temperature, hydrodynamics, presence of impurities and the properties of the established crystals. Thus, the K_R data obtained for each reactive crystallization isotherm of K_2SO_4 is used to estimate the secondary nucleation activation energy in the following.

3.5. Crystal Growth Rate of K_2SO_4

The crystal growth rate (G) of K_2SO_4 was simultaneously determined from the square weighted CLD data, according to Equation (14). Figure 7 shows the growth rate of K_2SO_4 crystals as a function of time during the reactive crystallization process for different S_0. At 5 °C (with S_0 = 5.15), G starts with a value of 231.06 µm/min determined at 4.01 min and reaches a minimum value of 4.94 µm/min after 20 min. The decrease in G is caused by the decrease in supersaturation due to crystallization. At 25 °C (with S_0 = 3.84), G has an initial value of 101.5 µm/min and decreases over time to 5.30 µm/min at ≈30 min.

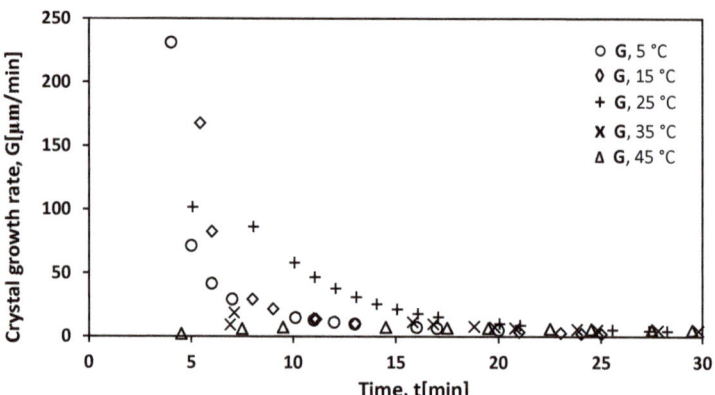

Figure 7. The growth rate of K_2SO_4 as a function of time at different S_0 generated by reactive crystallization isotherms at 5, 15, 25, 35 and 45 °C.

As seen in Figure 7, the crystal growth rate is low due to the predominance of 0th. moment values (μ_0) according to Equation (14). It is estimated that the nuclei are in a metastable state; therefore, the decrease in supersaturation is slight. Then, G becomes predominant, and S decreases.

As expected, the initial growth rates in the K_2SO_4 reactive crystallization are significantly higher than observed in the binary K_2SO_4-H_2O system. Bari and Pandit [24] reported a G of 8.82 µm/min at 49 °C studying K_2SO_4 growth by microscopy. Mohamed et al. [30] reported a maximum growth rate of 6 µm/min at 40 °C with seeding, based on population balance data measured in the multichannel Coulter Counter. Mullin and Gaska [27] reported single K_2SO_4 crystal growth rates of 4.68 and 5.16 µm/min for faces 100 and 001 at 20 °C, respectively. In the present report, a maximum growth rate of 231 µm/min was observed at 5 °C for the reciprocal quaternary system K^+, Mg^{2+}/Cl^-, SO_4^{2-}/H_2O using the FBRM probe in situ, without the need to take samples and without seeding.

With the G data for different degrees of S_0, the empirical fit parameters were determined according to Equation (15). As shown in Figure 7, the growth rates steadily decreased from an initial maximum to a minimum for all isotherms. However, for reaction isotherms 35 and 45 °C, G increased at the beginning and then diminished to the minimum due to depletion of supersaturation. Thus, the crystals grew, as seen in Figure 2a–c. To correlate G as a function of S_0, the G values of each isotherm were averaged to estimate the empirical parameters k_g and g. Figure 8 shows the correlation and adjustment of G versus S_0, obtaining a value of g = 4.64 and k_g = 0.028 (µm/min) with a coefficient of determination of R^2 = 0.761.

In reactive crystallization, the high supersaturation promotes both nucleation and crystal growth, leading to a fast [31] and relatively uncontrolled decrease in supersaturation. Thus, g is much higher due to the higher growth rate that implies higher S_0. However, in reactive crystallization processes for sparingly soluble salts, Taguchi et al. [31] correlated the G of $BaSO_4$ with the stirring speed (N) and S_0. They mention that the influence of N implies the occurrence of growth controlled by diffusion, while the order with respect to S_0 indicates that both diffusion and surface reaction exert some influence on growth rates. Tavare and Gaikar [47] correlated the growth rate of salicylic acid crystals with solution concentration and stirring speed. They report that due to the complexity of determining supersaturation, S was omitted in the correlation. Therefore, the adjusting exponents of both the concentrations of salicylic acid and the agitation speed prevented the attribution of any mechanism of influence on growth.

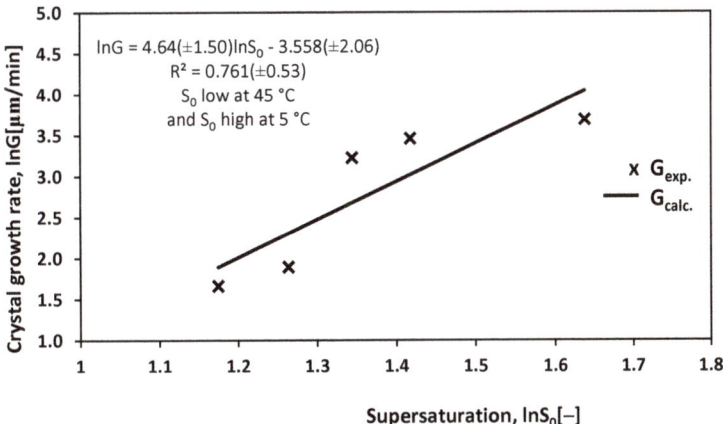

Figure 8. The fit of the empirical parameters g and k_g.

In this report, the value of g = 4.64 is strongly influenced by the initial supersaturation. When S_0 is higher for 5, 15, and 25 °C, a higher crystal growth rate is promoted. However, the growth rate is much lower for S_0 at 35 and 45 °C. Therefore, when adjusting G vs. S_0 (obtained for 5–45 °C) to an empirical Equation, the slope is greater than those known for the K_2SO_4-H_2O systems attributable to growth mechanisms. Thus, for future studies, it is proposed to develop a strategy that allows determining the growth mechanism of K_2SO_4, taking advantage of the generation of supersaturation by reaction and the initial presence of fine crystals so that a cubic cooling profile can be used to control supersaturation. As a result, a fairly narrow CSD and acceptable average crystal size [56,57] would be feasible.

On the other hand, there are numerous reports for the K_2SO_4-H_2O system for the isolated determination of the growth mechanisms in crystallization processes with low supersaturation controlled by a cooling technique. In these works, generally, the mechanism that controls the growth of K_2SO_4 is diffusion [21,25–27].

3.6. Crystal Suspension Density M_T

The variation of the suspension density with respect to time was determined by Equation (16), with the results presented in Figure S11 (SI). However, the absolute suspension density was identified via empirical Equation (17) to relate the amount by weight of crystals obtained gravimetrically at the end of reactive crystallization. The results allowed estimating the amount of potassium and sulfate ions in the solution during the reaction to later be used to determine the supersaturation profile.

In Figure 9, the absolute suspension densities at different reactive crystallization isotherms are shown as a function of time. At 5 °C (S_0 = 5.15), the suspension density increased linearly from 6.84 g of K_2SO_4/kg H_2O at 6 min to 108.45 g of K_2SO_4/kg H_2O at the end of the reactive crystallization at 20 min. At 25 °C (S_0 = 3.84), an increase in M_T from 3.95 g of K_2SO_4/kg H_2O at 10 min to 56.68 g of K_2SO_4/kg H_2O at the end of the reactive crystallization at 30 min is obtained. Thus, the higher S_0, the higher the suspension density reached in less time. The final M_T for the isotherms at 15, 35 and 45 °C, with the respective S_0 of 4.13, 3.54 and 3.24, were 67.41, 47.03 and 21.29 g of K_2SO_4/kg H_2O, respectively.

The suspension density (M_T) targeted by crystallization is limited by the solubility, and the S_0 reached in the system under study. It depends on the crystallization method and is further affected by the morphology and size distribution of the crystals, filtration demands, and the cocrystallization of an undesired compound. Mohamed et al. [30] reported an M_T between 58 and 95 (g of K_2SO_4 crystals/kg H_2O) by cooling from 63.5 to 24.6 °C, and 75 (g of K_2SO_4 crystals/kg H_2O) at 40 °C for a K_2SO_4-H_2O solution with seeding. In this report of reactive crystallization, a maximum M_T of 108.45 g of K_2SO_4/kg H_2O at 5 °C was obtained and was limited by the potential cocrystallization of another compound from the reciprocal salt pair.

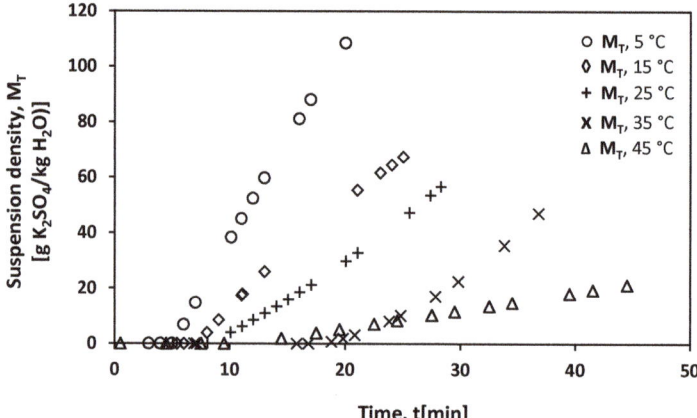

Figure 9. Suspension density of K_2SO_4 as a function of time at different S_0 generated by reactive crystallization isotherms at 5, 15, 25, 35 and 45 °C.

The product yield was estimated based on the suspension density data at the end of the reactive crystallization process. For the 5, 15, 25, 35, and 45 °C isotherms, the yields achieved were 72.6%, 45.2%, 37.9%, 31.5%, and 14.3% of K_2SO_4, respectively. As expected, the higher the S_0 in the system, the higher the yield and the less reactive crystallization time is required. The yield was determined by relating the amount of K_2SO_4 obtained experimentally with that calculated by stoichiometry. The mass balances of the reactive crystallization experiments established from the suspension density and the solute concentration (as obtained by Equations (17) and (18)) are depicted in Figure S12 (SI). While the solute concentration decreases, the suspension density increases due to the consumption of supersaturation by crystallization.

The solid K_2SO_4 obtained at 5 °C represents a solid concentration of 11 wt/wt% in the pulp, which is highly favourable for process monitoring by FBRM, as recommended by Senaputra et al. [58], who recommend a pulp concentration not greater than 20 wt/wt%.

3.7. Activation Energy E

As mentioned above, the dependence of the secondary nucleation rate constant K_R on the absolute temperature enables estimating the secondary nucleation activation energy E via the Arrhenius Equation (20). The results for the reactive crystallization process studied are presented in Figure 10, specifying the secondary nucleation activation energy as $E = 69.83$ kJ·mol^{-1}. However, to obtain E, the K_R values were considered as a function of the temperature that best fit. In addition, the Excel solver has been used under the restrictions of $j \geq 1$ for the isotherms of 5, 15, and 25 °C, also to $i \geq 1.5$ and 0.12 for the isotherms of 35 and 45 °C, respectively, which improves the determination coefficient ($R^2 = 0.92$) in the activation energy estimation. As seen in Figure 10, the secondary nucleation activation energy is higher in the 5–25 °C segment due to the higher S_0 promoting secondary nucleation, while S_0 is lower in the 25–45 °C range with the consequence of a lower E.

In general, the secondary nucleation activation energy of 69.83 kJ·mol^{-1} obtained for the reactive crystallization process of K_2SO_4 obeys the principle of positivity. However, it is believed that the activation energy for reactive crystallization primary nucleation of K_2SO_4 would be much lower due to the rate K_2SO_4 nucleated from a crystal-free solution. As there are no similar works, they cannot yet be compared with other values related to the reactive crystallization kinetics of soluble salts. However, for reference, Luo et al. [23] reported the activation energy of primary nucleation for cooling crystallization in the K_2SO_4-H_2O system to be 33.99 kJ·mol^{-1}, much less than obtained for the secondary nucleation of K_2SO_4

by reactive crystallization in this work. On the other hand, for reactive crystallization of a poorly soluble salt, Lu et al. [32] reported a nucleation activation energy of 73.049 kJ·mol^{-1} for Mg(OH)$_2$. As seen, when nucleation depends only on supersaturation, the activation energy of primary nucleation is much lower, implying that nucleation is faster than obtained in the present report, where the activation energy of secondary nucleation depends on the density of suspension.

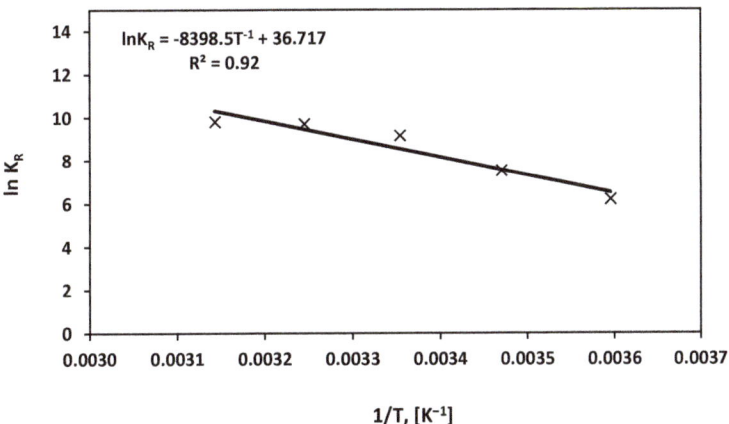

Figure 10. Dependence of the secondary nucleation rate constant K_R on the reciprocal reaction temperature.

3.8. K_2SO_4 Product Quality

The crystals of K_2SO_4 obtained as a final product in the reactive crystallization experiments were subjected to X-ray diffraction analysis to be compared with the reagents K_2SO_4, KCl, and $MgSO_4$ and synthesized picromerite. The diffractograms are compiled in Figure 11. In addition, the product crystals were analyzed regarding the presence of Magnesium as an impurity by ion chromatography. As a result, only very small contents of 0.51, 0.11, and 0.01 wt% of Magnesium were detected in the products of reaction temperatures 5, 15, and 25, 35, and 45 °C, respectively. The X-ray patterns in Figure 11 and the low residual Mg amounts in the crystals prove that the K_2SO_4 obtained is of high quality. The diffractogram obtained for the K_2SO_4 product at 5 °C exhibits several intense peaks that correspond to the patterns of K_2SO_4 and picromerite, verifying the somewhat higher Mg content in the respective product. K_2SO_4 is the majority phase, and no other crystalline phase is present, considering a detection limit of XRPD <1%. The presence of picromerite peaks might be attributed to mother liquor occluded between the K_2SO_4 crystals which, after filtration, crystallizes as picromerite when drying. In this work, the washing process of the K_2SO_4 product was omitted. Of course, when including washing after solid–liquid separation, a further increase of purity is possible but at the expense of yield. The washing process is essential to improve the quality of K_2SO_4, as accomplished in the crystallization process of K_2SO_4 from the KCl$_{(s)}$ picromerite$_{(s)}$-H$_2$O system reported by Fezei et al. [12–15].

It is important to mention the presence of the eutectic point in the system, which allows deriving the lowest feasible crystallization temperature. In this sense, despite the reactive crystallization of K_2SO_4 from the multicomponent system K$^+$, Mg^{2+}/Cl$^-$, SO$_4^{2-}$//H$_2$O, the remaining solution still contains salts like MgSO$_{4(aq)}$, KCl$_{(aq)}$, and MgCl$_{2(aq)}$ that could provide more K_2SO_4 at reaction temperatures below 5 °C. The eutectic points of the salts with solubilities are 7.29 g of K_2SO_4/100 g of saturated solution at −1.9 °C, 19 g of MgSO$_4$/100 g of saturated solution at −3.9 °C, 19.87 g of KCl/100 g of saturated solution at −10.8 °C, and 21.0 g of MgCl$_2$/100 g of saturated solution at −33.6 °C [45].

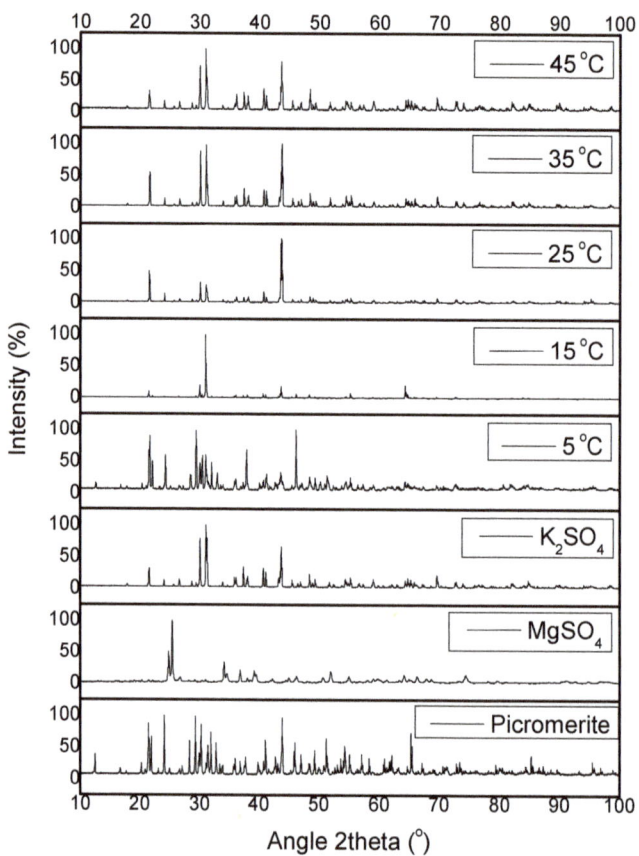

Figure 11. X-ray powder patterns of K_2SO_4, $MgSO_4$, picromerite and K_2SO_4 product obtained at reactive crystallization temperatures of 5, 15, 25, 35 and 45 °C (from bottom to top).

Based on the quality of the K_2SO_4 obtained by reactive crystallization at 5 °C and the eutectic points mentioned above, they allow working at reactive crystallization temperatures below 5 °C. In addition, this enables generating higher S_0 to improve the process performance since the eutectic point of K_2SO_4 is -1.9 °C, and it is estimated that this temperature is even much lower in the presence of K^+, Mg^{2+}, Cl^-, and SO_4^{2-} ions. In this regard, Song et al. [59] mention that the solubility of the precipitating compound in a reactive crystallization process increases in the presence of ions in the system; for example, for $CaSO_4$, the solubility increases in the presence of Cl^-, H^+, Ca^{2+}, and SO_4^{2-} ions at 60 °C.

4. Conclusions

Based on the research on the reactive crystallization kinetics of K_2SO_4 from $KCl_{(aq)}$ with $MgSO_{4(aq)}$ from picromerite, several conclusions can be drawn. The S_0 obtained is inversely proportional to the reactive crystallization temperature of K_2SO_4 and, the S_0 was sufficient to promote nucleation and crystal growth at all reactive crystallization conditions used. The CLD obtained at different S_0 is unimodal in the first minutes of reaction and bimodal in the final K_2SO_4 product. The bimodal CLD can be attributed to growth, secondary nucleation, and suspension density due to the higher S_0 generated by the reaction. The presence of bimodal CLD's is a reflection of the secondary nucleation effect and is unfavourable for subsequent processes such as filtration, drying, storage, etc. On the other hand, real time images captured during reactive crystallization evidenced the appearance and growth of crystals with pseudohexagonal morphology.

Online monitoring of the CLD and using square-weighted CLD counts shows that B_b, G, and M_T are directly proportional to S_0. The primary nucleation parameters were determined with the order $b = 3.61$ and constant $k_b = 83.68$ [#/min·kg H_2O] by correlation of B_b with S_0. The b-value indicates that the primary nucleation strongly depends on the supersaturation generated by the reaction in the K^+, Mg^{2+}/Cl^-, $SO_4^{2-}//H_2O$ system and the primary nucleation rate quantification method. In addition, G has been correlated with S_0 to estimate the empirical parameter $g = 4.61$. The allocation of one of the growth mechanisms, such as transport and surface reaction, to the mechanism controlling growth in the performed reactive crystallization of K_2SO_4 is generally challenging due to the rapid mass transfer of the solute to the solid phase in reactive crystallization processes. The K_2SO_4 crystals obtained were of high quality containing (unwashed) 0.01–0.51 wt% of magnesium as impurity under the conditions used.

In general, it can be concluded that it was possible to estimate the degree of S_0 and the trajectories of S with time in the K^+, Mg^{2+}/Cl^-, $SO_4^{2-}//H_2O$ system at 25 °C by directly varying the reaction temperature to produce soluble salt crystals like K_2SO_4. Furthermore, all the S_0 values obtained at different reaction isotherms were sufficient to promote the crystallization parameters. However, to improve the quality and performance of the crystals' CSD, it is suggested to apply a programmed cubic cooling profile to the studied system since the coexistence of crystals/solution, e.g., at 45 °C, could replace a seeding stage.

Supplementary Materials: The following are available online at https://www.mdpi.com/article/10.3390/cryst11121558/s1. Table S1: Empirical values to determine the ionic interaction parameters of KCl, Table S2. Empirical values to determine the ionic interaction parameters of $MgCl_2$, Table S3. Empirical values to determine the ionic interaction parameters of $MgSO_4$, Table S4. Solubility product for potassium sulfate at different temperatures based on solubility data, Table S5. Experimentation plan for reactive crystallization of potassium sulfate in batch crystallization mode[a], Table S6. Summary of empirical parameters of secondary nucleation rate and coefficient of determination at different temperatures, Figure S1. CLD and square-weighted count, for the final product of K_2SO_4 crystals at 5 °C, Figure S2. 0th moments μ_0(#/kg H_2O) as a function of time for different S_o generated at 5, 15, 25, 35 and 45 °C during reactive crystallization of K_2SO_4. Based on square-weighted counts, Figure S3. First moment, μ_1 (μm/kg H_2O) of K_2SO_4 as a function of time for different S_o generated at 5, 15, 25, 35 and 45 °C during reactive crystallization of K_2SO_4. Based on square-weighted counts, Figure S4. Second moment, μ_2(μm^2/kg H_2O) of K_2SO_4 as a function of time for different S_o generated at 5, 15, 25, 35 and 45 °C during reactive crystallization of K_2SO_4. Based on square-weighted counts, Figure S5. Third moment, μ_3(μm^3/kg H_2O) of K_2SO_4 as a function of time for different S_o generated at 5, 15, 25, 35 and 45 °C during reactive crystallization of K_2SO_4. Based on square-weighted counts, Figure S6. Dissolution process of picromerite in water at 25 °C. Figure S7. Absorbance profiles of the solution during reactive crystallization at different temperatures, Figure S8. Temperature profiles during reactive crystallization at different temperatures, Figure S9. Behavior of K^+, Mg^{2+}, Cl^- and SO_4^{2-} (g. ion/kg H_2O) as a function of time during the reactive crystallization process of potassium sulfate for S_0 at 5 °C, Figure S10. Supersaturation profiles S as a function of time for different S_0 generated at different reaction isotherms: (a) 5 °C, (b) 15 °C, (c) 25 °C, (d) 35 °C and (e) 45 °C, Figure S11. Suspension density rate of K_2SO_4 (dM_T/dt) as a function of time for different degrees of local initial supersaturation obtained at 5, 15, 25, 35 and 45 °C, Figure S12. Suspension density, M_T (grams K_2SO_4 crystals/kg H_2O) and solute concentration, C (grams solute/kg H_2O) as a function time for 5 to 45 °C isotherms.

Author Contributions: Conceptualization, A.A. and Y.P.J.; methodology, A.A., Y.P.J. and H.L.; formal analysis, A.A. and Y.P.J.; investigation, A.A.; writing—original draft preparation, A.A.; writing—review and editing, Y.P.J., T.A.G. and H.L.; visualization, A.A.; supervision, T.A.G. and H.L.; funding acquisition, A.A. and T.A.G.; All authors have read and agreed to the published version of the manuscript.

Funding: This research was funded by Universidad Autónoma Tomas Frías from Potosí Bolivia 2016–2019.

Data Availability Statement: Not applicable.

Acknowledgments: We are thankful to Universidad Autónoma Tomás Frías (U.A.T.F.) from Potosí Bolivia for declaring the first author in commission via Consejo Carrera de Química, Consejo Facultativo de Ciencias Puras, Comisión Académica and the Honorable Consejo Universitario de la U.A.T.F. 2016–2019. Further, we thank the Universidad de Antofagasta (U.A.) Chile, Programa Doctorado en Ingeniería de Procesos Minerales and the Max Planck Institute for Dynamics of Complex Technical Systems in Magdeburg/Germany for the support of the research internship.

Conflicts of Interest: The authors declare no conflict of interest.

References

1. Sanchez, J. Fertirrigación, Pricipios, Factores y Aplicaciones. In *Seminario de Fertirrigación Apukai-Comex*; Apukai-Comex: Lima, Perú, 2000; pp. 1–26.
2. Vergara-Edwards, L.; Parada-Frederick, N. Study of the Phase Chemistry of the Salar de Atacama Brines. In *Sixth International Symposium on Salt*; Salt Institute-Documentary: Santiago, Chile, 1983; Volume II, pp. 345–366.
3. Alonso, H. Origen De Los Componentes Y Balance Salino. *Rev. Geol. Chile.* **1996**, *23*, 113–122.
4. Risacher, F. *Estudio Econòmico Del Salar de Uyuni*; ORSTOM en Bolivie Mission de la Paz: La Paz, Bolvia, 1989.
5. Hidemaro, I.; Kenichi, H. Process of Producing a Potassium Sulfate Salt. U.S. Patent US4342737A, 3 August 1982.
6. Abu-eishah, S.I.; Bani-kananeh, A.A.; Allawzi, M.A. K_2SO_4 Production via the Double Decomposition Reaction of KCl and Phosphogypsum. *Chem. Eng. J.* **2000**, *76*, 197–207. [CrossRef]
7. Holdengraber, C.; Lampert, S. Process for Producing Potassium Sulfate from Potash and Sodium Sulfate. U.S. Patent US6143271A, 7 November 2000.
8. Grzmil, B.U.; Kic, B. Single-Stage Process for Manufacturing of Potassium Sulphate from Sodium Sulphate. *Chem. Pap.* **2005**, *59*, 476–480.
9. Mubarak, Y.A. Integrated Process for Potassium Sulfate and a Mixture of Ammonium Chloride/Potassium Sulfate Salts Production. *Int. J. Eng. Technol.* **2018**, *7*, 185–197. [CrossRef]
10. Marx, H.; Kaps, S.; Schultheis, B.; Pfänder, M. Potassium Sulfate—A Precious by-Product for Solar Salt Works. In Proceedings of the World Salt Symposium 2018, Park City, UT, USA, 11 June 2018.
11. Ryan, J.H.; Lukes, J.A.; Neitzel, U.E.; Duyster, H.P.; Great Salt Lake Minerals and Chemicals Corp. Method for the Production of Potassium Sulfate from Potassium-Containing Double Salts of Magnesium Sulfate. U.S. Patent US3634041A, 12 January 1972.
12. Jannet, D.B.; M'nif, A.; Rokbani, R. Natural Brine Valorisation: Application of the System K^+, Mg^{2+}/Cl^-, SO_4^{2-}/H_2O at 25 °C. *Desalination* **2004**, *167*, 319–326. [CrossRef]
13. Voigt, W. Solubility of Inorganic Salts and Their Industrial Importance. In *Developments and Applications in Solubility*; RSC Publishing: Freiberg, Germany, 2007; Chapter 24; pp. 390–406. [CrossRef]
14. Fezei, R.; Hammi, H.; Adel, M. Study of the Sylvite Transformation into Arcanite At 25 °C. *World J. Agric. Sci.* **2008**, *4*, 390–397.
15. Goncharik, I.I.; Shevchuk, V.V.; Krut, N.P.; Smychnik, A.D.; Kudina, O.A. Synthesis of Potassium Sulfate by Conversion of Potassium Chloride and Magnesium Sulfate. *Inorg. Synth. Ind. Inorg. Chem.* **2014**, *87*, 1804–1809. [CrossRef]
16. Jha, S.K.; Karthika, S.; Radhakrishnan, T.K. Modelling and Control of Crystallization Process. *Resour. Technol.* **2017**, *3*, 94–100. [CrossRef]
17. Togkalidou, T.; Tung, H.H.; Sun, Y.; Andrews, A.T.; Braatz, R.D. Parameter Estimation and Optimization of a Loosely Bound Aggregating Pharmaceutical Crystallization Using in Situ Infrared and Laser Backscattering measurements. *Ind. Eng. Chem. Res.* **2004**, *43*, 6168–6181. [CrossRef]
18. Shoji, M. Kinetic Studies for Industrial Crystallization to Improve Crystal Size Distribution and Crystal Shape. Available online: https://tuat.repo.nii.ac.jp/index.php?action=pages_view_main&active_action=repository_action_common_download&attribute_id=16&block_id=39&file_no=1&item_id=908&item_no=1&page_id=13 (accessed on 1 December 2012).
19. Nagy, Z.K.; Chew, J.W.; Fujiwara, M.; Braatz, R.D. Advances in the Modeling and Control of Batch Crystallizers. *IFAC Proc. Vol.* **2004**, *37*, 83–90. [CrossRef]
20. Tavare, N.S. Characterization of Crystallization Kinetics from Batch Experiments. *Sep. Purif. Rev.* **1993**, *22*, 93–210. [CrossRef]
21. Nemdili, L.; Koutchoukali, O.; Mameri, F.; Gouaou, I.; Koutchoukali, M.S.; Ulrich, J. Crystallization Study of Potassium Sulfate-Water System, Metastable Zone Width and Induction Time Measurements Using Ultrasonic, Turbidity and 3D-ORM Techniques. *J. Cryst. Growth* **2018**, *500*, 44–51. [CrossRef]
22. Bari, A.H.; Chawla, A.; Pandit, A.B. Ultrasonics Sonochemistry Sono-crystallization Kinetics of K_2SO_4: Estimation of Nucleation, Growth, Breakage and Agglomeration Kinetics. *Ultrason. Sonochem.* **2017**, *35*, 196–203. [CrossRef]
23. Luo, M.; Liu, C.; Xue, J.; Li, P.; Yu, J. Determination of Metastable Zone Width of Potassium Sulfate in Aqueous Solution by Ultrasonic Sensor and FBRM. *J. Cryst. Growth* **2017**, *469*, 144–153. [CrossRef]
24. Bari, A.H.; Pandit, A.B. Sequential Crystallization Parameter Estimation Method for Determination of Nucleation, Growth, Breakage, and Agglomeration Kinetics. *Ind. Eng. Chem. Res.* **2018**, *57*, 1370–1379. [CrossRef]
25. Gougazeh, M.; Omar, W.; Ulrich, J. Growth and Dissolution kKinetics of Potassium Sulfate in Pure Solutions and in the Presence of Cr^{3+} Ions. *Cryst. Res. Technol.* **2009**, *1210*, 1205–1210. [CrossRef]
26. Kubota, N.; Fukazawa, J.; Yashiro, H.; Mullin, J.W. Impurity Effect of Chromium (III) on the Growth and Dissolution Rates of Potassium Sulfate Crystals. *J. Cryst. Growth* **1995**, *149*, 113–119. [CrossRef]
27. Mullin, J.W.; Gaska, C. Potassium Sulfate Crystal Growth Rates in Aqueous Solution. *J. Chem. Eng. Data* **1973**, *18*, 217–220. [CrossRef]
28. Avila, S.T. *Control del Proceso de Cristalización Continuo No-isotérmico Empleando Lógica Difusa*; Celaya: Guanajuato, México, 2010.

29. Chianese, A.; Di Berardino, F.; Jones, A.G. On the Effect of Secondary Nucleation on the Crystal Size Distribution from a Seeded Batch Crystallizer. *Chem. Eng. Sci.* **1993**, *48*, 551–560. [CrossRef]
30. Mohamed-Kheir, A.K.M.; Tavare, N.S.; Garside, .J. *Crystallization Kinetics of Potassium Sulphate in a 1 M3 Batch Cooling Crystallizer*; Pergamon Press: Oxford, UK, 1987.
31. Taguchi, K.; Garside, J.; Tavare, N.S. Nucleation and Growth Kinetics of Barium Sulphate in Batch Precipitation. *J. Cryst. Growth* **1996**, *163*, 318–328. [CrossRef]
32. Lu, L.; Hua, Q.; Tang, J.; Liu, Y.; Liu, L.; Wang, B. Reactive Crystallization Kinetics of Magnesium Hydroxide in the Mg(NO$_3$)$_2$–NaOH System. *Cryst. Res. Technol.* **2018**, *53*, 1–10. [CrossRef]
33. Mignon, D.; Manth, T.; Offermann, H. Kinetic Modelling of Batch Precipitation Reactions. *Chem. Eng. Sci.* **1996**, *51*, 2565–2570. [CrossRef]
34. Pandit, A.V.; Vivek, V.R. Chord Length Distribution to Particle Sized Distribution. *AIChE J.* **2016**, *62*, 1–14. [CrossRef]
35. Heath, A.R.; Fawell, P.D.; Bahri, P.A.; Swift, J.D. Estimating Average Particle Size by Focused Beam Reflectance Measurement (FBRM). *Part. Part. Syst. Charact.* **2002**, *19*, 84–95. [CrossRef]
36. Trifkovic, M.; Sheikhzadeh, M.; Rohani, S. Kinetics Estimation and Single and Multi-Objective Optimization of a Seeded, Anti-solvent, Isothermal Batch Crystallizer. *Ind. Eng. Chem. Res.* **2008**, *47*, 1586–1595. [CrossRef]
37. Óciardhá, C.T.; Hutton, K.W.; Mitchell, N.A.; Frawley, P.J. Simultaneous Parameter Estimation and Optimization of a Seeded Antisolvent Crystallization. *Cryst. Growth Des.* **2012**, *12*, 5247–5261. [CrossRef]
38. Pitzer, K.S. *Activity Coefficients in Electrolyte Solutions*, 2nd ed.; Pitzer, K.S., Ed.; CRC Press Taylor & Francis Group: Berkeley, CA, USA, 2018. [CrossRef]
39. Harvie, C.E.; Møller, N.; Weare, J.H. The Prediction of Mineral Solubilities in Natural Waters: The Na-K-Mg-Ca-H-Cl-SO$_4$-OH-HCO$_3$-CO$_3$-CO$_2$-H$_2$O system to High Ionic Strengths at 25 °C. *Geochim. Cosmochim. Acta* **1984**, *48*, 723–751. [CrossRef]
40. Møller, N. The Prediction of Mineral Solubilities in Natural Waters: A Chemical Equilibrium Model for the Na-Ca-Cl-SO$_4$-H$_2$O System, to High Temperature and Concentration. *Geochim. Cosmochim. Acta* **1988**, *52*, 821–837. [CrossRef]
41. Marion, G.M.; Farren, R.E. Mineral Solubilities in the Na-K-Mg-Ca-Cl-SO$_4$-H$_2$O system: A re-evaluation of the Sulfate Chemistry in the Spencer-Moller-Weare Model. *Geochim. Cosmochim. Acta* **1999**, *63*, 1305–1318. [CrossRef]
42. Holmes, H.F.; Baes, C.F.; Mesmer, R.E. Isopiestic Studies of Aqueous Solutions at Elevated Temperatures. *J. Chem. Thermodyn.* **1978**, *10*, 983–996. [CrossRef]
43. De Lima, M.C.P.; Pitzer, K.S. Thermodynamics of Saturated Electrolyte Mixtures of NaCl with Na$_2$SO$_4$ and with MgCl$_2$. *J. Solut. Chem.* **1983**, *12*, 187–199. [CrossRef]
44. Phutela, R.C.; Pitzer, K.S. Heat Capacity and other Thermodynamic Properties of Aqueous Magnesium Sulfate to 473 K. *J. Phys. Chem.* **1986**, *90*, 895–901. [CrossRef]
45. Linke, W.F.; Seidell, A. *Solubilities of Inorganic and Metal Organic Compounds*, 4th ed.; Van Nostrand: Washington, DC, USA, 1958.
46. Jimenez, Y.P.; Taboada, M.E.; Galleguillos, H.R. Solid-Liquid Equilibrium of K$_2$SO$_4$ in Solvent Mixtures at Different Temperatures. *Fluid Phase Equilib.* **2009**, *284*, 114–117. [CrossRef]
47. Tavare, N.S.; Gaikar, G.V. Precipitation of Salicylic Acid: Hydrotropy and Reaction. *Ind. Eng. Chemisry Res.* **1991**, *30*, 722–728. [CrossRef]
48. Vollmer, U.; Raisch, J. Control of Batch Crystallization-A System Inversion Approach. *Chem. Eng. Process. Process. Intensif.* **2006**, *45*, 874–885. [CrossRef]
49. Lu, Y.; Wang, X.; Ching, C.B. Application of Preferential Crystallization for Different Types of Racemic Compounds. *Ind. Eng. Chem. Res.* **2009**, *48*, 7266–7275. [CrossRef]
50. Li, H.; Kawajiri, Y.; Grover, M.A.; Rousseau, R.W. Modeling of Nucleation and Growth Kinetics for Unseeded Batch Cooling Crystallization. *Ind. Eng. Chem. Res.* **2017**, *56*, 4060–4073. [CrossRef]
51. Lyczko, N.; Espitalier, F.; Louisnard, O.; Schwartzentruber, J. Effect of Ultrasound on the Induction Time and the Metastable Zone Widths of Potassium Sulphate. *Chem. Eng. J.* **2002**, *86*, 233–241. [CrossRef]
52. Garside, J.; Gibilaro, L.G.; Tavare, N.S. Evaluation of Crystal Growth Kinetics From a Desupersaturation Curve Using Initial Derivatives. *Chem. Eng. Sci.* **1982**, *37*, 1625–1628. [CrossRef]
53. Garside, J.; Gaska, C.; Mullin, J.W. Crystal Growth Rate Studies with Potassium Sulphate in a Fluidized Bed Crystallizer. *J. Cryst. Growth* **1972**, *13–14*, 510–516. [CrossRef]
54. Steyer, C.; Promotionskommission, M.; Lorenz, H.; Sundmacher, K.; Mangold, M. *Precipitation of Barium Sulfate in a Semi-Batch Stirred Tank Reactor: Influence of Feeding Policy on Particle Size and Morphology*; Universitätsbibl Otto von Guericke University Library: Magdeburg, Germany, 2012.
55. Jones, A.G.; Budz, J.; Mullin, J.W. Batch Crystallization and Solid-Liquid Separation of Potassium Sulphate. *Chem. Eng. Sci.* **1987**, *42*, 619–629. [CrossRef]
56. Jones, A.G.; Mullin, J.W. Programmed Cooling Crystallization of Potassium Sulphate Solutions. *Chem. Eng. Sci.* **1974**, *29*, 105–118. [CrossRef]
57. Jones, A.G.; Chianese, A. Fines Destruction During Batch Crystallization. *Chem. Eng. Commun.* **1987**, *62*, 5–16. [CrossRef]

58. Senaputra, A.; Jones, F.; Fawell, P.D.; Smith, P.G. Focused beam reflectance Measurement for Monitoring the Extent and Efficiency of Flocculation in Mineral Systems. *AIChE J.* **2014**, *60*, 251–265. [CrossRef]
59. Song, X.; Zhang, L.; Zhao, J.; Xu, Y.; Sun, Z.; Li, P.; Yu, J. Preparation of Calcium Sulfate Whiskers Using Waste Calcium Chloride by Reactive Crystallization. *Cryst. Res. Technol.* **2011**, *46*, 166–172. [CrossRef]

Article

Structural Properties of Bacterial Cellulose Film Obtained on a Substrate Containing Sweet Potato Waste

Izabela Betlej [1], Katarzyna Rybak [2], Małgorzata Nowacka [2], Andrzej Antczak [1], Sławomir Borysiak [3], Barbara Krochmal-Marczak [4], Karolina Lipska [5] and Piotr Boruszewski [5,*]

[1] Department of Wood Science and Wood Protection, Institute of Wood Sciences and Furniture, Warsaw University of Life Sciences–SGGW, 159 Nowoursynowska St., 02-776 Warsaw, Poland
[2] Department of Food Engineering and Process Management, Institute of Food Sciences, Warsaw University of Life Science–SGGW, 159C Nowoursynowska St., 02-776 Warsaw, Poland
[3] Institute of Chemical Technology and Engineering, Faculty of Chemical Technology, Poznan University of Technology, Berdychowo 4, 60-965 Poznan, Poland
[4] Department of Plant Production and Food Safety, Carpathian State College in Krosno, 38-400 Krosno, Poland
[5] Department of Technology and Entrepreneurship in Wood Industry, Institute of Wood Sciences and Furniture, Warsaw University of Life Sciences–SGGW, 159 Nowoursynowska St., 02-776 Warsaw, Poland
* Correspondence: piotr_boruszewski@sggw.edu.pl

Abstract: The paper presents the results of research on the microstructure of bacterial cellulose (BC-SP) obtained on a medium containing sweet potato peel, which was compared to cellulose obtained on a synthetic medium containing sucrose and peptone (BC-N). The properties of cellulose were analyzed using the methods: size exclusion chromatography (SEC), X-ray diffraction (XRD), scanning electron microscope (SEM), and computer microtomograph (X-ray micro-CT). BC-SP was characterized by a higher degree of polymerization (5680) and a lower porosity (1.45%) than BC-N (4879, 3.27%). These properties give great opportunities to cellulose for various applications, e.g., the production of paper or pulp. At the same time, for BC-SP, a low value of relative crystallinity was found, which is an important feature from the point of view of the mechanical properties of the polymer. Nevertheless, these studies are important and constitute an important source of knowledge on the possibility of using cheap waste plant materials as potential microbiological substrates for the cultivation of cellulose-synthesizing micro-organisms with specific properties.

Keywords: bacterial cellulose; crystallinity; polymerization degree; porosity; sweet potato waste

1. Introduction

Bacterial cellulose (BC) is a polymer with great application potential, synthesized by aerobic micro-organisms. Due to its high mechanical strength, high crystallinity, and a much greater degree of polymerization than plant cellulose, it has become a promising polymer for use in various technical fields, and even in medicine.

The main quality parameters of cellulose, determining its desired properties, is the crystallinity and the degree of polymerization. Allomorph Iα is dominant in the bacterial polymer. Aleshina et al. [1] indicate that it may constitute from 70 to 100% of the morphological composition, and additionally the quality and composition of the culture medium on which cellulose is synthesized affects its level in cellulose. Skiba et al. [2] reported that the synthesis of cellulose on unconventional substrates from plant materials causes a reduction in crystallinity and a decrease in the content of Iα in the polymer. The same authors, referring to the works of other authors, indicated that cellulose synthesized on a substrate from agricultural waste in the form of grape bagasse is characterized by a content of allomorph Iα from 70 to 56%. Another important morphological parameter influencing the high tensile strength of cellulose is the degree of crystallinity. In addition, this parameter may vary depending on the method of culturing cellulose-synthesizing micro-organisms [3], the types of carbon source and other components of the medium [4], or the procedure and method of drying [5]. 6. Illa et al. [6] showed that in the case of conventional drying, the degree of crystallinity of bacterial cellulose was slightly higher than during drying by lyophilization. Particular attention is paid to the influence of the composition of the culture medium on the degree of crystallinity of the cellulose. Because the development of low-cost culture media, on which it will be possible to obtain high-quality polymer, additionally with high efficiency, can guarantee its commercial application. Xu et al. [7], using a substrate of sweet potatoes, obtained cellulose with a crystallinity ranging from 83 to 87%. Other authors report that cellulose obtained on substrates containing agricultural waste in the form of oil palm leaf juice [8] or sweet sorghum leaves [9] was characterized by much lower crystallinity.

The properties of bacterial cellulose are also inextricably linked to its degree of polymerization, which is much higher than that of its plant-based counterpart and can be up to 20,000 [10]. Like crystallinity, the degree of polymerization of cellulose can be influenced by various external factors accompanying the synthesis process by micro-organisms. Surma-Ślusarska et al. [11] obtained cellulose on a substrate with glucose and mannitol with a degree of polymerization of approximately 1700, while Betlej et al. [12] obtained a cellulose polymerization degree of 6080 on a substrate with sucrose and peptone.

The conditions for culturing cellulose-synthesizing micro-organisms, including the composition of the culture medium, have a significant impact on the structural features of cellulose, which will reflect its properties. One of the key features of a bacterial polymer, determining its potential utility, is tensile strength and porosity. Porosity seems to be of particular importance in the case of the use of cellulose in the form of medical dressings, being gas-permeable and thus preventing the growth of anaerobic bacteria in places protected by it [13].

However, it should be remembered that the guarantee of the production volume of bacterial cellulose and its global demand is the reduction of production costs, while maintaining excellent physical and mechanical properties. According to Rivas et al. [14], the cost of cultivation on standard microbiological media may account for approximately 30% of the total cost of the process, therefore, efforts should be made to search for alternative sources of nutrients in the processes of microbial cultivation. It seems that a good alternative to synthetic substrates may be waste from plant production, which are rich in sugars, proteins, vitamins, and microelements necessary for the development of cellulose-synthesizing micro-organisms. At the same time, the management and reuse of plant waste can bring many benefits, including by reducing the costs of exportation and disposal or the production of new products.

The aim of the study was to investigate the structural features of bacterial cellulose, such as crystallinity, degree of polymerization and porosity, obtained on the culture medium from sweet potato peel and to compare them to the characteristics of cellulose obtained on a semi-synthetic medium containing sucrose and peptone. The indirect goal of the study was therefore to determine the suitability of plant waste materials, grown in many countries on a large scale, as a low-cost substrate for the production of high-quality polymer for various applications. In this way, we indicate environmentally friendly methods of bacterial cellulose production, which can be used in many industrial areas.

2. Materials and Methods

Bacterial cellulose (BC) was synthesized by micro-organisms known as Symbiotic Culture of Bacteria and Yeast (SCOBY) grown on two types of media. SCOBY were obtained from the organic farm Wolanin (Wolanin, Szczawnik, Poland). According to literature data, the dominant bacterial cultures are the species *Acetobacter xylium*, *A.pasteurianus*, *A. aceti*, and *Gluconobacter oxydans* [15], among the fungi yeasts belonging to *Saccharomyces*, *Saccharomycodes*, *Schizosaccharomyces*, or *Zygosaccharomyces* [16] are the those that are dominant. The test cultures were stored on agar slants containing 0.03% peptone (Biomaxima SA, Lublin, Poland), 0.05% yeast extract (Biomaxima SA, Lublin, Poland), 2.5% glucose (PPF

HASCO-LEK S.A., Wrocław, Poland), and 2.5% agar (AphaVit, Biała Podlaska, Poland). Before starting the experiment, an inoculum of micro-organisms was taken and introduced into 100 cm^3 of a liquid medium containing peptone, yeast extract, and glucose and cultured for 14 days in a heat incubator. During this time, the formation of bacterial cellulose on the surface of the medium was checked. Cultures were carried out in glass beakers with a diameter of 5 cm. The test culture was homogenized and used for inoculation of the media used in the test.

The reference medium contained 10% sucrose (Krajowa Spółka Cukrowa SA, Toruń, Poland) and 0.03% peptone (Biomaxima SA, Lublin, Poland). The second type of medium was based on ingredients of vegetable origin (sweet potato peel), treated as waste. The sweet potato tubers were stored at 4 °C before the start of the study. To prepare a broth medium based on plant material: 200 g of sweet potato peel, varieties 'Carmen Rubin', 'Purple' and 'Beauregard', grown in the field in Żyznów (49°49′ N 21°50′ E, Poland) on the soil of the defective wheat complex, with a slightly acidic reaction (pH = 6.1, in 1N KCl), suspended in 500 cm^3 of water and ground with a blender, model MMBM401W (Bosch, Gerlingen, Germany). Thus, a homogeneous homogenate was prepared. The individual sweet potato homogenates were combined and then mixed. The homogenate was then filtered through the filter paper using a water pump, separating the clear solution from the solids. A clear solution was used as a microbiological broth medium, divided into equal portions, and sterilized in a steam autoclave (Spółdzielnia Mechaników SMS, Warsaw, Poland) for 20 min at 121 °C. A total of 1 cm^3 of the inoculum was sterile added to both types of media. Cultures were incubated in a heat incubator (J.P. Selecta Laboratory Equipment Manufacturer, Barcelona, Spain) for a period of 14 days. The incubation temperature was 26 ± 2 °C. After the end of the cultivation time, the cellulose was purified according to the procedure described by Betlej et al. [17]. Both the cellulose obtained on the standard medium (BC-N) and the cellulose obtained on the sweet potato peel medium (BC-SP) were washed several times with distilled water, then rinsed in 0.1% NaOH solutions (Avantor Performance materials Poland SA, Gliwice, Poland) and 0.1% citric acid (Avantor Performance Materials Poland SA, Gliwice, Poland). Distilled water was always used between uses of the individual alkali and acid solutions and at the end of the rinsing process. The polymer thus prepared was dried at a temperature of 24 ± 2 °C in a laboratory dryer (J.P. Selecta Laboratory Equipment Manufacturer, Barcelona, Spain) until obtaining the constant mass of the polymer. The total sugar content in individual sweet potato varieties was presented and described by Krochmal-Marczak et al. [18] in earlier studies (Table 1). Krochmal-Marczak et al. [19] in other studies reported that the average protein content in dry matter in the raw material used is 1.35 g 100 g^{-1}, the average content of vitamin C is 22.86 mg 100 g^{-1}, and macroelements (P, K, Ca, Mg, Na) are 0.26, 2.12, 0.51, 0.13, and 0.19 mg 100 g^{-1}, respectively

Table 1. Total sugars in sweet potato with peel based on studies by Krochmal-Marczak et al. [16].

Cultivars	Total Sugars (Average) g kg^{-1} FM *
'Purple'	4.90
'Beauregard'	7.42
'Carmen Rubin'	9.43

* FM—fresh matter.

2.1. Polymerization Degree and Crystallinity of Bacterial Cellulose

The degree of polymerization of bacterial cellulose was determined by the size exclusion chromatography (SEC) method [20]. The degree of polymerization of bacterial cellulose were determined according to the methodology described by Antczak et al. [21] and Waliszewska et al. [22], with changes described by Betlej et al. [12]

The crystallinity of polymer was analyzed using a TUR M-62 X-ray diffractometer (Carl Zeiss AG, Jena, Germany) with the method described by Betlej et al. [12]. On the basis of XRD tests, the structural parameters of cellulose were determined:

- Crystallite size was calculated using the Scherrer equation (Equation (1)):

$$D = \frac{k \cdot \lambda}{\beta \cdot \cos\theta} \qquad (1)$$

where D is the crystallite size perpendicular to the plane; k-Scherrer constant; λ is the X-ray wavelength; β is the full-width at half-maximum in radians; and θ is the Bragg angle.

- The crystallinity of bacterial cellulose by comparison of the areas under crystalline peaks and the amorphous curve was determined. Deconvolution of peaks was performed by the method proposed by Hindeleh and Johnson [23].

After the separation of X-ray diffraction lines, the relative crystallinity was determined by comparing the areas under crystalline peaks and the amorphous curve. Relative crystallinity (%) was calculated using Equation (2).

$$\text{Relative crystallinity} = \frac{\text{crystalline area}}{(\text{crystalline} + \text{amorphous})\text{area}} \qquad (2)$$

2.2. Microstructure of Bacterial Cellulose

The microstructure of bacterial cellulose was examined using a Hitachi scanning electron microscope, (TM-3000, Hitachi Ltd., Tokyo, Japan). Gold was used as a sputter (Cressington 108 auto sputter coater, Netherlands). The cross-section was observed. The photos of the samples at accelerating voltages equal to 15 kV were taken with 500 and 1000 magnification, and the record was saved using SEM software (TM3000, Hitachi Ltd., Tokyo, Japan).

2.3. Porosity Analysis

To examine the porosity of bacterial cellulose, samples were analyzed using X-ray micro-CT Skyscan 1272 system (Bruker, Kontich, Belgium). The parameters of the process carried out were as follows: X-ray source, voltage at 40 kV, and 193 μA current. Scans were done with a rotation step of 0.3° and a resolution of 25 μm. NRecon software (Bruker, Kontich, Belgium) was used to reconstruct cross-section images from μCT projection into 3D images. The determination of porosity was done with the application of CTAnn software (Bruker). Raw images were binarized at a threshold value of 25–255, and custom processing with internal plugins (despeckle, ROI shrink-wrap, 3D analysis) were applied for the selected volume of interest. The images were binarized by means of assigning pixels with lower intensity as background (air, pores) and pixels with higher intensity as matter. Two samples of each experimental variant were scanned.

2.4. Statistical Analysis

TIBCO company software (STATISTICA program, version 13, Palo Alto, CA, USA) was used to conduct the ANOVA analysis. The samples of bacterial cellulose film were divided into homogenous groups with the use of Tukey's test ($\alpha = 0.05$).

3. Results

3.1. Characteristics of the Crystallinity and Degree of Polymerization of Bacterial Cellulose

Bacterial cellulose is a polymer characterized by high crystallinity, which is a decisive feature influencing the mechanical and physical properties of the polymer. XRD analysis is a key method for imaging crystallinity to verify the effect of various nutrient media on the crystallization properties of BC. X-ray patterns of the BC-N and BC-SP polymers presented in Figure 1 show significant differences in the heights as well as the widths of the diffraction peaks, which proves some changes in the supermolecular structure. In the case of BC-N obtained on a standard medium, typical diffraction maxima originating from the polymorphic variety of cellulose I were observed (Figure 1). The recorded diffraction peaks at the diffraction angles of 2θ corresponded to the crystal planes (100), (010), (110) of cellulose type I_α [1]. On the basis of the performed calculations, it has been shown that for bacterial cellulose from standard medium the value of the degree of crystallinity is 65%, which is close to the crystallinity value obtained on Hestrin–Schramm substrates, so far considered as reference substrates for cellulose-synthesizing micro-organisms [24]. The crystallinity of the cellulose obtained on the sweet potato medium was relatively low at 27%. Fan et al. [25] also observed lower crystallinity of cellulose obtained on media containing plant components.

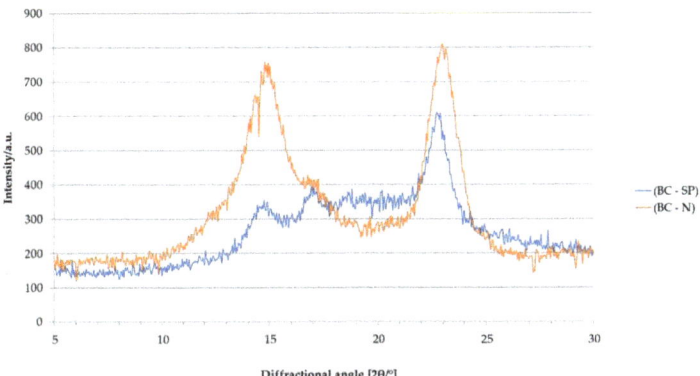

Figure 1. XRD bacterial cellulose obtained from different medium. BC-N—bacterial cellulose from standard medium, BC-SP—bacterial cellulose from sweet potato peel medium.

The conducted research also showed significant differences in the determined sizes of crystallites in individual types of cellulose. It can be noted (Table 2) that bacterial cellulose from sweet potato peel medium is characterized by a much larger crystallite size (70–94 Å depending on the plane) compared to BC-N cellulose, where the crystallite size is in the range of approximately 44–56 Å) (Table 2). The reason for this phenomenon can also be seen as BC-SP is not a pure cellulose. On the subject, information can be found that bacterial cellulose contains up to 90–95% pure cellulose, the remaining components may be fractions of other polysaccharides, such as levane [26].

Table 2. Based structural properties of bacterial cellulose obtained from different medium broth.

Parameter	BC-N	BC-SP
Crystallite size (D) of (100) plane (Å)	44.4	70.3
Crystallite size (D) of (010) plane (Å)	56.4	94.1
Crystallite size (D) of (110) plane (Å)	50.2	77.6
Relative crystallinity (%)	65	27
Parameter	**BC-N (SD) ***	**BC-SP (SD) ***
Molar mass:		
Number average M_n (kg/mol)	266 a (29)	336 b (11)
Weight average M_w (kg/mol)	791 a (14)	920 a (55)
Molar mass dispersity Đ	3.01 a (0.41)	2.75 a (0.28)
Polymerization degree DP_w	4879 a (98)	5680 b (383)

* SD—standard deviations in parentheses. The different lowercase letters in row show different homogeneous groups with the use of Tukey's HSD test with α = 0.05.

Despite its low crystallinity, BC-SP is characterized by a higher degree of polymerization compared to BC-N (Table 2). The reason for this can be seen in the greater availability of saccharides in the sweet potato medium than in the standard medium containing only sucrose. Sweet potatoes are a rich source of sugars, both mono and polysaccharides [27], and the latter can be broken down by enzyme into simple sugars, which are then used by micro-organisms not only for energy purposes but also in the process of polymer synthesis. In addition, the medium based on plant ingredients is rich in compounds such as vitamins, minerals, and enzymes, which can additionally regulate cellular processes or affect complex enzyme complexes involved in the biosynthesis of the polymer [28].

3.2. Microstructure Identification Using SEM

Figure 2 illustrates the surface cross-sections of bacterial cellulose. The cross-section of the polymer obtained on the sweet potato peel medium differs significantly from that obtained on the standard substrate. BC-N has a clearly layered structure in which the individual layers are significantly folded and clearly visibly separated from each other. Void spaces between the layers are observed. The cross-section of the BC-SP is completely different. The individual layers of the polymer clearly adhere to each other, creating a uniform structure. The cross-section structure is not folded, and the individual layers are flat and firmly integrated with each other.

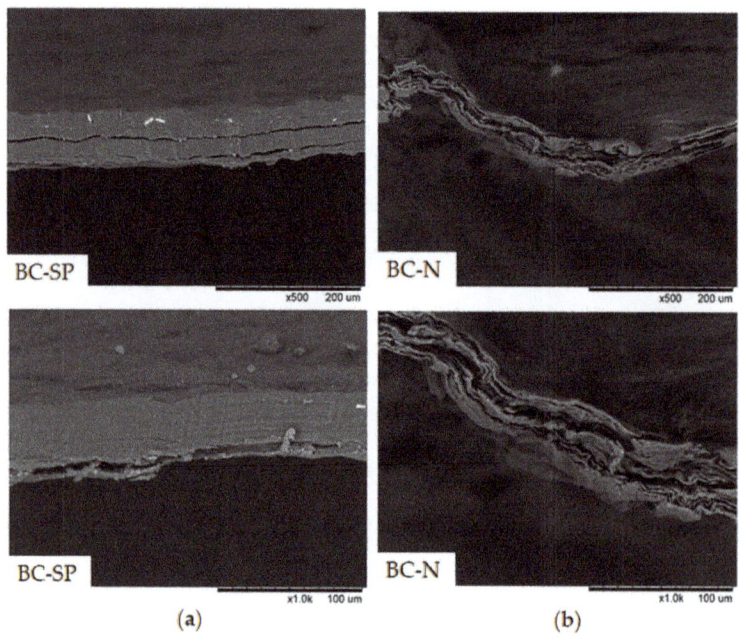

Figure 2. Cross-section of bacterial cellulose imaging by SEM with ×500 and ×1000 magnification: (**a**) BC-SP; (**b**) BC-N.

3.3. Porosity of Bacterial Cellulose

Porosity is one of the most important morphological parameters of materials. It is particularly important for the application of bacterial cellulose in papermaking [29] or as a medical product [30]. Tang et al. [31] showed that the porosity of cellulose depends not only on the conditions and method of cultivation but also on the polymer drying process. BC-SP cellulose was characterized by a smaller number of pores than BC-N cellulose, which may correlate with a greater degree of polymerization and thus a greater amount of microfibers and a more compact structure, which was confirmed by SEM tests. The tested polymers were characterized by exceptionally low porosity (Table 3). The low porosity of the two types of polymers obtained may also be due to the mild drying conditions. Moreover, as reported by Tang et al. [31], the carbon sources in the medium also have an effect on the porous structure of cellulose. The observed morphological changes may be a consequence of the use of microbiological media with a specific composition. Molina-Ramírez et al. [32] by examining the different composition of the substrate on the morphology of the synthesized cellulose using SEM scanning microscopy, showed that the nutrients contained in the microbiological substrate affect the degree of porosity, which results from the density of the cellulose nanofiber network. Studies by other authors have shown that the synthesis of bacterial cellulose on various types of substrates does not affect the size of the produced nanofibers, but with some types of substrates a polymer is obtained with a larger amount of micro- and nanofibrils [33]. The same authors also report that crystallinity is inversely related to porosity. The larger the crystal size, the smaller the number of pores, which is consistent with the results of this study.

Table 3. Porosity of bacterial cellulose.

Parameter	BC-N (SD) *	BC-SP (SD) *
Total volume of pore space	0.43 a (0.09)	0.18 b (0.08)
Total porosity (%)	3.27 a (0.11)	1.45 b (0.22)

* SD—standard deviations in parentheses. The different lowercase letters in row show different homogeneous groups with the use of Tukey's HSD test with $\alpha = 0.05$.

In this study, the authors showed that cellulose obtained on the sucrose medium broth was characterized by a greater porosity than the polymer synthesized on the medium with sweet potato medium.

4. Conclusions

Culture media play a key role in the economic viability of bacterial cellulose synthesis. Striving to lower the costs of cellulose production on a large scale, readily available and cheap sources of carbon and nitrogen are sought. It seems that waste plant raw materials can successfully replace commercial microbiological substrates, while significantly reducing the costs of cellulose production for various applications. The sweet potato peel medium has proven to be suitable for the synthesis of cellulose with specific quality features. The research results presented in the paper show that the use of a microbiological medium broth based on plant-based ingredients as a medium for the synthesis of bacterial cellulose has an impact on the structural parameters of the polymer. In terms of polymer characteristics, such as degree of polymerization or porosity, it seems that this type of support is better than the standard, which is based solely on sucrose and peptone. The obtained polymer was characterized by a higher degree of polymerization, lower porosity, and a more compact structure. The degree of polymerization of SP-BC was over 14% higher than BC-N, and the percentage of porosity of cellulose obtained on the sweet potato substrate was over two times lower than BC-N. At the same time, from the point of view of crystallinity, the use of a microbiological medium based on sweet-potato peel gives worse results than on a sucrose and peptone based microbiological medium, which was only 27%. It can be concluded that the usefulness of the microbiological medium based on sweet potatoes is desirable, especially for applications of cellulose that should be characterized by a high degree of polymerization, and in this direction, it should intensify the process of polymer synthesis.

Author Contributions: Conceptualization, I.B.; methodology, I.B., A.A., K.R., K.L. and S.B.; validation, P.B. and I.B.; investigation, I.B. and B.K.-M.; writing—original draft preparation, I.B.; writing—review and editing, I.B., P.B., A.A., S.B., M.N. and B.K.-M.; formal analysis, I.B. and P.B.; supervision, I.B. and P.B. All authors have read and agreed to the published version of the manuscript.

Funding: This research was performed by using research equipment purchased as part of the "Food and Nutrition Centre—modernisation of the WULS campus to create a Food and Nutrition Research and Development Centre (CŻiŻ)" co-financed by the European Union from the European Regional Development Fund under the Regional Operational Programme of the Mazowieckie Voivodeship for 2014–2020 (Project No. RPMA.01.01.00-14-8276/17).

Acknowledgments: The research was carried out thanks to funding from the Warsaw University of Life Sciences—SGGW.

Conflicts of Interest: The authors declare no conflict of interest.

References

1. Aleshina, L.A.; Gladysheva, E.K.; Budaeva, V.V.; Mironova, G.F.; Skiba, E.A.; Sakovich, G.V. X-ray Diffraction Data on the Bacterial Nanocellulose Synthesized by *Komagataeibacter xylinus* B-12429 and B-12431 Microbial Producers in Miscanthus- and Oat Hull-Derived Enzymatic Hydrolyzates. *Crystallogr. Rep.* **2022**, *67*, 391–397. [CrossRef]
2. Skiba, E.A.; Gladysheva, E.K.; Budaeva, V.V.; Aleshina, L.A.; Sakovich, G.V. Yield and quality of bacterial cellulose from agricultural waste. *Cellulose* **2022**, *29*, 1543–1555. [CrossRef]
3. Czaja, W.; Romanovicz, D.; Brown, R.M. Structural investigations of microbial cellulose produced in stationary and agitated culture. *Cellulose* **2004**, *11*, 403–411. [CrossRef]
4. Yim, S.M.; Song, J.E.; Kim, H.R. Production and characterization of bacterial cellulose fabrics by nitrogen sources of tea and carbon sources of sugar. *Process Biochem.* **2017**, *59*, 26–36. [CrossRef]
5. Stanisławska, A.; Staroszczyk, H.; Szkodo, M. The effect of dehydration/rehydration of bacterial nanocellulose on its tensile strength and physicochemical properties. *Carbohydr. Polym.* **2020**, *236*, 116023. [CrossRef] [PubMed]
6. Illa, M.P.; Sharma, C.S.; Khandelwal, M. Tuning the physiochemical properties of bacterial cellulose: Effect of drying conditions. *J. Mater. Sci.* **2019**, *54*, 12024–12035. [CrossRef]
7. Xu, S.; Xu, S.; Ge, X.; Tan, L.; Liu, T. Low-cost and highly efficient production of bacterial cellulose from sweet potato residues: Optimization, characterization, and application. *Int. J. Biol. Macromol.* **2022**, *196*, 172–179. [CrossRef]
8. Azmi, S.N.N.S.; Fabli, S.N.N.F.M.; Aris, F.A.F.; Samsu, Z.A.; Asnawi, A.S.F.M.; Yusof, Y.M.; Ariffin, H.; Abdullah, S.S.S. Fresh oil palm frond juice as a novel and alternative fermentation medium for bacterial cellulose production. *Mater. Today Proc.* **2021**, *42*, 101–106. [CrossRef]
9. Wang, Q.; Nnanna, P.C.; Shen, F.; Huang, M.; Tian, D.; Hu, J.; Zeng, Y.; Yang, G.; Deng, S. Full utilization of sweet sorghum for bacterial cellulose production: A concept of material crop. *Ind. Crop. Prod.* **2021**, *162*, 113256. [CrossRef]
10. Bednarczyk, D.; Betlej, I.; Boruszewski, P. Bacterial cellulose—Characteristics, synthesis, properties. *Bull. Res. Dev. Cent. Wood-Based Panels Czarna Woda* **2021**, *3–4*, 122–138. [CrossRef]
11. Surma-Ślusarska, B.; Presler, S.; Danielewicz, D. Characteristics of bacterial cellulose obtained from *Acetobacter xylinum* culture for application in papermaking. *Fibres Text. East. Eur.* **2008**, *16*, 108–111.
12. Betlej, I.; Antczak, A.; Szadkowski, J.; Drożdżek, M.; Krajewski, K.; Radomski, A.; Zawadzki, J.; Borysiak, S. Evaluation of the Hydrolysis Efficiency of Bacterial Cellulose Gel Film after the Liquid Hot Water and Steam Explosion Pretreatments. *Polymers* **2022**, *14*, 2032. [CrossRef] [PubMed]
13. Swingler, S.; Gupta, A.; Gibson, H.; Kowalczuk, M.; Heaselgrave, W.; Radecka, I. Recent Advances and Applications of Bacterial Cellulose in Biomedicine. *Polymers* **2021**, *13*, 412. [CrossRef]
14. Rivas, B.; Moldes, A.B.; Domínguez, J.M.; Parajó, J.C. Development of culture media containing spent yeast cells of Debaryomyces hansenii and corn steep liquor for lactic acid production with *Lactobacillus rhamnosus*. *Int. J. Food Microbiol.* **2004**, *97*, 93–98. [CrossRef] [PubMed]
15. Liu, C.-H.; Hsu, W.-H.; Lee, F.-L.; Liao, C.-C. The isolation and identification of microbes from a fermented tea beverage, Haipao, and their interactions during Haipao fermentation. *Food Microbiol.* **1996**, *13*, 407–415. [CrossRef]
16. Jayabalan, R.; Malbaša, R.V.; Lončar, E.S.; Vitas, J.S.; Sathishkumar, M. A Review on Kombucha Tea-Microbiology, Composition, Fermentation, Beneficial Effects, Toxicity, and Tea Fungus. *Compr. Rev. Food Sci. Food Saf.* **2014**, *13*, 538–550. [CrossRef]
17. Betlej, I.; Salerno-Kochan, R.; Krajewski, K.J.; Zawadzki, J.; Boruszewski, P. The influence of culture medium components on the physical and mechanical properties of cellulose synthesized by kombucha microorganisms. *BioResources* **2020**, *15*, 3125–3135. [CrossRef]
18. Krochmal-Marczak, B.; Sawicka, B.; Krzysztofik, B.; Danilčenko, H.; Jariene, E. The Effects of Temperature on the Quality and Storage Stalibity of Sweet Potato (*Ipomoea batatas* L. [Lam]) Grown in Central Europe. *Agronomy* **2020**, *10*, 1665. [CrossRef]
19. Krochmal-Marczak, B.; Sawicka, B.; Słupski, J.; Cebulak, T.; Paradowska, K. Nutrition value of the sweet potato (*Ipomoea batatas* (L.) Lam) cultivated in south—eastern Polish conditions. *Int. J. Agron. Agric. Res.* **2014**, *4*, 169–178.
20. Bikova, T.; Treimanis, A. Problems of the MMD analysis of cellulose by SEC using DMA/LiCl: A review. *Carbohydr. Polym.* **2002**, *48*, 23–28. [CrossRef]
21. Antczak, A.; Radomski, A.; Drożdżek, M.; Zawadzki, J.; Zielenkiewicz, T. Thermal ageing of cellulose with natural and syn-thetic antioxidants under various conditions. *Drewno* **2016**, *59*, 139–152. [CrossRef]
22. Waliszewska, H.; Waliszewska, B.; Zborowska, M.; Borysiak, S.; Antczak, A.; Czekała, W. Transformation of Miscanthus and Sorghum cellulose during methane fermentation. *Cellulose* **2018**, *25*, 1207–1216. [CrossRef]
23. Hindeleh, A.M.; Johnson, D.J. The resolution of multipeak data in fibre science. *J. Phys. D Appl. Phys.* **1971**, *4*, 259–263. [CrossRef]
24. Barshan, S.; Rezazadeh-Bari, M.; Almasi, H.; Amiri, S. Optimization and characterization of bacterial cellulose produced by *Komagatacibacter xylinus* PTCC 1734 using vinasse as a cheap cultivation medium. *Int. J. Biol. Macromol.* **2019**, *136*, 1188–1195. [CrossRef]
25. Fan, X.; Gao, Y.; He, W.; Hu, H.; Tian, M.; Wang, K.; Pan, S. Production of nano bacterial cellulose from beverage industrial waste of citrus peel and pomace using *Komagataeibacter xylinus*. *Carbohydr. Polym.* **2016**, *151*, 1068–1072. [CrossRef] [PubMed]
26. Bae, S.O.; Shoda, M. Production of bacterial cellulose by Acetobacter xylinum BPR2001 using molasses medium in a jar fermentor. *Appl. Microbiol. Biotechnol.* **2005**, *67*, 45–51. [CrossRef]

27. Lai, Y.-C.; Huang, C.-L.; Chan, C.-F.; Lien, C.-Y.; Liao, W.C. Studies of sugar composition and starch morphology of baked sweet potatoes (*Ipomoea batatas* (L.) Lam). *J. Food Sci. Technol.* **2011**, *50*, 1193–1199. [CrossRef]
28. Ul-Islam, M.; Khan, S.; Ullah, M.W.; Park, J.K. Bacterial cellulose composites: Synthetic strategies and multiple applications in bio-medical and electro-conductive fields. *Biotechnol. J.* **2015**, *10*, 1847–1861. [CrossRef]
29. Mautner, A.; Bismarck, A. Bacterial nanocellulose papers with high porosity for optimized permeance and rejection of nm-sized pollutants. *Carbohydr. Polym.* **2021**, *251*, 117130. [CrossRef]
30. Huang, Y.; Wang, J.; Yang, F.; Shao, Y.; Zhang, X.; Dai, K. Modification and evaluation of micro-nano structured porous bacterial cellulose scaffold for bone tissue engineering. *Mater. Sci. Eng. C* **2017**, *75*, 1034–1041. [CrossRef]
31. Tang, W.; Jia, S.; Jia, Y.; Yang, H. The influence of fermentation conditions and post-treatment methods on porosity of bacterial cellulose membrane. *World J. Microbiol. Biotechnol.* **2010**, *26*, 125–131. [CrossRef]
32. Molina-Ramírez, C.; Castro, C.; Zuluaga, R.; Gañán, P. Physical Characterization of Bacterial Cellulose Produced by *Komagataeibacter medellinensis* Using Food Supply Chain Waste and Agricultural By-Products as Alternative Low-Cost Feedstocks. *J. Polym. Environ.* **2018**, *26*, 830–837. [CrossRef]
33. Costa, A.F.S.; Almeida, F.C.G.; Vinhas, G.M.; Sarubbo, L.A. Production of Bacterial Cellulose by *Gluconacetobacter hansenii* Using Corn Steep Liquor As Nutrient Sources. *Front. Microbiol.* **2017**, *8*, 2027. [CrossRef] [PubMed]

Article

Gypsum Seeding to Prevent Scaling

Taona Malvin Chagwedera, Jemitias Chivavava and Alison Emslie Lewis *

Crystallization and Precipitation Research Unit, University of Cape Town, Rondebosch, Cape Town 7700, South Africa; taona.chagwedera@alumni.uct.ac.za (T.M.C.); jemitias.chivavava@uct.ac.za (J.C.)
* Correspondence: alison.lewis@uct.ac.za

Abstract: Eutectic freeze crystallization (EFC) is a novel separation technique that can be applied to treat brine solutions such as reverse osmosis retentates. These are often a mixture of different inorganic solutes. The treatment of calcium sulphate-rich brines using EFC often results in gypsum crystallization before any other species. This results in gypsum scaling on the cooled surfaces of the crystallizer, which is undesirable as it retards heat transfer rates and hence reduces the yield of other products. The aim of this study was to investigate and understand gypsum crystallization and gypsum scaling in the presence of gypsum seeds. Synthetic brine solutions were used in this research because they allowed an in-depth understanding of the gypsum bulk crystallization process and scaling tendency without the complexity of industrial brines. A cooled, U-shaped stainless-steel tube suspended in the saturated solution was employed as the scaling surface. This was because a tube-shaped surface enabled the introduction of a constant temperature cold surface in the saturated solution and most industrial EFC crystallizers are constructed from stainless steel. Gypsum seeding was effective in decreasing the mass of scale formed on the heat transfer surface. The most effective seed loading was 0.25 g/L, which reduced scale growth rate by 43%. Importantly, this seed loading is six times the theoretical critical seed loading. The seeding strategy also increased the gypsum crystallization kinetics in the bulk solution, which resulted in an increase in the mass of gypsum product. These findings are relevant for the operability and control of EFC processes, which suffer from scaling problems. By using an appropriate seeding strategy, two problems can be alleviated. Firstly, scaling on the heat transfer surface is minimised and, secondly, seeding increases the crystallization kinetics in the bulk solution, which is advantageous for product yield and recovery. It was also recommended that the use of silica as a seed material to prevent gypsum scaling should be investigated in future studies.

Keywords: gypsum; scaling; seeding; eutectic freeze crystallization; brine

Citation: Chagwedera, T.M.; Chivavava, J.; Lewis, A.E. Gypsum Seeding to Prevent Scaling. *Crystals* **2022**, *12*, 342. https://doi.org/10.3390/cryst12030342

Academic Editor: Linda Pastero

Received: 29 October 2021
Accepted: 24 February 2022
Published: 2 March 2022

Publisher's Note: MDPI stays neutral with regard to jurisdictional claims in published maps and institutional affiliations.

Copyright: © 2022 by the authors. Licensee MDPI, Basel, Switzerland. This article is an open access article distributed under the terms and conditions of the Creative Commons Attribution (CC BY) license (https://creativecommons.org/licenses/by/4.0/).

1. Introduction

South Africa is an industrialized semi-arid country [1] that produces numerous saline solutions. Reverse osmosis (RO) is an economical and energy efficient way of treating these saline solutions. However, a highly concentrated brine stream (reverse osmosis retentate) is produced in the process, which must be treated before disposal. The brine production in South Africa is forecast to reach a peak daily production of 17,000 m^3/day in 2030 compared to approximately 3000 m^3/day in 2010 [2,3].

Conventional brine disposal methods in South Africa include discharging the brine into lined evaporation ponds, the use of mechanical evaporators, and injecting the brine into deep wells [3]. The main limitation of these methods is their inability to fully separate the brine into reusable products. As an example, evaporative methods result in the formation of a sludge, which is a mixture of salts that needs another disposal method [4]. In contrast, Eutectic Freeze Crystallization (EFC) is theoretically able to fully separate the brine into its constituents, thus having an advantage compared to evaporative methods.

Eutectic Freeze Crystallization (EFC) is a novel brine treatment process for separating the salts from water by cooling the brine to sub-eutectic temperatures. This results in the co-crystallization of ice and salts. The ice naturally floats to the top, because it is less dense than the brine, and the salts sink, because they are denser than the brine, making the products separable [5]. The salts produced can be sold depending on their purity and production quantities.

The treatment of calcium sulphate-rich brines, such as reverse osmosis retentates, using EFC, results in the formation of calcium sulphate dihydrate (gypsum) scale deposits on the cooled surfaces of the crystallizers and surfaces of ancillary equipment. This is due to the sparingly soluble nature of gypsum in water. Gypsum scaling is undesirable because the scale forms an insulating layer on the crystallizer heat exchange surfaces, thus retarding heat transfer rates and thereby lowering yields. Gypsum scaling also results in frequent stoppages to clean the scale layer.

Scaling or crystallization fouling is a process in which a deposit forms on a surface. This is due to either bulk crystallization followed by adhesion onto the surface or heterogeneous nucleation and growth on the surface [6]. Gypsum scaling due to adhesion is common for membrane processes [7–9]. Gypsum scaling on hot surfaces is a result of heterogeneous nucleation and growth [10,11]. There is no literature available for gypsum scaling mechanisms under cooling or freeze crystallization conditions, as previous studies [11–15] were conducted under heating crystallization conditions due to the recurrence of gypsum scaling in the handling of geothermal brines for energy production and water distillation.

Heterogeneous nucleation is a form of primary nucleation induced by foreign surfaces such as dust and vessel walls [16]. The nucleation energy barrier for heterogenous nucleation is higher than that for secondary nucleation. Secondary nucleation occurs in the presence of crystals of the material to be crystallized [17,18]. Seeding with parent crystals of the solute in a supersaturated solution lowers the nucleation energy barrier for the dissolved solute particles to crystallize [18].

Seed quality, seed surface area, and seed loading influence the effectiveness of a seeding protocol. Characteristics such as surface smoothness of seed crystals and the structural integrity of the seed crystals constitute the quality of the seeds. Jagadesh and co-workers [19] observed that precipitated potassium seeds were the most effective seed type to precipitate potassium alum from its solution compared to ground and commercial potassium seeds. This may have been due to the precipitated seed crystals having fewer strains in their crystal lattice, which are usually induced through milling. The strains in the crystal lattice are known to dampen the ability of crystals to grow [20].

Seed loading is a measure of the mass of seeds per unit volume of the supersaturated solution. The critical seed loading refers to the minimum amount of seeds required to promote growth without prior nucleation [21]. Doki and co-workers [22] give two correlations that can be used to determine the critical seed loading for a system. Equation (1) is used to determine the critical seed loading ratio using the mean seed crystal size, L_s.

$$C_R^* = 2.17 \times 10^{-6} L_s^2 \qquad (1)$$

where C_R^* = critical seed loading ratio;
L_s = mean seed crystal size (µm).

Equation (2) is then used to determine the critical seed loading of the system, using the critical seed loading ratio determined above, as well as the theoretical yield and the volume of the solution.

$$C_S^* = C_R^* \times \frac{W_T}{V} \qquad (2)$$

where C_S^* = critical seed loading (g/L)
W_T = theoretical yield of the salt (g)
V = volume of the solution (L)

It has been found that specific seed surface area plays an important role. Wang and co-workers [23] showed that 25 µm seed crystals were the most effective in enhancing bulk crystallization compared to larger crystals; 48 µm and 75 µm.

Higher seed loading introduces more surface area for nucleation and growth in the system and thus increases the crystallization rate of the target material. Liu and Nancollas [24] observed that the induction time for gypsum crystallization was shortened by increasing the seed loading from 0.42 to 1.89 g/L. However, the addition of an excessive number of seeds above the critical seed loading was observed to have no pronounced effect on gypsum crystallization kinetics [24].

Seeding has been employed in batch crystallization systems to control the crystal size distribution of the product crystals [19,21,22,25]. It is also an established method to enhance bulk crystallization of the target salt or ice in the EFC context [26–28]. Bulk crystallization of gypsum from a reverse osmosis brine was increased significantly when the brine was seeded with gypsum crystals [29].

A few studies on the use of seeding as a method to prevent scaling have been published. Adams and Papangelakis [30] observed that introducing gypsum seed crystals at 10 g/L in a laboratory scale neutralization reactor resulted in a 50% decrease in the mass of scale formed at 70 °C. Wang and co-workers [23] established that seeding was more effective in preventing scaling in brine transportation pipes compared to brine dilution. Gainey et al. [31] reported that seeding in evaporators resulted in the elimination of the calcium sulphate scale at the Rosewell laboratory and pilot plants. The actual details of the seed characteristics and seed loading were not published.

In this work, seeding was tested as a method to prevent scaling under cooling crystallization conditions. The aim of the study was to investigate and understand the interaction of gypsum crystallization in the bulk and gypsum scaling on the crystallizer surfaces in the presence of gypsum seeds.

2. Materials and Methods

2.1. Experimental Equipment

The experiments were conducted using the apparatus shown in Figure 1. A jacketed and insulated glass crystallizer with a working volume of 1.25 L was used. A U-shaped stainless-steel tube, 290 mm long with an outer diameter of 3.18 mm, was suspended from the lid into the supersaturated solution. The tube was maintained at 0.0 °C by a Lauda Proline PP855 thermostatic unit (Lauda, Germany), which circulated polydimethylphenylsiloxane (Kryo 51™) through it to cool the solution from 22.3 to 3 °C.

The jacket of the crystallizer was maintained at 2.5 °C by a Lauda ECO RE1050G thermostatic unit (Lauda, Königshofen, Germany), which circulated polydimethylphenylsiloxane (Kryo 51™, Lauda, Königshofen, Germany) through it. The temperatures of the bulk solution, coolant into and out of the tube, and coolant into and out of the jacket of the crystallizer were measured to an accuracy of ±0.01°C, at 3-s intervals, using platinum resistance thermometers (Pt100) (Tempcontrol, Nootdorp, The Netherlands). The thermometers were connected to a CTR5000 precision bridge (ASL, Horsham, UK), which communicated with the computer via the ULog software (Ulog V6, ASL WIKA, Manchester, UK). A 4-blade pitched-blade impeller, attached to an overhead stirrer, was used to agitate the solution inside the crystallizer.

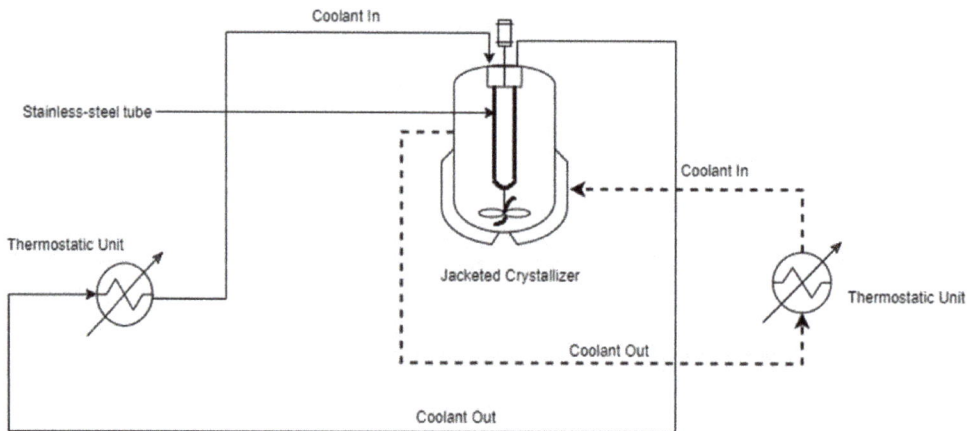

Figure 1. Batch crystallizer with cooled stainless-steel tubing suspended from the lid.

2.2. Feed Solution Preparation

The brine solution was prepared by reacting equal quantities of 0.11 M Ca(OH)$_2$ (Merck, Modderfontein, South Africa) and 0.11 M H$_2$SO$_4$ (Sigma-Aldrich, Modderfontein, South Africa) in order to prepare a supersaturated calcium sulphate–water solution as illustrated by the reactions in Equation (3). The average concentration of the feed solution was 6.13 g/L CaSO$_4$, as shown in Figure 7, resulting in an average starting supersaturation, S, of 7.71 that was calculated using the Debye–Huckel theory.

$$Ca(OH)_2(aq) + H_2SO_4 = CaSO_4(aq) + 2H_2O \quad (3)$$

Feed solution preparation was not possible through dissolving reagent grade gypsum powder in de-ionised water because of the sparingly soluble nature of gypsum. The suspension formed from the reaction was filtered through a 0.22 µm cellulose acetate membrane (Kimix Chemical and Lab Supplies, Cape Town, South Africa) held by a 250 mL Merck Millipore glass holder connected to a vacuum pump at room temperature (23.5 °C). However, this filtration step does not completely eliminate nano fraction particles as determined by Oshchepkov and co-workers [32]. The filtrate was used as feed solution due to technological limitations to further remove nano-sized particles.

2.3. Seeds Preparation

Gypsum seeds were precipitated by mixing equal quantities of aqueous 0.6 M sodium sulphate solution (Merck, Modderfontein, South Africa) and 0.6 M calcium chloride solution (Merck, Modderfontein, South Africa) as illustrated by Equation (4). The resistivity of deionised water used to prepare both solutions was 10.9 MΩ-cm.

$$Na_2SO_4(aq) + CaCl_2(aq) + 2H_2O = CaSO_4 \cdot 2H_2O(s) + 2NaCl(aq) \quad (4)$$

Calcium chloride solution was added one drop at a time to sodium sulphate solution at 70 °C to allow slow distribution of the supersaturation and precipitation of needle-type gypsum crystals. This method was adapted from Liu and Nancollas [24]. The suspension formed was filtered through a 0.22 µm cellulose acetate membrane held by a 250 mL Merck Millipore glass holder connected to a vacuum pump. Gypsum crystals were repeatedly washed with 0.50 L of deionised water to remove sodium chloride before they were dried.

2.4. Experimental Procedure

Briefly, 1.25 L of the feed solution was measured and transferred into the crystallizer. The overhead stirrer was set to 450 rpm, which is equivalent to a Reynolds number, (Re) of 4.21 × 10^5, and the thermostatic units were switched on to start the experiment. Seed crystals with a mean size of 58 µm were added into the crystallizer at the start of the experiments in which seeding was employed.

At the end of the experiment, the thermostatic units and the overhead stirrer were switched off. The tube was removed from the lid and allowed to dry before it was weighed. The suspension in the crystallizer was filtered using the same apparatus as above and the filtrate was analysed for sulphate ion concentration.

2.5. Measurement/Analytical Techniques

The sulphate concentration for the feed solution and spent solution was analysed using the turbidimetric method. In this method, the sulphate ion is converted to barium sulphate through addition of barium chloride dihydrate (Merck, Modderfontein, South Africa) and the turbidity of the suspension is measured. A photometer (Merck Spectroquant Nova 60, Merck, Modderfontein, South Africa) set at a wavelength of 410 nm was used.

The mass of the scale was determined arithmetically from the difference between the mass of the scaled tube and the mass of the clean tube, which were both measured using a digital scale (Mettler™ Toledo ML204, Greinfensee, Switzerland) with an accuracy of ±0.0003 g.

Crystal size and morphology of the scaled tube were analysed using a Scanning Electron Microscope (Tescan™ MIRA3 Rise, TESCAN, Brno-Kohoutovice, Czech Republic).

3. Results and Discussion

3.1. Seed Crystals

Figure 2 shows SEM micrographs of the seeds. The seed crystals were a mixture of the needle-type habit and prisms.

Figure 2. Micrographs of the seed crystals. Scale bar = 100 µm in 4.2 (**a**) and 50 µm in 4.2 (**b**).

The crystal size distribution of the seed crystals is presented in Figure 3. An average of three samples was taken and most of the seed crystals (61%) were below 55 µm in size. The mean size of the seed crystals was 57 µm with a modal size of 26 µm.

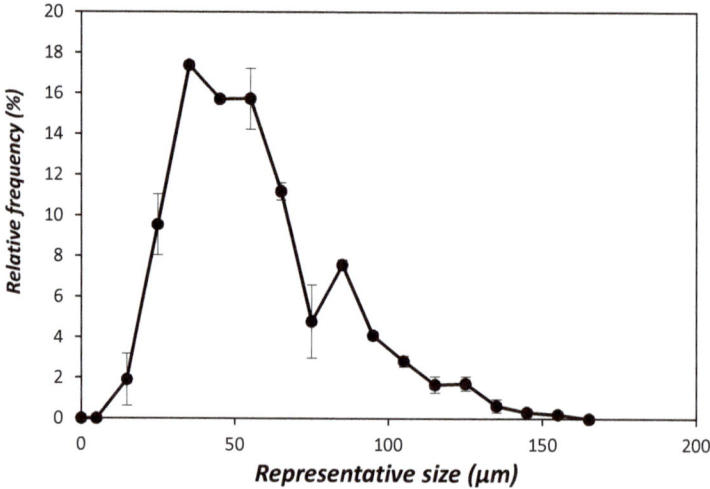

Figure 3. Crystal size distribution of gypsum seed crystals.

3.2. Preliminary Experiments

It was established that the required experiment run time was 4 h for a measurable mass of gypsum scale to be deposited on the stainless-steel tube. The mass of gypsum scale recorded was 0.045 g. The bulk solution temperature was 3 °C at the time of stopping the experiment. There was negligible mass of gypsum scale on the inner wall of the crystallizer. A temperature of 3 °C was maintained in all experiments as this allowed the study of gypsum scaling, testing the effectiveness of gypsum seeding, without the complexity caused by ice formation.

Figure 4 shows the micrographs of the scaled tube after running the experiment. The lighter phase represents the stainless-steel tube surface and the darker phase represents gypsum crystals. There was also a very thin layer of finely grained crystals, which could have been due to the adhesion of gypsum crystals when the tube was left to dry.

Figure 4. Micrographs of a scaled tube after running the experiment for 4 h. Scale bar = 100 μm (**a**), 50 μm (**b**) and 20 μm (**c**) respectively.

The micrographs presented in Figure 4 show that the tube was not fully covered with scale after running the experiment for 4 h. Needle-type crystals grew normal to the plane of the tube into the bulk solution; although, the expectation was that they would grow along the cold tube surface, which provided high local supersaturation conditions. This was due to the difference in the crystallographic structure of stainless steel and gypsum, which inhibited the growth of crystals along the plane of the tube. It is also possible that the integration of gypsum lattice units into the scale crystals that crystallized first on the tube could have caused the gypsum crystals to grow into the bulk solution. There were tiny crystals that were lying parallel to the tube, possibly due to the adhesion of crystals precipitated in the bulk solution or shearing of crystals by fluid motion. The fluid motion around the tube was turbulent (Re = 4.21×10^5).

The experiment duration was increased further by 6 h to develop an understanding of how the scale crystals grew. The mass of scale deposited on the tube was 0.080 g and the final temperature of the solution was 3 °C. The longer experiment time did not change the predominant habit of crystals, with needle-type crystals of varying lengths constituting the scale layer. Growth of the crystals was also into the bulk solution, which resulted in small 'islands' of the tube that were not fully covered with gypsum scale.

An increase in the duration of the experiment to 24 h resulted in an increase in the mass of gypsum scale that deposited on the stainless-steel tube. However, the increase was not linear as was the case when the experiment duration was further increased to 48 h from 24 h. This was due to the decrease in the supersaturation of the system with time. Choi and co-workers [33] asserted that gypsum crystallization rates decrease in batch tests as the calcium ion concentration decreases. The mass of gypsum scale deposited on the tube after 24 h and 48 h was 0.179 g and 0.260 g, respectively. Figure 5 shows the increase in the mass of scale deposited on the tube as the run time was increased.

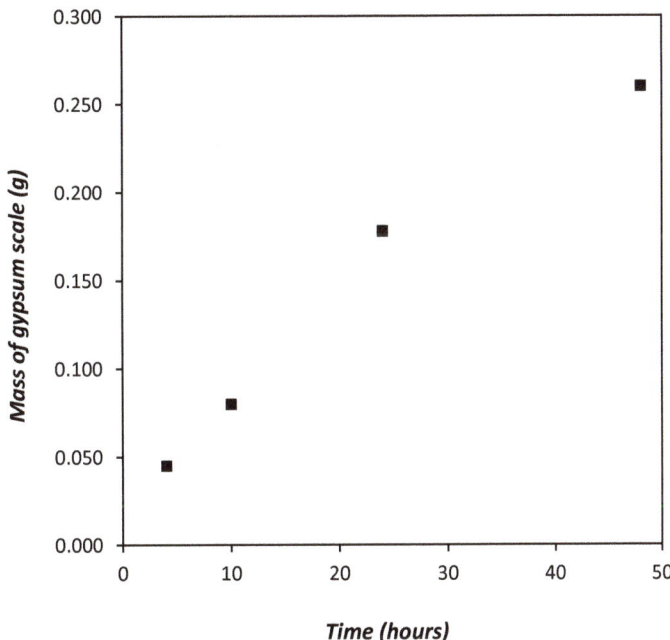

Figure 5. Mass of gypsum scale deposited on the stainless-steel tube as a function of time.

Figure 6 shows the micrographs of the scaled tube after 48 h. In the micrographs, the darker phase represents the stainless-steel tube surface, and the lighter phase represents gypsum crystals.

Figure 6. Micrographs of a scaled tube after running the experiment for 48 h. Scale bar = 100 μm (**a**), 50 μm (**b**) and 20 μm (**c**) respectively.

Figure 6a shows that the scale layer was predominantly composed of needle-type crystals varying in length between 40 and 100 μm. Most of the smaller sized crystals were on the top surface of the scale layer while the larger sized crystals were underneath. This was possibly because the underlying crystals crystallized first and had more time to grow and, hence, became larger than the top surface crystals.

Figure 6c shows attachment of smaller sized crystals to the larger crystals. This could have been due to the underlying crystals serving as growth sites for subsequent scale crystals. There was crystal twinning during scale layer growth as depicted in Figure 6a,b (white circles). The twinning may have resulted from the combination of moderate supersaturation conditions at the start of the experiment, prolonged growth time, and close contact with the cold tube surface where heat transfer was the highest.

Analysis of the micrographs of the scaled tube acquired after each preliminary experiment enabled the formulation of a possible mechanism of gypsum scaling on the tube, even though it was not conclusive. Gypsum scale layer was predominantly composed of needle-type crystals showing that the stainless-steel tube did not alter its habit under moderate supersaturation conditions present in the system. This is similar to what was observed by Amjad [10], although on a brass tube. The phenomenon would support the notion of gypsum scaling through adhesion. However, the plausible reasons for gypsum scaling through adhesion were outweighed by those for heterogeneous nucleation and supported by the micrographs.

It was proposed that gypsum scaling on the stainless-steel tube most likely proceeded via heterogeneous nucleation followed by growth. The growth of the scale layer crystals was into the bulk solution. Gypsum scaling was found to begin between 0 and 30 min. Based on this, it was decided that gypsum seeds would be added at the beginning of the experiment.

3.3. Effect of Increasing Gypsum Seed Loading on Gypsum Scale Formation

Synthetic gypsum seeds of the type described earlier were used. At the time the experiments were stopped, the calcium sulphate concentration was on average 5.33 g/L, which is above the thermodynamic equilibrium concentration of 2.27 g/L at 3 °C [34]. Since the calcium sulphate concentration in the spent solutions was double the equilibrium concentration, more gypsum may have theoretically crystallized from the solution if the experiments were run for longer. Gypsum crystallization kinetics were generally slow. Figure 7 is a graphical representation of the changes in solution concentration from feed to spent solution for 4-h run times plotted on the same axis for the different experiments.

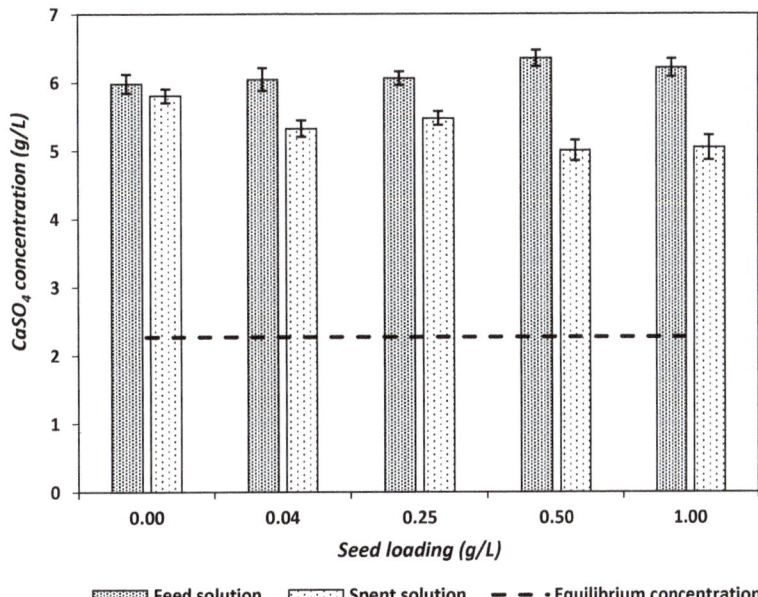

Figure 7. CaSO$_4$ concentration in solutions as a function of gypsum seed loading.

The graph shows that the average change in concentration between the feed solution and the spent solution for all seed loadings was 1.0 g/L CaSO$_4$. At seed loadings less than 0.50 g/L, the least average change in concentration of 0.50 g/L CaSO$_4$ was recorded while at 1.0 g/L this change in concentration was 1.76 g/L CaSO$_4$. The significant concentration change at higher seed loading was a result of faster gypsum crystallization kinetics.

Figure 8 shows the mass of gypsum scale that deposited on the tube and the mass of gypsum that crystallized in the bulk solution as a function of gypsum seed loading. The x-axis is from a minimum value of −0.2 to show the data points at 0.0 g/L. Figure 8 shows that the mass of gypsum that deposited on the stainless-steel tube was several orders of magnitude less than the mass of gypsum that crystallized in the bulk solution.

The mass of gypsum that crystallized in the bulk solution increased rapidly as gypsum seed loading was increased due to faster gypsum crystallization kinetics. The increase in seed loading increased the available surface area with favourable energetics for gypsum growth to occur. In addition, the abundance of gypsum crystals in suspension increased crystal–crystal, crystal–impeller, and crystal–crystallizer surface collisions. These collisions increased the rate of secondary nucleation, which requires the lowest activation energy; thus, crystallization kinetics increased. The observed increase in crystallization rates as the seed loading was increased corroborated the results found by Choi and co-workers [33], where they observed that the induction time shortened in the presence of seeds compared to unseeded solutions.

Although the tube was the heat transfer surface area and the coldest part of the apparatus, causing high local supersaturation, less mass of gypsum crystallized on it than in the bulk solution. This is because the surface area provided by the tube (28.9 cm^2) was very small compared to that provided by the bulk solution (1415 cm^2) and the seed material. Surface area is a key determinant of crystallization rate processes. The surface area provided by the bulk solution was calculated using the internal dimensions of the crystallizer. It was difficult to quantify the surface area provided by gypsum seeds at the different seed loadings due to technological limitations. In addition, the surface area provided by the tube had poor energetics for gypsum nucleation and growth compared to the gypsum seeds.

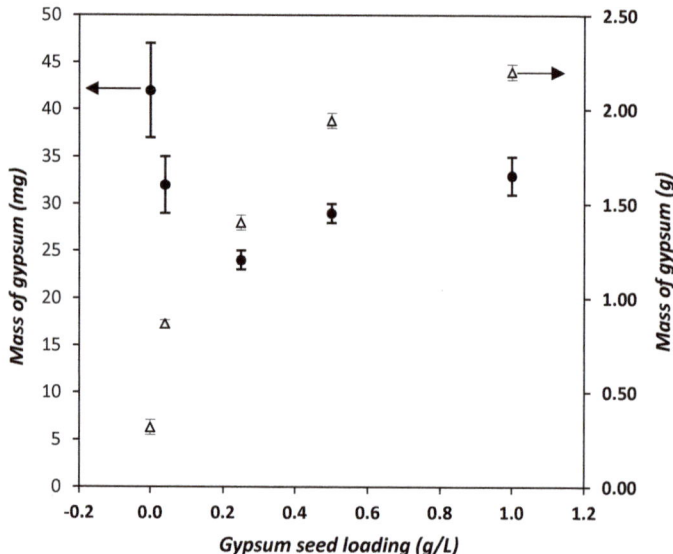

● Mass of gypsum scale on tube   △ Mass of gypsum crystallized in the bulk

Figure 8. Seed loading against mass of gypsum scale deposited on the tube and mass of gypsum crystallized in the bulk solution.

A zoomed view of Figure 8 showing the change in the mass of gypsum scale with increase in gypsum seed loading is presented in Figure 9. The x-axis has a minimum value of −0.2 to show the data point at 0.0 g/L.

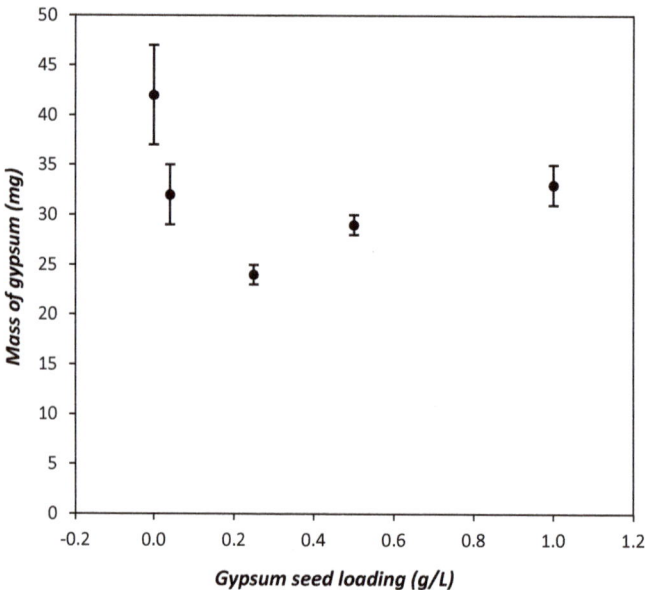

Figure 9. Seed loading against mass of gypsum scale deposited on the tube.

Figure 9 shows that the highest mass of gypsum scale deposited on the tube in the control experiment. The mass of scale deposited on the tube decreased in the presence of gypsum seeds because the added seeds consumed some of the available supersaturation to sufficiently low levels to decrease the rate of heterogenous nucleation on the tube, but still promoting secondary nucleation in the bulk solution. An increase in gypsum seed loading (specific surface area) decreased the mass of scale up to the seed loading of 0.25 g/L. Beyond the seed loading of 0.25 g/L, a further increase in gypsum seeds resulted in an increase in the mass of gypsum scale deposited. Although the mass of gypsum scale deposited on the tube increased at seed loadings greater than 0.25 g/L, it was still less than the amount deposited in the control experiment without seeding.

Contrary to expectation, the mass of gypsum scale deposited on the tube when the critical seed loading was employed was not the lowest. It was anticipated that the mass of gypsum scale deposited on the tube would be the least at the critical seed loading because this seed loading is associated with growth without any prior nucleation. Hence, at seed loadings greater than the critical seed loading the surface area provided by the seeds would have been in excess compared to the available supersaturation. Instead, the lowest mass of scale deposited on the tube was realised when a seed loading approximately six times higher than the critical seed loading (0.25 g/L) was employed. This deviation could have been because the surface area provided by the critical seed loading was too small to sufficiently reduce nucleation on the stainless-steel tube surface any further.

In addition, the calculated contact angle for gypsum nuclei to form on a stainless-steel surface was small. The contact angle calculation was done using Equation (A1) provided in the Appendix A together with the values from literature which were used (Table A1). A range of the contact angle was determined since the dispersive component of gypsum surface free energy was found as a range. The contact angle range found was 16° to 50°. The lower limit of the contact angle range implies the degree of wetting was high, thus heterogeneous nucleation of gypsum on stainless-steel occurred easily. The respective surface energy reduction factors, $f(\emptyset)$, for the contact angles calculated using Equation (A2) (see Appendix A) were 0.001 and 0.08. This shows that the nucleation work on the stainless-steel tube which needed to be overcame by the dissolved gypsum molecules was low.

The relative ease of gypsum to heterogeneously nucleate [17,18] on the stainless-steel tube as determined from the contact angle calculations may have hampered the ability of gypsum seed crystals to sufficiently reduce heterogeneous nucleation. Figure 9 shows that increasing gypsum seed loading six times from 0.04 to 0.25 g/L only resulted in a further 25% reduction in scale mass.

At seed loadings greater than 0.25 g/L, the specific surface area provided by the seeds could have been in excess for this system since the contact angle calculations showed that the degree of wetting on stainless-steel tube was relatively high, resulting in some of the seed crystals possibly adhering onto the tube surface. It should be noted that the scaling mechanism postulated for these experiments in which gypsum seeding was employed is different to the one for the preliminary experiments where there was no seeding. This is because the presence of gypsum seeds in relatively high quantities (0.50 and 1.0 g/L) made adhesion a possibility. However, this may not have been to a great extent since the mass of scale deposited on the tube in these experiments remained lower than that deposited in the control experiment.

The total amount of gypsum crystallized from the experiment was computed as the sum of the mass of gypsum scale and the mass of gypsum crystallized in the bulk solution. Figure 10 shows the total mass of gypsum crystallized as a function of gypsum seed loading.

The graph shows that the total mass of gypsum crystallized from the solution was much less than the theoretical yield expected. Theoretical yield was calculated using the feed solution concentration and the thermodynamic equilibrium concentration at 3 °C. This may have been due to slow gypsum crystallization kinetics stated earlier. Preliminary experiments, which were ran for 48 h, did not yield a spent solution concentration that is comparable to the thermodynamic equilibrium concentration.

Figure 11 shows the micrographs of the scaled tube at different seed loadings. The light phase represents the stainless-steel surface while the dark phase represents gypsum crystals.

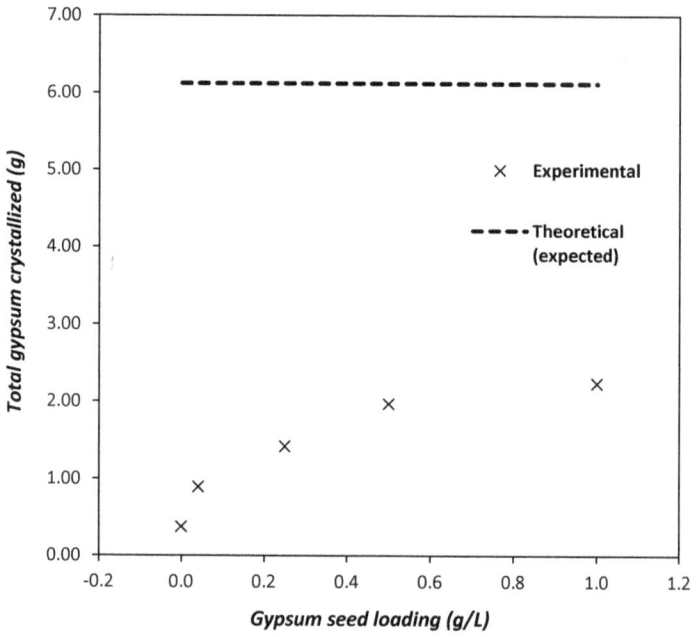

Figure 10. Total mass of gypsum crystallized as a function of gypsum seed loading.

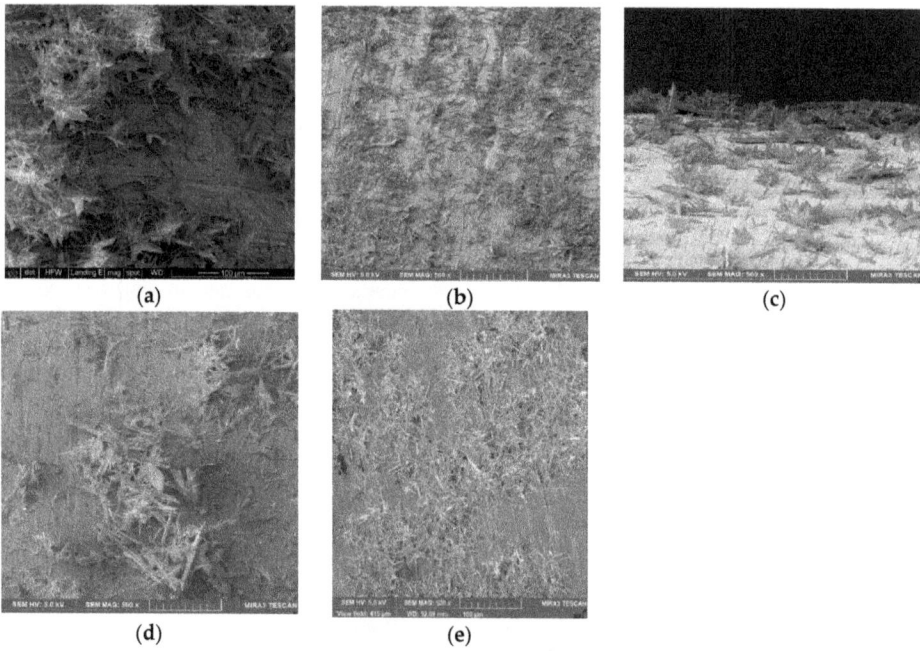

Figure 11. Micrographs of the scaled tube (**a**) control experiment, (**b**) C_S = 0.04 g/L, (**c**) C_S = 0.25 g/L, (**d**) C_S = 0.50 g/L, (**e**) C_S = 1.0 g/L. Scale bar = 100 µm.

The micrographs show that the predominant habit of the crystals that formed the scale layer was needles. An increase in gypsum seed loading led to fewer scale layer crystals per unit area of the tube because some of the available supersaturation was consumed by the seeds, leaving less available for heterogeneous nucleation and growth on the tube. Additionally, more gypsum seed crystals meant fewer prism-shaped crystals in the scale layer, as some of the supersaturation for growth of needle-type crystals into prisms was consumed by the gypsum seeds.

The growth direction of the crystals that formed the scale layer was comparable to that which was observed in the preliminary experiments at different durations. Figure 12 shows the normalized growth rate of the scale layer as a function of seed loading. The minimum on the x-axis (−0.2) was chosen to ensure the data point at 0.0 g/L would show clearly.

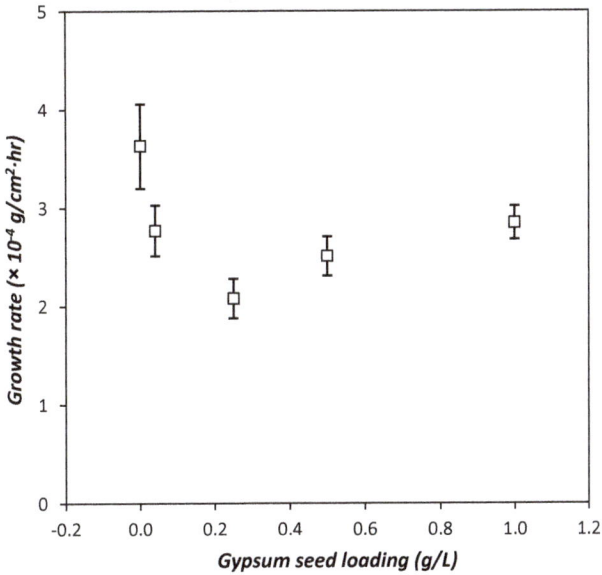

Figure 12. Normalized gypsum scale growth rate as a function of seed loading.

The graph shows that the normalised scale growth rate followed the same trend as was observed for the mass of scale deposited on the tube (Figure 9). The normalised scale growth rate was calculated by dividing the mass of gypsum scale by the product of the experiment duration and the tube surface area (same divisor). The experiment duration and the tube surface area were constants, hence the similarity in the trends.

The micrographs of the crystals recovered from the suspension at the end of each experiment are presented in Figure 13. The light phase represents the gypsum crystals and the dark phase represents the mounting glue.

The micrographs show that in the control experiment (Figure 13a), the crystals in the bulk solution were composed predominantly of needle-type crystals. There was evidence of some crystal twinning (white circles). The addition of 0.04 g/L seed crystals to the system decreased the proportion of needle-type crystals and the degree of twinning (Figure 13b). As the seed loading was increased, the habit of the crystals transformed from being predominantly needle-type to prisms. This was because in the absence of seeds, the supersaturation was relatively high and numerous crystallites were birthed. The available supersaturation was distributed among the crystallites for their growth, which resulted in needle-type habit. The presence of seeds and increase thereof possibly reduced the degree of nucleation and promoted crystal growth resulting in the formation of prisms.

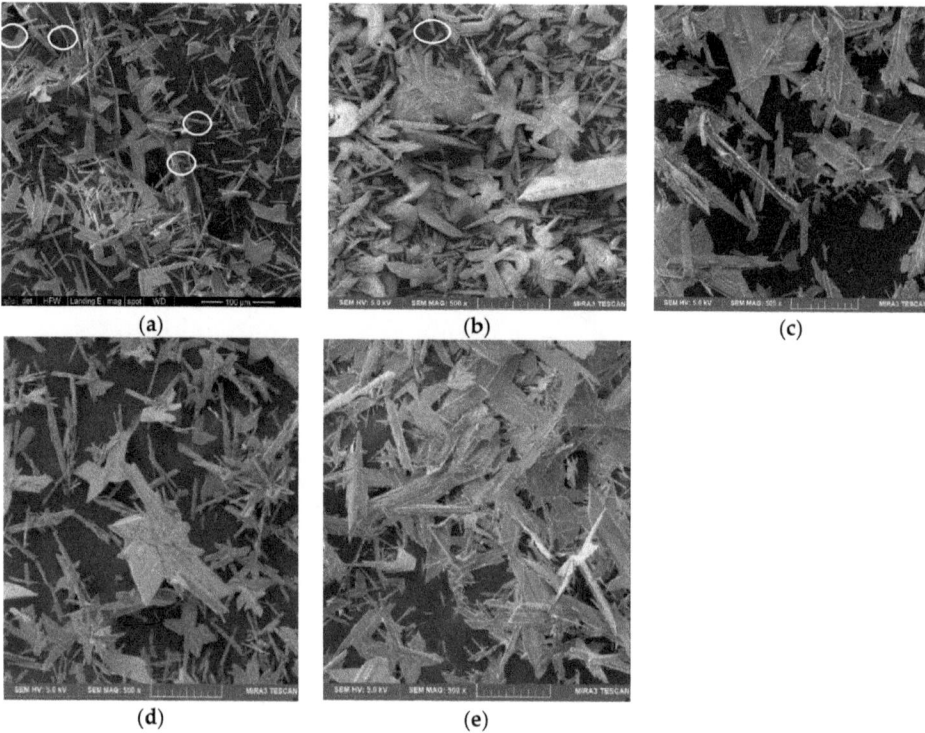

Figure 13. Micrographs of the solid suspension at the end of the experiment (**a**) control experiment, (**b**) C_S = 0.04 g/L, (**c**) C_S = 0.25 g/L (**d**) C_S = 0.50 g/L, and (**e**) C_S = 2.01 g/L. Scale bar = 100 μm.

4. Conclusions

Gypsum scale formation on the cooled stainless-steel tube was most likely a result of heterogenous nucleation and growth. The micrographs of the scaled tube showed that the rough patches on the stainless-steel tube were nucleation sites for gypsum scale.

Gypsum seeding was effective in decreasing the mass of gypsum scale deposited on the stainless-steel tube. This was attributed to the gypsum seeds providing a surface area that had favourable energetics for gypsum crystallization compared to the tube surface. The most effective seed loading was 0.25 g/L.

The amount of gypsum crystallized in the bulk solution increased as gypsum seed loading was increased. This was due to the increase in specific surface area that had growth sites on which gypsum dissolved in solution could crystallize.

These findings are relevant for the operability and control of EFC processes, which suffer from scaling problems. By using an appropriate seeding strategy, two problems can be alleviated. Firstly, scaling on the heat transfer surface is minimised and, secondly, seeding increases the crystallization kinetics in the bulk solution, which is advantageous for product yield and recovery.

This is of great importance towards scaling-up EFC for industrial applications. Beyond that, seeding to prevent scaling has potential applicability in other processes where the scale is regarded a product and/or purity is of importance, thus making addition of polymeric scale inhibitors undesirable.

5. Recommendation

There is need to investigate the effect of silica as a seed material to prevent gypsum scale formation. The gypsum crystallized in the bulk solution was still to a greater extent composed of fines, which poses separation problems in EFC. Silica is a robust and inert material that can ideally maintain its structural integrity throughout the experiment. If gypsum dissolved in solution can crystallize on silica, then there is a possibility of yielding coarser silica-gypsum crystals that may be relatively easy to separate from ice during EFC.

Author Contributions: Data curation, A.E.L.; Investigation, T.M.C.; Methodology, J.C. All authors have read and agreed to the published version of the manuscript.

Funding: This research was funded by Julian Baring Scholarship and University of Cape Town grant number. The APC was funded by University of Cape Town.

Institutional Review Board Statement: Not applicable.

Informed Consent Statement: Not applicable.

Data Availability Statement: The data presented in this study are available on request from the corresponding author.

Acknowledgments: The authors would like to acknowledge and thank Gerda van Rosmalen, who gave us some very valuable input at the early stages of this work. The authors would also like to thank the Julian Baring Scholarship Fund, all members of the Crystallization and Precipitation Research Unit, and the Mechanical Workshop in the Chemical Engineering Department at the University of Cape Town. Special mention goes to Miranda Waldron, Electron Microscope Unit.

Conflicts of Interest: The authors declare no conflict of interest.

Appendix A

$$\gamma_{sl} = \gamma_{cs} + \gamma_{cl}\cos\theta \tag{A1}$$

$$\Delta G'_{Heterogeneous} = f(\varnothing)\Delta G'_{Homogeneous} \tag{A2}$$

Table A1. Parameters used to calculate contact angle.

Parameter.	Value/Range (mJ/m^2)	Source
Stainless-steel surface free energy, γ_s	37	[35]
Dispersive component of stainless-steel surface free energy, γ_s^d	33.72	[36]
Water surface tension, γ_l	72.8	[37]
Dispersive component of water surface tension, γ_l^d	21.8	[37]
Gypsum surface free energy, γ_g	37	[18]
Dispersive component of gypsum surface free energy, γ_g^d	25.7–47.1	[37]

References

1. Fakir, S. The Industrialisation Debate in SA—What Are the Lessons? Available online: https://www.engineeringnews.co.za/article/the-industrialisation-debate-in-sa-what-are-the-lessons-2018-09-21/rep_id:4136 (accessed on 27 October 2020).
2. Claassen, M.; Masangane, W. *The Current State and Future Priorities of Brine Research in South Africa: Workshop Proceedings*; KV338/15; W.R. Commission: Pretoria, South Africa, 2015.
3. Van der Merwe, I.W.; Lourens, A.; Waygood, C. *An Investigation of Innovative Approaches to Brine Handling*; 1669/1/09; W.R. Commission: Pretoria, South Africa, 2010.
4. Macedonio, F.; Katzir, L.; Geisma, N.; Simone, S.; Drioli, E.; Gilron, J. Wind-Aided Intensified eVaporation (WAIV) and Membrane Crystallizer (MCr) integrated brackish water desalination process: Advantages and drawbacks. *Desalination* **2011**, *273*, 127–135. [CrossRef]
5. Randall, D.G.; Nathoo, J.; Lewis, A.E. A case study for treating a reverse osmosis brine using Eutectic Freeze Crystallization—Approaching a zero waste process. *Desalination* **2011**, *266*, 256–262. [CrossRef]
6. Helalizadeh, A.; Müller-Steinhagen, H.; Jamialahmadi, M. Mixed salt crystallisation fouling. *Chem. Eng. Process. Process Intensif.* **2000**, *39*, 29–43. [CrossRef]

7. Benecke, J.; Haas, M.; Baur, F.; Ernst, M. Investigating the development and reproducibility of heterogeneous gypsum scaling on reverse osmosis membranes using real-time membrane surface imaging. *Desalination* **2018**, *428*, 161–171. [CrossRef]
8. Jaramillo, H.; Boo, C.; Hashmi, S.M.; Elimelech, M. Zwitterionic coating on thin-film composite membranes to delay gypsum scaling in reverse osmosis. *J. Membr. Sci.* **2021**, *618*, 118568. [CrossRef]
9. Matin, A.; Rahman, F.; Shafi, H.Z.; Zubair, S.M. Scaling of reverse osmosis membranes used in water desalination: Phenomena, impact, and control; future directions. *Desalination* **2019**, *455*, 135–157. [CrossRef]
10. Amjad, Z. Calcium sulfate dihydrate (gypsum) scale formation on heat exchanger surfaces: The influence of scale inhibitors. *J. Colloid Interface Sci.* **1988**, *123*, 523–536. [CrossRef]
11. Hasson, D.; Avriel, M.; Resnick, W.; Rozenman, T.; Windreich, S. Mechanism of Calcium Carbonate Scale Deposition on Heat-Transfer Surfaces. *Ind. Eng. Chem. Fundam.* **1968**, *7*, 59–65. [CrossRef]
12. Gill, J.S.; Nancollas, G.H. Kinetics of growth of calcium sulfate crystals at heated metal surfaces. *J. Cryst. Growth* **1980**, *48*, 34–40. [CrossRef]
13. Linnikov, O.D. Investigation of the initial period of sulphate scale formation Part 1. Kinetics and mechanism of calcium sulphate surface nucleation at its crystallization on a heat-exchange surface. *Desalination* **1999**, *122*, 1–14. [CrossRef]
14. Linnikov, O.D. Investigation of the initial period of sulphate scale formation Part 2. Kinetics of calcium sulphate crystal growth at its crystallization on a heat-exchange surface. *Desalination* **2000**, *128*, 35–46. [CrossRef]
15. Mwaba, M.G.; Rindt, C.C.M.; Van Steenhoven, A.A.; Vorstman, M.A.G. Experimental investigation of CaSO4 crystallization on a flat Plate. *Heat Transf. Eng.* **2006**, *27*, 42–54. [CrossRef]
16. Kashchiev, D. On the relation between nucleation work, nucleus size, and nucleation rate. *J. Chem. Phys.* **1982**, *76*, 5098–5102. [CrossRef]
17. Mersmann, A.; Eble, A.; Heyer, C. Crystal Growth. In *Crystallization Technology Handbook*; Mersmann, A., Ed.; Marcel Dekker, Inc.: New York, NY, USA, 2001.
18. Mullin, J.W. *Crystallization*, 4th ed.; Butterworth-Heinemann: Oxford, UK, 2001.
19. Jagadesh, D.; Kubota, N.; Yokota, M.; Sato, A.; Tavare, N.S. Large and mono-sized product crystals from natural cooling mode batch crystallizer. *J. Chem. Eng. Jpn.* **1996**, *29*, 865–873. [CrossRef]
20. Lewis, A.E.; Seckler, M.; Kramer, H.J.M.; Rosmalen, G. *Industrial Crystallization: Fundamentals and Applications*; Cambridge University Press: Cambridge, UK, 2015.
21. Doki, N.; Kubota, N.; Sato, A.; Yokota, M.; Hamada, O.; Masumi, F. Scaleup experiments on seeded batch cooling crystallization of potassium alum. *AIChE J.* **1999**, *45*, 2527–2533. [CrossRef]
22. Doki, N.; Kubota, N.; Yokota, M.; Chianese, A. Determination of critical seed loading ratio for the production of crystals of uni-modal size distribution in batch cooling crystallization of potassium alum. *J. Chem. Eng. Jpn.* **2002**, *35*, 670–676. [CrossRef]
23. Wang, H.; Wang, D.; Li, Z.; Demopoulos, G.P. Solubility and scale prevention of gypsum in transportation pipes of well brine with salinities up to 5 M at temperature range of 278–298 K. *Desalination Water Treat.* **2012**, *22*, 78–86. [CrossRef]
24. Liu, S.T.; Nancollas, G.H. The kinetics of crystal growth of calcium sulfate dihydrate. *J. Cryst. Growth* **1970**, *6*, 281–289. [CrossRef]
25. Doki, N.; Kubota, N.; Sato, A.; Yokota, M. Effect of cooling mode on product crystal size in seeded batch crystallization of potassium alum. *Chem. Eng. J.* **2001**, *81*, 313–316. [CrossRef]
26. Aspeling, B.J.; Chivavava, J.; Lewis, A.E. Selective salt crystallization from a seeded ternary eutectic system in Eutectic Freeze crystallization. *Sep. Purif. Technol.* **2020**, *248*, 117019. [CrossRef]
27. Randall, D.G.; Nathoo, J.; Lewis, A.E. Seeding for selective salt recovery during eutectic freeze crystallization. In Proceedings of the International Mine Water Conference, Pretoria, South Africa, 19–23 October 2009.
28. Randall, D.G. Development of a Brine Treatment Protocol Using Eutectic Freeze Crystallization. Ph.D. Thesis, University of Cape Town, Cape Town, South Africa, 2010.
29. Randall, D.G.; Mohamed, R.; Nathoo, J.; Rossenrode, H.; Lewis, A.E. Improved calcium sulfate recovery from a reverse osmosis retentate using eutectic freeze crystallization. *Water Sci. Technol.* **2013**, *67*, 139–146. [CrossRef] [PubMed]
30. Adams, J.F.; Papangelakis, V.G. Gypsum scale formation in continuous neutralization reactors. *Can. Metall. Q.* **2013**, *39*, 421–432. [CrossRef]
31. Gainey, R.J.; Thorp, C.A.; Cadwallader, E.A. CaSO4 seeding prevents CaSO4 scale. *Ind. Eng. Chem.* **1963**, *55*, 39–43. [CrossRef]
32. Oshchepkov, M.; Popov, K.; Kovalenko, A.; Redchuk, A.; Dikareva, J.; Pochitalkina, I. Initial stages of gypsum nucleation: The role of "nano/microdust". *Minerals* **2020**, *10*, 1083. [CrossRef]
33. Choi, J.Y.; Lee, T.; Cheng, Y.; Cohen, Y. Observed crystallization induction time in seeded gypsum crystallization. *Ind. Eng. Chem. Res.* **2019**, *58*, 23359–23365. [CrossRef]
34. OLI Systems Inc. *OLI Studio 10.0 User Guide*; OLI Systems Inc.: Morris Plains, NJ, USA, 2020.
35. MacAdam, J.; Parsons, S.A. Calcium carbonate scale formation and control. *Environ. Sci. Bio Technol.* **2004**, *3*, 11. [CrossRef]
36. Ozbay, S.; Erbil, H.Y. Ice accretion by spraying supercooled droplets is not dependent on wettability and surface free energy of substrates. *Colloids Surf. A Physicochem. Eng. Asp.* **2016**, *504*, 210–218. [CrossRef]
37. Nikoo, A.H.; Kalantariasl, A.; Malayeri, M.R. Propensity of gypsum precipitation using surface energy approach. *J. Mol. Liq.* **2020**, *300*, 112320. [CrossRef]

Article

Continuous Isolation of Particles with Varying Aspect Ratios up to Thin Needles Achieving Free-Flowing Products

Claas Steenweg, Jonas Habicht and Kerstin Wohlgemuth *

Laboratory of Plant and Process Design, Department of Biochemical and Chemical Engineering, TU Dortmund University, D-44227 Dortmund, Germany; claas.steenweg@tu-dortmund.de (C.S.); jonas.habicht@tu-dortmund.de (J.H.)
* Correspondence: kerstin.wohlgemuth@tu-dortmund.de; Tel.: +49-231-755-3020

Abstract: The continuous vacuum screw filter (CVSF) for small-scale continuous product isolation of suspensions was operated for the first time with cuboid-shaped and needle-shaped particles. These high aspect ratio particles are very common in pharmaceutical manufacturing processes and provide challenges in filtration, washing, and drying processes. Moreover, the flowability decreases and undesired secondary processes of attrition, breakage, and agglomeration may occur intensively. Nevertheless, in this study, it is shown that even cuboid and needle-shaped particles (L-alanine) can be processed within the CVSF preserving the product quality in terms of particle size distribution (PSD) and preventing breakage or attrition effects. A dynamic image analysis-based approach combining axis length distributions (ALDs) with a kernel-density estimator was used for evaluation. This approach was extended with a quantification of the center of mass of the density-weighted ALDs, providing a measure to analyze the preservation of the inlet PSD statistically. Moreover, a targeted residual moisture below 1% could be achieved by adding a drying module (T_{dry} = 60 °C) to the modular setup of the CVSF.

Keywords: continuous manufacturing; pharmaceutical manufacturing; continuous crystallization; continuous particle isolation; filtration; washing; needles; drying

1. Introduction

The small-scale production of active pharmaceutical ingredients (APIs) has been traditionally performed in batch mode, including all process steps from raw material treatment to the final drug formulation [1–3]. Much effort has been made to mirror the progress achieved in continuous pharmaceutical reaction technology in the development of continuous crystallization apparatuses by realizing various concepts [1,4–8]. Since crystallization steps are included in more than 90% of API manufacturing processes, a high potential is seen in reliable continuous end-to-end manufacturing processes [5,9,10]. The major benefits of these integrated processes are the elimination of batch-to-batch variability, an increase in capacity either through parallelization of units or simply through longer plant run times, and a shorter time-to-market [5,6,11–14]. The latter arises with the opportunity to directly use the equipment designed in the research and development of a small-scale apparatus concept. Moreover, high-purity continuous crystallization is nowadays seen as the first downstream step to selectively determine the final product properties regarding the particle size distribution (PSD) and particle shape [1,4]. Thus, preserving the produced quality attributes is a crucial requirement for all subsequent isolation steps, namely, filtration, washing, and drying to achieve free-flowing particles usable for secondary processing.

Various research groups using different apparatuses showed recent advances in small-scale continuous product isolation. Alconbury Weston Ltd. launched the continuous filter carousel (CFC) to the market. A multitude of current publications shows the potential of automated, cyclic batch filtration in the CFC [9,15–22]. A second concept, the continuous

rotary plate filter (CRPF) was invented as part of an integrated continuous manufacturing pilot plant of Continuus Pharmaceuticals. Recent studies showed the potential of combined filtration and washing in the CRPF as a part of an end-to-end continuous manufacturing process [10,23–26].

We developed the modular continuous vacuum screw filter (CVSF) in our research group and patented this innovative apparatus in 2021 [27]. The CVSF is used to efficiently perform all necessary product isolation steps after a preceding small-scale continuous crystallization in a fully continuous operation [27–31]. It provides salient features as its modularity allows a rapid adaption to varying material systems and throughputs. The CVSF is designed for a typical API production rate of 250–1000 kg per year, which can be transferred to an approximated volume flow rate of 10–100 mL min^{-1} of feed suspension [28]. It can be used in flexible modular connections of filtration, washing, and drying modules. The first characterization studies were performed with a well filterable model system of L-alanine/water resulting in bipyramidal particles [28,29]. For this system, suitable residual moistures of approx. 1% with a preserved PSD were obtained using two washing modules without an extra drying module [29]. Moreover, a narrow residence time distribution of the solid phase (RTD$_s$) was defined as a fundamental requirement and thus, systematically proven in previous works [28,29]. A narrow RTD$_s$ is thereby explicitly demanded by the American Food and Drug Administration to ensure material traceability [32]. Furthermore, stable steady-state operation for more than one hour was demonstrated and product loss was negligible [29].

It is well known that particle bulk porosity, which is mainly defined by the particle shape and the width of the PSD, highly influences the processability with regard to filterability and solids handling [33]. To illustrate the dependence of particle shape on bulk properties, an analysis of five parameters (porosity, flowability, flow rate, tensile strength, and angle of internal friction) was performed by Pudasaini et al. using six different crystal habits of acetylic salicylic acid [34]. In general, higher aspect ratio shapes result in higher porosity and lower flowability. On the one hand, a higher porosity may influence the filterability of the crystals resulting in the potential need for additional modules to reach targeted residual moistures. On the other hand, a lower flowability may lead to breakage issues in the CVSF owing to increased compression.

Thereby, solid–liquid separation using filtration is highly dependent on the properties of the particles used. Besides device-specific constants (e.g., filter area, filter material, and pressure difference), the dehumidification challenges vary with particle size, cake porosity, and particle morphology [33]. Regarding the 'crystal process chain', these component-specific parameters are defined mainly by the preceding crystallization step. Thus, CVSF operability with various particle properties is an essential requirement for widespread industrial application. The above-mentioned bipyramidal L-alanine particles showed good filterability and a low breakage tendency due to their approximate spherical structure. However, for pharmaceutical crystals, smaller particle sizes between 50–500 µm and a mean particle size of around 80 µm is targeted to enable direct compression while still enabling dissolution in the human body [35]. Additionally, more anisotropic structures such as cuboid-shaped or needles-shaped particles are common for APIs [34,36,37]. The smaller particle size and varying shape result in a more challenging separation task defined by the filter cake resistance [33,38]. Referring to the Carman–Kozeny equation, higher residual moisture at the end of the filtration and washing process is expected for decreasing particle size, due to an increased filter cake resistance [33]. Therefore, a longer deliquoring time may be necessary. Moreover, for increasing aspect ratios the porosity often increases because particles with high aspect ratios tend to create high-porosity bulk solids [33]. This leads to a higher residual moisture at a constant liquid saturation degree after filtration or washing and therefore more pore liquid (higher free volume due to increased porosity) has to be evaporated during the subsequent drying process compared with spherical particles. Furthermore, for more elongated particles, major risks are particle breakage or increased attrition and agglomeration [39].

Especially for high aspect ratio particles, challenges regarding the measurement and evaluation are still remaining, since the calculation of the equivalent diameter d_{eq} as one-dimensional size descriptor is not suitable [39,40]. Therefore, various approaches for image-analysis-based measurements and simulations were developed in the past [38,40–43]. Especially in the group of Mazzotti [40–42], a multitude of scientific papers were published introducing the axis length distribution (ALD) as a two-dimensional approach using automated image analysis. ALDs show the corresponding particles with their respective major and minor axis lengths in a 2D illustration. From this, a sole qualitative comparison is possible and statements regarding changes in particle sizes are possible. Nevertheless, to the best of our knowledge, corresponding quantitative values as known from spherical particles (e. g., d_{50}, d_{90}–d_{10}) have not yet been determined. However, this is an important prerequisite to being able to provide a quantitative statement about possible changes of the particle size before and after particle isolation.

The objective of this study is to demonstrate the broad applicability of CVSF for particles with varying aspect ratios and mean particle sizes while maintaining PSD and reducing residual moisture to free-flowing particles. As a first step, a quantitative measure is introduced enabling comparison of ALDs using 2D image analysis. Afterwards, crystal suspensions of L-alanine crystallized from aqueous solution with different L-glutamic acid concentrations were filtered. L-glutamic acid acts as a habit modifier and inhibits crystal growth selectively so that pure L-alanine crystals with varying shapes and sizes result. The following requirements are addressed:

1. Maintaining ALD through particle isolation process
2. Reducing residual moisture also for higher aspect ratios

2. Materials and Methods

2.1. Substances Used

As model material system L-alanine/water is chosen. It presents an intensively investigated system at our group and is well-known regarding multiple properties such as aqueous solubility, crystal morphology, solids handling, and cooling crystallization behavior. As solely used in previous works regarding the CVSF, this model system is further used to increase the comparability to the results obtained. L-alanine is purchased from Evonik Industries AG, Essen, Germany (purity 99.7%). As solvent ultrapure, deionized and bacteria-free filtered water (Milli-Q® Advantage A10, 0.05 µS cm^{-1}, Merck KGaA, Darmstadt, Germany) is used. The solubility curve c^* of L-alanine/water can by approximated by Equation (1) [44].

$$c^* \left(g_{ala} \, g_{solution}^{-1} \right) = 0.11238 \cdot \exp\left(9.0849 \cdot 10^{-3} \cdot \vartheta^* (^\circ C)\right) \tag{1}$$

The density of a saturated L-alanine/water solution at ambient conditions (23 °C) is 1042 kg m^{-3} and the corresponding particle density of pure L-alanine amounts to 1420 kg m^{-3}. To get L-alanine particles with varying particle shapes, L-glutamic acid (Acros Organics™, purity 99%, Fisher Scientific GmbH, Schwerte, Germany) is added to the binary system which acts as a habit modifier.

2.2. Experimental CVSF Setup

The experimental setup is schematically shown in Figure 1. The setup is the same as in our previous study [29] and therefore just briefly summarized here. From a 450 mL borosilicate glass suspension vessel equipped with a stirrer, vertical baffles, and a conical-shaped bottom (all out of borosilicate glass) developed by Lührmann et al. [45], the suspension is pumped (60 mL min^{-1}) to the CVSFs inlet using a peristaltic pump (Ismatec, Wertheim, Germany, ISM597D, 2.79 mm Saint Gobain Tygon Tubing). The crystal suspension is discharged through a vertically adjustable outlet tube. Its opening is located 0.5 cm above the profiled bottom. The distance between the vessel and the pump, as well as the distance between the pump and the CVSF is 20 cm, respectively. The CVSF is mounted into an aluminum frame and consists of two basic elements: tubular, cylindrical modules (A, B1, and B2) and a rotating custom-made polytetrafluoroethylene (PTFE) screw (outer diameter 25 mm, distance to glass wall 0.2 mm, screw pitch 11 mm, and 33 screw coils for two-stage washing—11 screw coils per module) for axial transport of the particles in the inner part. All modules have an inner diameter of 25.4 mm (DN25), a length of 13 cm, and are made out of borosilicate glass. Porous glass frits (porosity 2, 40–100 µm, ROBU GmbH, Hattert, Germany) are installed as filter media and vacuum can be applied using vacuum pumps (Vacuubrand PC2004 Vario and Vacuubrand PC3002 Vario select, Wertheim, Germany). For axial transport of the particles, the PTFE screw can be rotated using a motor (Osmtec Nema Co., Ltd., 173 Nm, Nanjing, China). For washing in modules B1 and B2, a low-pressure flat-fan nozzle (Lechler, 610.145, Metzingen, Germany), adapted to fit into the respective inlets, is integrated at the top. The outer jackets of all modules are separated to prevent back-mixing of the withdrawn liquids and thus simplify solvent recovery (not conducted in this work).

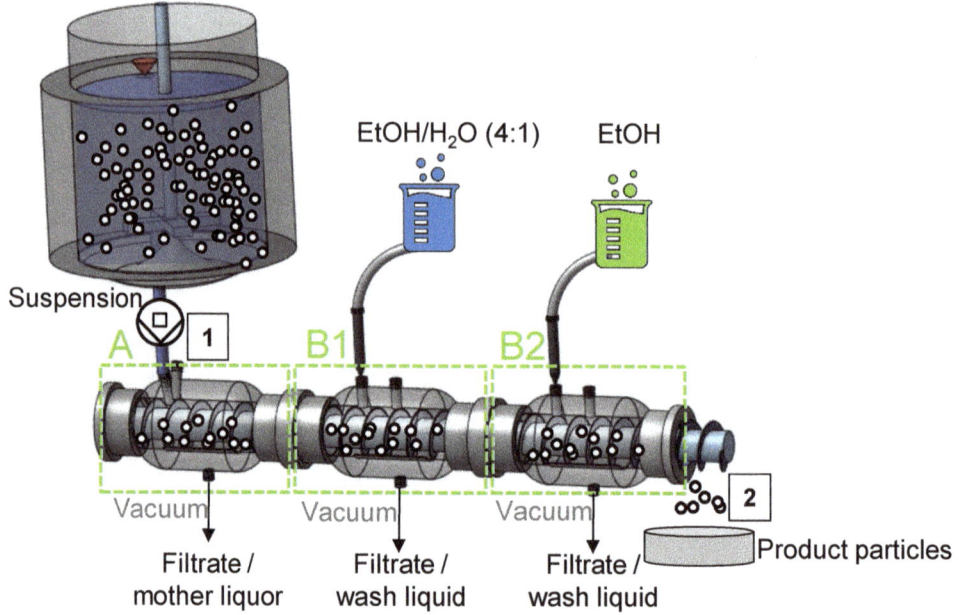

Figure 1. Experimental setup of the experiments with varying particle shapes. The CVSF consists of the module A (filtration), module B1 (1st washing), and module B2 (2nd washing).

2.3. Experimental Procedures

All experiments with varying particle shape and size are carried out at the same process conditions, only changing the particle shapes. Therefore, first the production of the particles with different shapes is described and then the procedure of the CVSF experiments is explained.

2.3.1. Procedure for Preparation of Inlet Suspension with Particles of Varying Shape

The preparation of L-alanine particles with varying shapes used for preparation of the inlet suspension includes five steps: crystallization, filtration, washing, drying, and sample dividing, all of them are performed in batch mode. Batch cooling crystallization of L-alanine from water is carried out in a 10 L vessel. First, a saturated solution of L-alanine is prepared at 50 °C according to Equation (1) (9000 g water and 1935.5 g L-alanine). To ensure a crystal-free liquid with a defined temperature, the solution is heated up to 60 °C. Having stirred the solution at constant temperature for 60 min, a cooling rate of 0.45 K min^{-1} is set to the system and cooled down to 20 °C. Due to the primary homogenous nucleation, high cooling rates create a high local supersaturation, resulting in a preferentially high number of small crystals. The product yield of a single batch crystallization is approx. 450–500 g. The same procedure is basically chosen to obtain other crystal shapes (cuboids and needles), but the appropriate amount of L-glutamic acid is added before starting the cooling ramp from 60 °C to 20 °C. According to previous work by Saal, L-glutamic acid selectively influences the crystal growth of L-alanine during batch cooling crystallization [46]. Due to the zwitterionic character of L-alanine, the additive adsorbs preferentially at a defined crystal surface, selectively influencing crystal growth. Thus, crystal growth at this particular surface can be either inhibited or promoted. In the case of L-glutamic acid, selective growth is more and more inhibited with increasing L-glutamic acid concentration resulting in smaller crystals with higher aspect ratios, as seen in Figure 2 [47].

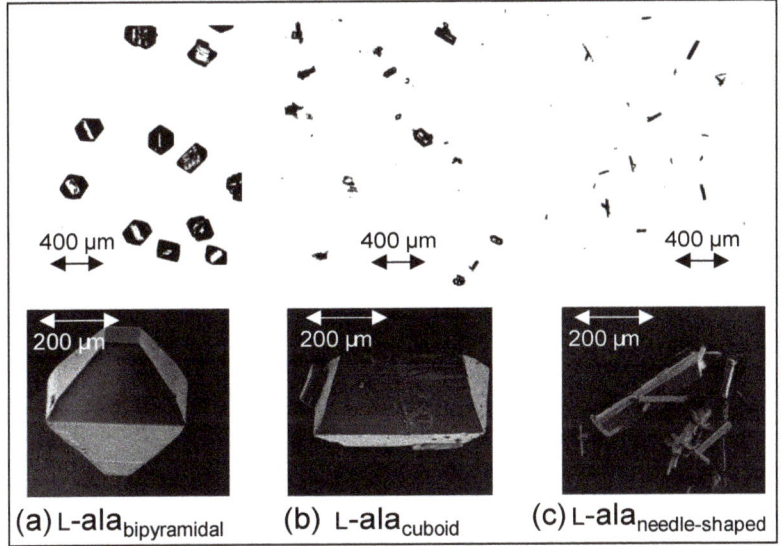

Figure 2. Upper half: exemplary QICPIC binary images of (**a**) bipyramidal—no $_{L-GA}$ impurity, (**b**) cuboid—2.9 g$_{L-GA}$ kg^{-1} $_{water}$, and (**c**) needle-shaped 5.8 g$_{L-GA}$ kg^{-1} $_{water}$. Lower half: corresponding scanning electron microscope images.

In the case of the L-alanine cuboid habit, the concentration of L-glutamic acid was chosen to 2.9 g$_{L-GA}$ kg^{-1} $_{water}$. In contrast, for the L-alanine needle-shaped habit, an additive concentration of 5.8 g$_{L-GA}$ kg^{-1} $_{water}$ was used. Additive concentrations were selected similarly to the works of Heisel et al. and Saal [46,47]. Besides the addition of L-glutamic acid, all other crystallization parameters were held constant. The resulting crystals are harvested, filtrated, washed, and dried so that statistically uniform samples can subsequently be prepared using a sample divider (described below). Product-isolation realized by filtration is performed with a sintered glass filter frit by ROBU GmbH, Hattert, Germany. The porosity of four samples according to ISO 4793-80 with pore sizes from 10–16 µm is used to guarantee retention of the crystals of targeted size fraction (>25 µm). To remove the mother liquor, a vacuum Nutsche system is used containing the glass filter and a vacuum pump (Vacuubrand PC2004 Vario, Wertheim, Germany). Retained crystals are then washed according to the procedure described by Terdenge and Wohlgemuth [48] in order to minimize the agglomeration of the seed crystals. Therefore, a two-stage washing procedure is applied. First, a volumetric mixture of 4/1 of ethanol (EtOH) absolute (99.9%, VWR, Darmstadt, Germany) and water is used as wash liquid. This ratio is defined as the optimal ratio to avoid anti-solvent crystallization by maintaining displacement washing [48]. Next, the filter cake is washed with pure EtOH to improve the washing efficiency. For an evenly distributed washing film, the solvent is sprayed with the use of an atomizer in both cases. The amount of washing solvent is chosen in relation to the amount of crystals processed. Approximately 40 mL of solvent are used per 100 g of crystals. Washed crystals are then transferred into a drying oven at ambient pressure and 50 °C for 24 h. Then, the pre-dried crystals are further deliquefied in a vacuum oven (Thermo Heraeus, type Vacutherm VT 6060M, Schwete, Germany) at the same temperature for at least 48 h to ensure complete solvent evaporation. In a final step, four samples of around 120 g each are created using a rotary sample divider (Fritsch, Laborette 27, Idar-Oberstein, Germany) to enable statistical comparability. Due to the usage of the rotary divider, each sample is assumed to contain particles of the same statistical population. The corresponding inlet suspension is then prepared by filling saturated solution (see Equation (1)) into the suspension vessel and adding the L-alanine particles of the prepared sample separately.

2.3.2. Procedures of CVSF Experiments

The CVSF is operated with the parameters shown in Table 1, which presents a well-known operational point with a narrow RTD$_S$ [28].

Table 1. Operating parameters for experiments in this study. Q_{susp}: suspension volume flow rate; w_s: feed solid mass fraction; n_{screw}: rotational speed of the PTFE screw; Δp_{set}: applied vacuum pressure difference; and Q_{wash}: wash liquid flow rate.

Operating Parameters	Value
Q_{susp} [mL min^{-1}]	60
w_s [wt.%, g g$_{sol}^{-1}$]	6
n_{screw} [rpm]	3
Δp_{set} [mbar]	400
Q_{wash} [mL min^{-1}]	35

The CVSF is set up with three modules to enable two-stage washing as shown in Figure 1 Assuming an ideal residence time distribution of the solid phase (RTD$_S$), the ideal solids residence time $\tau_{CVSF,id}$, is solely dependent on the operational parameter n_{screw}. Thereby, the ideal residence time is derived from the number of helical mounts in the apparatus N_{coil} and the rotational screw speed n_{screw}. For the modular design, this can be specified in more detail using the number of helical mounts per module $N_{coil,mod}$ and the corresponding number of modules consisting of $N_{mod,filtration}$ and $N_{mod,washing}$ according to Equation (2) [29].

$$\tau_{CVSF,id} = \frac{N_{coil}}{n_{screw}} = \frac{N_{coil,mod} \cdot \left(N_{mod,filtration} + N_{mod,washing}\right)}{n_{screw}} \quad (2)$$

For the CVSF used, $N_{coil,mod}$ is 11 [29]. Having reached a steady-state in sole filtration mode after a run time of 10.5 min (referring to the ideal solids residence time $\tau_{CVSF,id}$ after two-stage washing), QICPIC (PSD measurement) and residual moisture samples are taken. Then, the first washing stage (4/1 EtOH/water) is activated in module B1. After a run time of 7 min, one-stage washed filter cake is discharged and samples are withdrawn. Following the same procedure, two-stage washing (EtOH) is applied, and after a run time of 3.5 min the last samples are taken. Through this experimental procedure, the material consumption is minimized that only around 120 g of particles are necessary to perform a single experiment. With the number of particles available, a three-fold determination of the residual moisture is performed per experiment.

2.4. Analytics

The dynamic image analysis system QICPIC (resolution 1024 × 1024 pixel, module M6, Sympatec GmbH, Clausthal-Zellerfeld, Germany), equipped with the liquid dispersion system LIXELL (Sympatec GmbH, Clausthal-Zellerfeld, Germany) is used for measurement of particle size at the inlet and outlet of the CVSF. At the respective sampling locations, suspension (after feed pump) or particles (product particles after CVSF) are collected (>10,000 particles) in excess of tempered saturated aqueous solution to sufficiently dilute potential local supersaturation. Each sample is transferred to a stirred vessel, which is directly connected to the QICPIC measuring cell. The image acquisition and quantification of shape descriptors are executed by artificial neural networks implemented in MATLAB [47,49]. The axis length distribution of the inlet suspension (ALD$_{IN}$) and axis length distribution of the product particles (ALD$_{OUT}$) are measured for needle-shaped and cuboid L-alanine particles, and possible changes are quantified, as described in Section 3.

Particles leaving the CVSF are analyzed regarding the residual moisture by a procedure based on loss-on-drying [28,29]. The residual moisture φ_{RM} is determined using a pre-drying step in a drying oven for 24 h (50 °C) and further drying in a vacuum drying oven (Thermo Heraeus, type Vacutherm VT 6060M, Schwerte, Germany) at 50 °C for 72 h to ensure gentle solids handling and prevent stirring up of the particles. For that, at least 2 g of wet sample was collected in a weighed screw cap container. φ_{RM} is calculated by the mass of the evaporated liquid m_l divided by the corresponding mass of the wet sample, consisting of the mass of solids m_s and mass of liquid m_l as shown in Equation (3) [29].

$$\varphi_{RM} \left[\text{g g}_{\text{wet}}^{-1} \right] = \frac{m_l}{m_s + m_l} \qquad (3)$$

3. Extension of ALD Approach of Non-Spherical Particles

The traditional approach of dynamic image analysis (DIA) for bipyramidal L-alanine evaluates particles by calculating an equivalent diameter d$_{eq}$ for each particle. This one-dimensional size descriptor is only valid for approximate spherical particles because the DIA is taken out based on a 2D-projection of the particles [50,51]. With increasing aspect ratio, the orientation of particles to the camera in DIA highly influences the 2D-projection and accordingly the size measured by equivalent diameters. This phenomenon is schematically depicted in Figure 3 for nearly spherical particles and needle-shaped particles with a high aspect ratio. The grey surfaces represent the projection area measured by DIA for two different observation angles each. It can be noted that in the case of non-spherical particles, these areas may differ significantly.

According to this, depending on its orientation to the imaging device, even one single, non-spherical particle results in different projection areas measured. Therefore, to accurately measure non-bipyramidal particles by 2D-DIA, two-dimensional size descriptors are needed [40]. During this work, major and minor axis lengths of the best fitting ellipsoid are used as size descriptors for each particle analyzed. Initially, to calculate the major and minor axes, the area and centroid of each 2D-projection are directly calculated by the same DIA-combined MATLAB routine used in our previous studies [28,29] as introduced by Heisel et al. [47,49]. An ellipsoid with an equal moment of inertia is fitted to the 2D-projection based on this data. The corresponding values of major and minor axis length generate the measured data for each particle based on its 2D-projection [41]. The detailed procedure of ellipsoid fitting is given in [40,52]. The aspect ratio (AR) is then defined as the quotient of the major axis length (MAL) and the minor axis length (MIL) according to Equation (4).

$$AR = \frac{\text{MAL}}{\text{MIL}} \qquad (4)$$

The result of all ellipsoids measured is the ALD. The major advantage over different two-dimensional size descriptors such as maximal/minimal Feret diameter is the compensation of irregularities in the 2D-projection or blurred particles [40,41]. It can be stated that even for a relatively low aspect ratio of two, a variety of measurements can occur during DIA. A measurable ALD represents the 2D-projection of the unknown PSD overlapped by the Gaussian-distributed orientation of each particle in the flow-through cell, if flow-induced preferential orientations are neglected. To reduce the influence of random particle orientation, it is recommendable to analyze a high number of particles for each measurement [53].

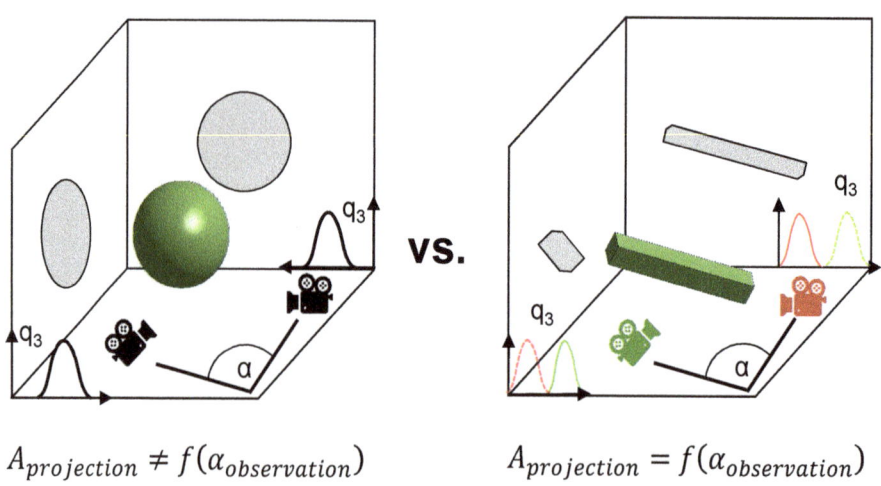

Figure 3. Challenges occurring with 2D-image analysis of non-spherical particles. The grey surfaces show the measured areas by dynamic image analysis for two different angles of $\alpha_{observation}$. The differences in the resulting PSDs are schematically shown in the left and right corners of the cubic analytical space.

Although the exact size of particles is not measurable by 2D-DIA, it shows high potential for application in the analysis of the inlet particle size distribution (PSD_{IN}) and the outlet particle size distribution (PSD_{OUT}) of CVSF since the main challenge in filtration is not reaching a targeted particle size but maintaining the PSD_{IN} during critical process steps, which is a crucial requirement for the performance behavior characterization of the apparatus. Having depicted the challenges of size determination by DIA for non-spherical particles, this work aims to develop a shape-independent, PSD-related, analytic routine to evaluate and quantify changes in the particle size distribution during CVSF operation. Therefore, an ALD-based approach containing **three** steps is applied, which is schematically depicted in Figure 4. The different steps are described in detail, for an exemplary data set containing needle-shaped L-alanine particles. In the first two steps, the approach of the group of Mazzotti is applied [40,41]. In the third step, a new quantitative measure is introduced so that quantitative comparison of two ALDs can be obtained.

In **Step 1,** the scatter plot is obtained by illustrating DIA raw data. The resolution limit of the QICPIC used was included in the ALD construction. Based on the equivalent diameter, lines with equal area regarding an ellipsoid were calculated and all particles below the resolution limit are excluded. For the QICPIC used, a minimal equivalent diameter of 25 µm is required for the imaging module to ensure image analysis with ISO-standard and adequate size analysis [54]. Each point represents one measured particle of all 38,702 particles (>25 µm) within the sample located by its major and minor axis length. The right-side half-plane is empty by definition since the MAL is always greater than (ellipsoid) or equal (circle) to its associated MIL. Since the measurement points are highly overlapping, a suitable raw data treatment is needed to illustrate and further analyze DIA-measurements.

In **Step 2** of the routine, a Gaussian kernel-density-estimator is used, which is implemented as the "density dots" function provided by the Origin Pro 2021 software [55]. Of course, this algorithm could also be applied with many other software solutions. First, for each data point (X_i, Y_i) out of a population of N points, a bivariate Gaussian distribution Φ_i according to Equation (5) is created.

$$\Phi_i = \frac{1}{2\pi b_x b_y} \cdot \exp\left(-\frac{(x-X_i)^2}{2b_x{}^2} - \frac{(y-Y_i)^2}{2b_y^2}\right) \tag{5}$$

The location parameters X_i and Y_i can, in the case of ALD measurements, be defined as minor axis length and major axis length, respectively. The so-called bandwidths b_x and b_y are defined as the scales of the Gaussian distribution. The algorithm used in the "density dots" function relies on minimizing the mean integrated squared error between the estimator and the underlying distribution. The exact procedure of bandwidth calculation is based on linear diffusion processes [56]. Having calculated Φ_i for each data point i, each Gaussian distribution is scaled by factor N^{-1} and summed up to create the estimator E of the population investigated according to Equation (6) [56]. Values of E represent the corresponding kernel-density of a point defined by its major/minor axis length (x, y).

$$E(x, y, b_x, b_y) = \frac{1}{N} \cdot \sum_{i=1}^{N} \Phi_i(x, y, b_x, b_y) \tag{6}$$

This scaling gives an estimator normed in sample size, and thus a comparability of estimator values for different populations investigated. This characteristic is a crucial requirement for applying kernel-density estimators in the analysis of ALDs since the sample size of DIA-measurements varies inevitably. Having a suitable estimator as a mathematic tool at hand, E is evaluated at an arbitrary number of k orthogonal grid points. These grid points allocate a dimensionless kernel-density value to each measurement point (X_i, Y_i) by linear interpolation. High values of E thereby are formed when multiple measurements occur in a sufficiently close neighborhood. Finally, the 4th root of kernel-density values is applied to fit the density scale to a color scale. ALD measurements for cuboid-shaped and needle-shaped L-alanine particles were performed in a maximum window of 400 µm of major axis length in this work. Thereupon, grid points were chosen to a number of 400 for each dimension to create an equally spaced grid. Related to the 2D-projection measurement of particles, the calculated kernel-density can be interpreted as an indicator for the probability to measure a certain point in the ALD. Since particle orientation to the imaging device is assumed to be Gaussian distributed, deviations in the probability distribution (kernel density) are mainly formed by the corresponding PSD if the number of sampled particles is sufficiently high. In other words, the more particles of a certain size fraction and shape exist, the more probable a corresponding area in the ALD is. Thus, combining a kernel-density estimator and an ALD offers the potential to decouple the linkage of Gaussian particle orientation and the underlying PSD. However, despite the progress, the scope of the ALD is limited to a qualitative comparison between different ALDs.

This challenge is intended to be overcome in **Step 3** of the approach by introducing a characteristic value to the ALDs. To enable statistical comparability between ALDs, a characteristic value, which represents the qualitative information of the ALD, is needed, similar to a d_{50}-value in one-dimensional particle size analysis. Due to the initially described interpretation as measurement probability in DIA, the k grid points calculated by "density dots" can be seen as a curved surface. Accordingly, a "center of mass" of the curved surface can be calculated based on the grid matrix weighted by the belonging measurement probability E_{ij}. The center of mass is composed of the characteristic major and minor axis length (MAL_{ch}, MIL_{ch}) as "center of mass coordinates" at the corresponding axes as shown by the red arrows in Figure 4. MAL_{ch} and MIL_{ch} are determined according to Equations (7) and (8).

$$MAL_{ch} = \frac{\sum_{j=1}^{k} \sum_{i=1}^{k} y_i \cdot E_{ij}(y_i, x_j)}{\sum_{j=1}^{k} \sum_{i=1}^{k} E_{ij}(y_i, x_j)} \qquad (7)$$

$$MIL_{ch} = \frac{\sum_{j=1}^{k} \sum_{i=1}^{k} x_i \cdot E_{ij}(y_i, x_j)}{\sum_{j=1}^{k} \sum_{i=1}^{k} E_{ij}(y_i, x_j)} \qquad (8)$$

This approach presents a powerful tool to add "center of mass" to the ALDs, which allows statistical comparison of ALDs. Hereby, MAL_{ch} and MIL_{ch} as coordinates can be compared by a two-sample t-test for different experiments and different process steps [57].

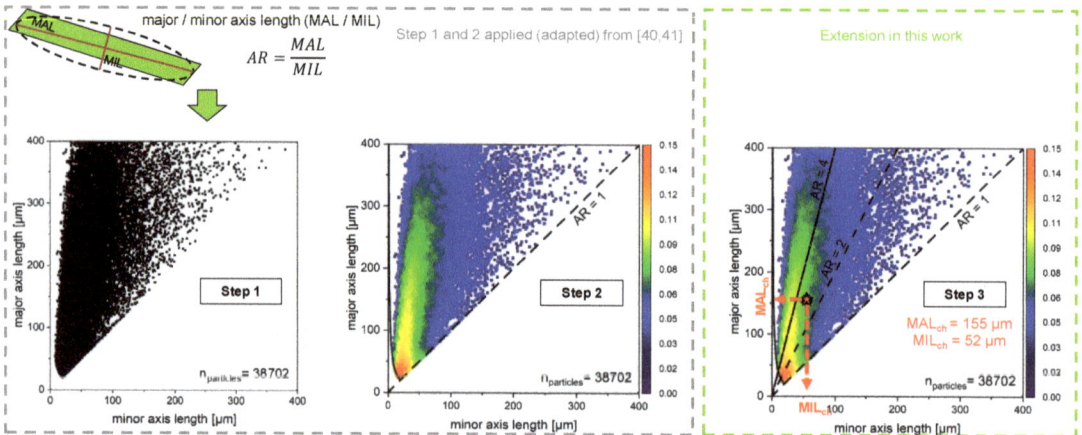

Figure 4. Three step process of the extended approach to analyze ALDs of non-spherical particles using needle-shaped L-alanine particles as an example. Step 1: Scatter plot of the measured MIL and MAL values of the sample with 38,702 particles. Step 2: Calculation of probability values with the function "density dots" of OriginPro 2021 for measured MIL and MAL. Step 3: Determination of the "center of mass" (red star), quantified by the coordinates of MAL_{ch} and MIL_{ch} at the corresponding axes, enabling comparability between ALDs.

4. Results and Discussion

4.1. Maintaining ALD through Integrated Process

Figure 5 shows the results of the ALDs of the inlet suspension (ALD_{IN}) and of the product particles (ALD_{OUT}) for cuboid-shaped and needle-shaped L-alanine particles. A comparably constant ALD ($ALD_{IN} = ALD_{OUT}$) is achieved for both particle shapes (compare Figure 5 left and right) processed in the CVSF, which hypothesizes no particle attrition, breakage, and agglomeration during CVSF operation. Thus, a full maintenance of ALD is always ensured.

The differentiation between different particle shapes can be made directly by detecting the areas with high measurement probabilities (red–yellow–green). Attrition or breakage would become visible by a shift of these areas in the direction of the bisecting line. The cuboid-shaped particles have an average AR of approximately two (see Figure 5 upper half) and the needle-shaped particles have an average AR of three to four (see Figure 5 lower half), which can be additionally observed for the MAL_{ch}/MIL_{ch}-values (red stars) and are in good accordance with the SEM images analyzed (see Figure 2) Low aspect ratio data points, which indicate a spherical shape, could be caused by complex agglomerates. These large agglomerates are marginally seen in QICPIC-videos and SEM-images and may be found predominantly in the upper right corner of the ALD. Due to the low measurement probability indicated by the blue dyeing in the ALD, these values are negligible in terms of

MAL$_{ch}$/MIL$_{ch}$ determination. To statistically justify the maintenance of ALDs, a two-sided two-sample *t*-test is performed to compare the coordinates MAL$_{ch}$ and MIL$_{ch}$ with the center of mass [57]. Here, no statistically significant (95% confidence interval) change of MAL$_{ch}$ and MIL$_{ch}$ for both cuboid-shaped and needle-shaped particles is observed. This demonstrates both visually (ALD) and quantitatively (MAL$_{ch}$/MIL$_{ch}$) that the CVSF is also suitable for product isolation of higher aspect ratio particles while maintaining product quality in terms of particle size.

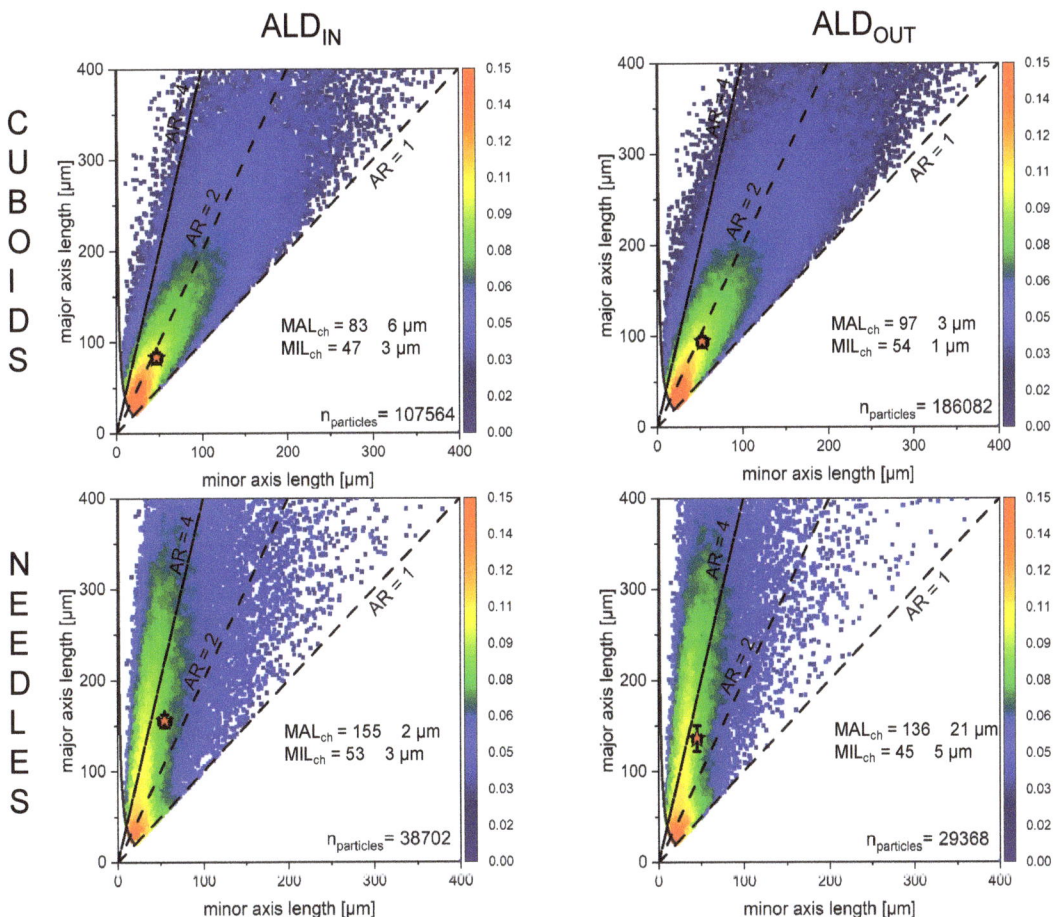

Figure 5. Axis length distribution of the inlet suspension (ALD$_{IN}$, left) and washed product particles (ALD$_{OUT}$, right) after two-stage washing: upper half—for cuboid-shaped particles with an average aspect ratio of 2; lower half—for needle-shaped particles with an average aspect ratio of 3–4.

4.2. Residual Moisture

The residual moistures measured are shown in Figure 6 for the different operating stages. As a benchmark, the results from our previous study of bipyramidal L-alanine particles are added [29].

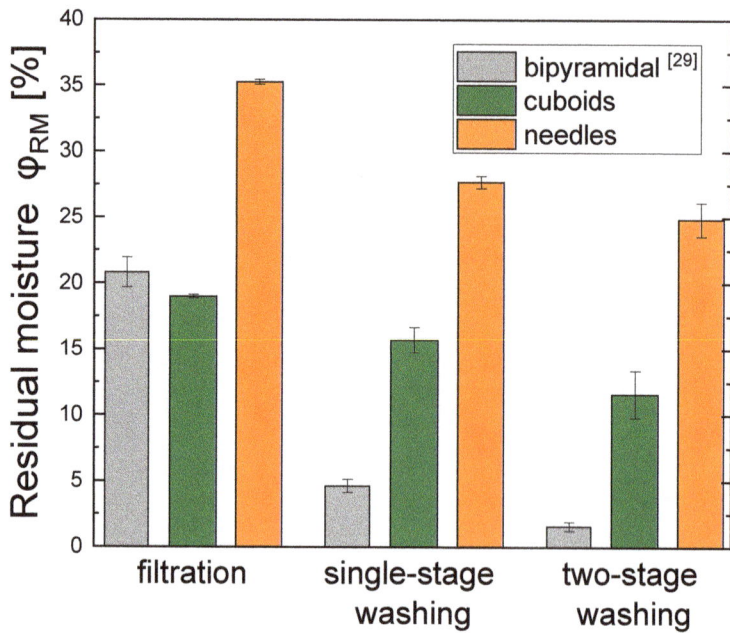

Figure 6. Averaged residual moistures of the product particles after CVSF operation of bipyramidal (grey), cuboids (green), and needles (orange).

Based on this figure, different conclusions can be made. First, a general decrease in residual moisture for all particle shapes investigated can be stated for single-stage washing and two-stage washing, respectively. Here, the possibility to use the same wash liquid for all three particle habits shows its full potential. Due to the varying shape by maintaining the material system, high comparability of the experiments is given. Additionally, all other influence factors on washing efficiency such as enthalpy of evaporation, solvent density, solvent surface tension, solvent viscosities, or solubility curves can be set constant. Second, regarding the sole filtration in CVSF, a high increase in aspect ratio (needles, AR = 3–4) leads to an increase in the residual moisture compared with bipyramidal particles (AR≈1). Therefore, in filtration without washing, the higher porosity could induce a change in residual moisture. Regarding the washing steps, this assumption cannot be expanded. Here, for the bipyramidal particles, a drastic decrease in residual moisture is measured, which leads to residual moistures around 1% after two-stage washing, which is sufficient for further processing in the secondary continuous manufacturing process. In contrast, the smaller, non-spherical particles (cuboids and needles) remain at a high level of residual moisture (11.6 ± 1.8% for cuboids, 24.9 ± 1.3% for needles). However, the residual moistures obtained for non-bipyramidal particles are too high for further processing and the particles do not leave the apparatus in a free-flowing manner. Instead, attached wet particle collectives are leaving the CVSF as shown in Video S1 for the cuboid-shaped particles exemplarily (Solid-Discharge_withoutDrying) in the Supplementary Materials. Therefore, an additional drying module is required for further deliquoring and will be introduced in the following section.

4.3. Operation with Added Drying Module

Figure 7 shows the experimental setup of the CVSF with the extended drying module C. The designed drying module consists of a double-jacketed DN25 glass tube without a filter medium and has the same length as the other modules A, B1, and B2 (13 cm). The resulting heat exchanger area is 53.2 cm². The module's outer jacket is connected to a thermostat (Huber CC-K6, water) via two tangentially mounted GL18 ports. The temperature is set to 60 °C by controlling the tempering medium at the outlet of the module C. The same operating conditions as before (see Table 1) are used to characterize the drying behavior. As the rotational screw speed n_{screw} remains at 3 rpm, the resulting mean drying time is approx. 3.5 min.

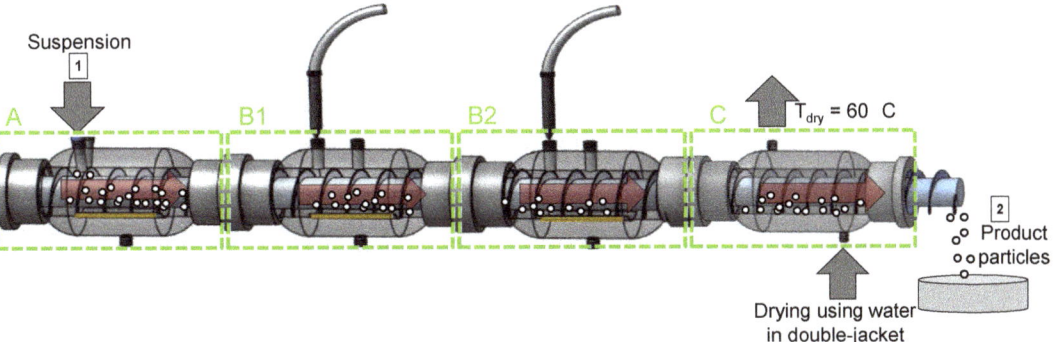

Figure 7. Experimental setup of the experiments with varying particle shapes.

For the cuboid particles, the residual moisture is $\varphi_{RM} = 0.56 \pm 0.30\%$ and for the needle-shaped particles, the residual moisture is $\varphi_{RM} = 0.81 \pm 0.11\%$. This result clearly proves the assumption of a drastic decrease in residual moisture to a sufficient value below 1%, resulting in free-flowing particles after the drying step. The difference in residual moisture is shown for the cuboid-shaped particles exemplarily in the Video S2 (Solid-Discharge_withDrying) in the Supplementary Materials. This shows, again, the major advantage of the modular design of the CVSF. Depending on the particle shape, it may not be necessary to use the additional drying module C, thus saving time and operating costs. For more challenging drying tasks, however, it is possible to simply add one or more drying modules C.

Further investigation of the product particles related to the ALD is necessary, to prove the preservation of the ALD_{IN} with the added drying module. The same experimental conditions and analytical evaluations, solely adding the module C, were used for this purpose. Figure 8 shows the experimental results of the experiments of the ALDs comparing the inlet suspension with the dried product particles at the CVSF outlet. It can be seen that the respective ALD are very similar to each other. The respective color ranges can be identified in the same sections of the ALDs. Nevertheless, slight tendencies of abrasion and agglomeration are recognizable. Abrasion or disaggregation can be seen in the MAL_{ch} values, which decreases from 98 μm to 87 μm (cuboids) and from 129 μm to 121 μm (needles). Therefore, it seems to be attrition, since particle breakage would probably show a stronger reduction in MAL_{ch}. The MIL_{ch} values remain at a constant level of approx. 50 μm for both particle shapes. In addition, the stronger appearance of the blue areas of the ALD_{OUT} (ALDs at the right side) is a possible indication of agglomeration. Even if the wash liquid is selected in such a way that agglomeration can be avoided as much as possible during the whole crystal process chain, this cannot be completely avoided, especially during the drying process. For the first washing step, a mixture of ethanol absolute and water 4:1 (solubility of L-alanine at 25 °C: 2.83 g kg^{-1} [48]) and for the second washing step pure ethanol absolute (solubility of L-alanine at 25 °C: 0.18 g kg^{-1} [48]) were utilized as described by Terdenge and Wohlgemuth [48]. Even if the solubility is already very low after the second washing step, the remaining solute can cause increased agglomeration during drying due to bridging of single particles to agglomerates. A more detailed investigation for agglomeration of smaller particles (<100 μm) is challenging based on Figure 8, since many particles are present in this region (see red–yellow areas of Figure 8) and an increased agglomeration would only be recognizable by a color gradient shift in the direction of the bisecting line or an increase in the corresponding MAL_{ch} /MIL_{ch} values. However, this is not the case here for either cuboids or needles.

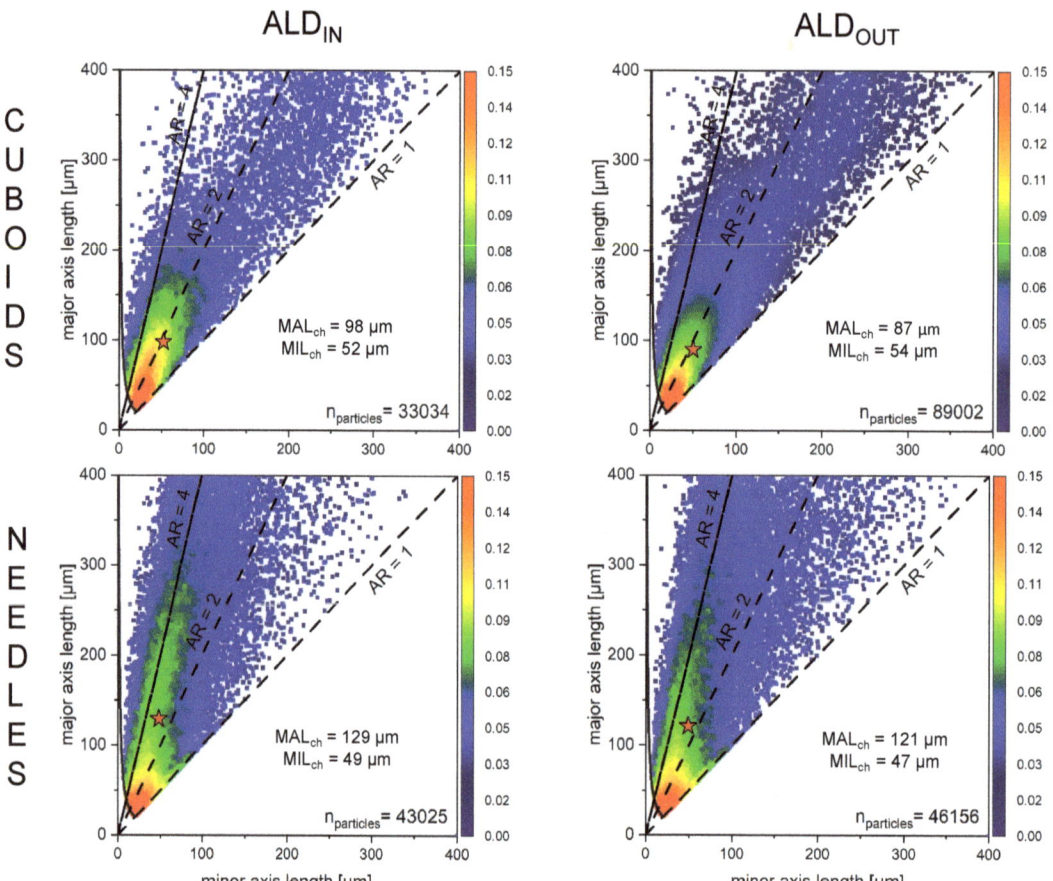

Figure 8. Axis length distribution in the inlet suspension (ALD_{IN}, left) and washed product particles after drying (ALD_{OUT}, right): upper half—for cuboid-shaped particles; lower half—for needle-shaped particles.

While at first impression there is an increased influence of attrition compared with agglomeration (reduction in MAL_{ch} values as described above), this needs to be further investigated in future work. For this purpose, corresponding ALDs consisting of subpopulations (single particles and agglomerates separately) would be useful. First preliminary work on this has already been developed by Heisel et al. [47,49] and could be used for this purpose in the future with appropriate adaptations for smaller agglomerates.

This requirement can also be confirmed by comparing the results to published literature data. Attrition and agglomeration phenomena occur quite frequently during the drying processes [58,59] because there is no lubricant, which prevents direct particle–particle collision. Needles in particular tend to a simultaneous occurrence of abrasion and agglomeration, as already shown by Lekhal et al. for similar particles (L-threonine) during agitated drying [39].

In order to investigate this in more detail for CVSF operation, exemplary binary images are shown in Figure 9. Although no detailed quantitative statement can be made, a representative overview is possible. Abrasion, breakage, and agglomeration cannot be fully excluded, but a maintenance of the product quality in terms of particle shape and size distribution is clearly recognizable. However, in order to be able to characterize this statistically in the future for the drying process in the CVSF, systematic experiments should be carried out with material systems that have a stronger tendency to abrasion and agglomeration during drying.

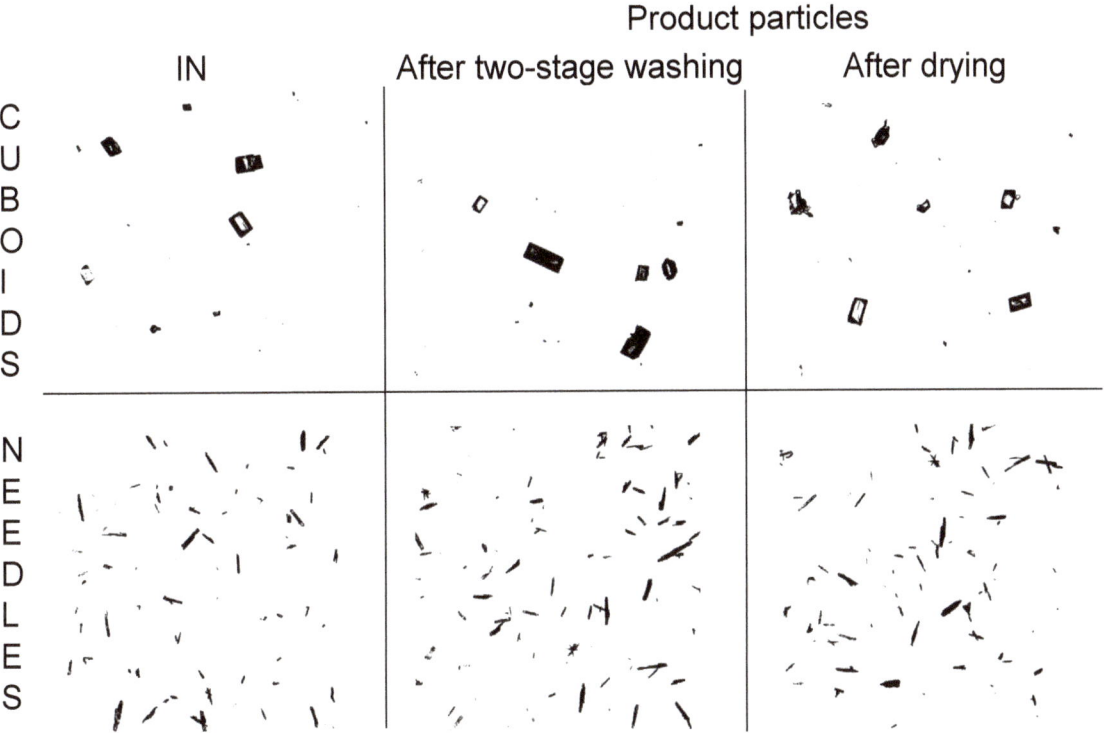

Figure 9. Exemplary QIPCIC binary images of the inlet suspension (left), product particles after two-stage washing (middle, compare ALD$_{OUT}$ of Figure 5), and product particles after washing and drying (right, compare ALD$_{OUT}$ of Figure 8): upper half—for cuboid-shaped particles; lower half—for needle-shaped particles.

Comparing the CVSF with industrial drying processes, the major advantage of the CVSF is the product-friendly drying. Whereas particle–particle collision may occur in conventional fluidized-bed drying, thus promoting agglomeration and breakage, this can be prevented to a large extent inside the CVSF. Using the drying module, a combined/mixed drying process takes place (contact drying by heated double-jacket walls and convection drying by drawn air of the vacuum pump). At the same time, the particles are conveyed evenly through the apparatus, whereby on the one hand the drying time is uniform and on the other hand particle–particle collision can most likely be avoided.

5. Conclusions and Outlook

In this study, full continuous particle isolation, from suspension to dried product particles, was performed for the first time for particles with varying shapes using the continuous vacuum screw filter (CVSF). The former calculation of the equivalent diameter d_{eq} as one-dimensional size descriptor is not suitable for the high aspect ratio particles processed within the CVSF (cuboid-shaped and needle-shaped L-alanine). Thus, an extension of the dynamic-image analysis approach was made using the axis length distribution (ALD) to access potential changes in product quality. To enable a statistical validation of different ALDs, the center of mass of the probability-weighted ALDs as a characteristic value was introduced. The corresponding coordinates MAL_{ch} and MIL_{ch} were used for comparison of ALDs. With this tool at hand, reasonable analysis of potential changes in the ALD for non-spherical particles was possible for the first time.

Even for needle-shaped particles with a high breakage risk, no significant change in the ALD was shown during filtration and two-stage washing. Due to the decreased particle sizes caused by the corresponding selective growth-inhibiting batch crystallization process, an analysis of the agglomeration degree was not possible. Since agglomeration processes may influence the particle properties of discharged filter cake, a detailed analysis of the agglomeration degree should be performed in future work.

Regarding the residual moisture, the assumption of higher filter cake resistance resulting in higher residual moistures was confirmed. For cuboid-shaped and needle-shaped particles, residual moistures of >10% (cuboids) and >20% (needles) were obtained after two-stage washing. Thus, the necessity to add a drying module was shown and residual moistures below 1%, to produce free-flowing particles, was proven for these particles. During drying processes, slight indications of attrition, disaggregation, and agglomeration were visible, which still need to be proven in future work. With the systematic experimental investigations in this study, a remarkable contribution towards full continuous primary manufacturing processes was made. The product quality preservation was also shown for smaller needle-shaped particles and thus enables a significant achievement towards the industrial applicability of CVSF for small-scale continuous particle isolation, as many APIs have a needle-shaped morphology. Due to the modular design of the CVSF, it is possible to adapt very flexibly to the varying requirements of different material systems, thus ensuring gentle particle isolation at all times.

6. Patents

The CVSF was patented in 2021 as "Rotating-Screw Drying Reactor" as WO2021/148108 A1.

Supplementary Materials: The following are available online at https://www.mdpi.com/article/10.3390/cryst12020137/s1, Video S1: Solid-Discharge_withoutDrying.mp4, Video S2: Solid-Discharge_withDrying.mp4.

Author Contributions: Conceptualization, C.S. and K.W.; methodology, C.S., J.H. and K.W.; investigation, J.H. and C.S.; writing—original draft preparation, C.S. and J.H.; writing—review and editing, K.W.; visualization, C.S. and J.H.; supervision, K.W.; project administration, K.W.; funding acquisition, C.S. and K.W. All authors have read and agreed to the published version of the manuscript.

Funding: This research work was funded by the state of North-Rhine Westphalia (NRW) and the European Regional Development Fund (EFRE), Project "NRW Patent Validation Program" (Grant EFRE-0400357).

Institutional Review Board Statement: Not applicable.

Informed Consent Statement: Not applicable.

Data Availability Statement: Not applicable.

Acknowledgments: The authors express their special thanks to Daniela Ermeling for the technical support during the experiments.

Conflicts of Interest: The authors declare no conflict of interest. The funders had no role in the design of the study; in the collection, analyses, or interpretation of data; in the writing of the manuscript; or in the decision to publish the results.

References

1. Cote, A.; Erdemir, D.; Girard, K.P.; Green, D.A.; Lovette, M.A.; Sirota, E.; Nere, N.K. Perspectives on the current state, challenges, and opportunities in pharmaceutical crystallization process development. *Cryst. Growth Des.* **2020**, *20*, 7568–7581. [CrossRef]
2. Hofmann, G. *Kristallisation in der Industriellen Praxis*; Wiley-VCH: Weinheim, Germany, 2005.
3. Beckmann, W. *Crystallization: Basic Concepts and Industrial Applications*; Wiley-VCH: Weinheim, Germany, 2013.
4. Wood, B.; Girard, K.; Polster, C.S.; Croker, D. Progress to date in the design and operation of continuous crystallization processes for pharmaceutical applications. *Org. Process Res. Dev.* **2019**, *23*, 122–144. [CrossRef]
5. Orehek, J.; Teslić, D.; Likozar, B. Continuous crystallization processes in pharmaceutical manufacturing: A review. *Org. Process Res. Dev.* **2021**, *25*, 16–42. [CrossRef]
6. Zhang, D.; Xu, S.; Du, S.; Wang, J.; Gong, J. Progress of pharmaceutical continuous crystallization. *Engineering* **2017**, *3*, 354–364. [CrossRef]
7. Wang, T.; Lu, H.; Wang, J.; Xiao, Y.; Zhou, Y.; Bao, Y.; Hao, H. Recent progress of continuous crystallization. *J. Ind. Eng. Chem.* **2017**, *54*, 14–29. [CrossRef]
8. Kleinebudde, P.; Khinast, H.; Rantanen, J. *Continuous Manufacturing of Pharmaceuticals*; John Wiley & Sons, Ltd.: Chichester, UK, 2017.
9. Acevedo, D.; Peña, R.; Yang, Y.; Barton, A.; Firth, P.; Nagy, Z.K. Evaluation of mixed suspension mixed product removal crystallization processes coupled with a continuous filtration system. *Chem. Eng. Process. Process. Intensif.* **2016**, *108*, 212–219. [CrossRef]
10. Mascia, S.; Heider, P.L.; Zhang, H.; Lakerveld, R.; Benyahia, B.; Barton, P.I.; Braatz, R.D.; Cooney, C.L.; Evans, J.M.B.; Jamison, T.F.; et al. End-to-end continuous manufacturing of pharmaceuticals: Integrated synthesis, purification, and final dosage formation. *Angew. Chem. Int. Ed.* **2013**, *52*, 12359–12363. [CrossRef] [PubMed]
11. Bourcier, D.; Féraud, J.; Colson, D.; Mandrick, K.; Ode, D.; Brackx, E.; Puel, F. Influence of particle size and shape properties on cake resistance and compressibility during pressure filtration. *Chem. Eng. Sci.* **2016**, *144*, 176–187. [CrossRef]
12. Ma, Y.; Wu, S.; Macaringue, E.; Zhang, T.; Gong, J.; Wang, J. Recent Progress in Continuous Crystallization of Pharmaceutical Products: Precise Preparation and Control. *Org. Process Res. Dev.* **2021**, *24*, 1785–1801. [CrossRef]
13. Domokos, A.; Nagy, B.; Szilágyi, B.; Marosi, G.; Nagy, Z.K. Integrated continuous pharmaceutical technologies—A review. *Org. Process Res. Dev.* **2021**, *25*, 721–739. [CrossRef]
14. Johnson, M.D.; Burcham, C.L.; May, S.A.; Calvin, J.R.; Groh, J.M.; Myers, S.S.; Webster, L.P.; Roberts, J.C.; Reddy, V.R.; Luciani, C.V.; et al. API continuous cooling and antisolvent crystallization for kinetic impurity rejection in cGMP manufacturing. *Org. Process Res. Dev.* **2021**, *25*, 1284–1351. [CrossRef]
15. Nagy, B.; Szilágyi, B.; Domokos, A.; Tacsi, K.; Pataki, H.; Marosi, G.; Nagy, Z.K.; Nagy, Z.K. Modeling of pharmaceutical filtration and continuous integrated crystallization-filtration processes. *Chem. Eng. J.* **2021**, *413*, 127566. [CrossRef]
16. Domokos, A.; Nagy, B.; Gyürkés, M.; Farkas, A.; Tacsi, K.; Pataki, H.; Liu, Y.C.; Balogh, A.; Firth, P.; Szilágyi, B.; et al. End-to-end continuous manufacturing of conventional compressed tablets: From flow synthesis to tableting through integrated crystallization and filtration. *Int. J. Pharm.* **2020**, *581*, 119297. [CrossRef]
17. Liu, Y.C.; Domokos, A.; Coleman, S.; Firth, P.; Nagy, Z.K. Development of continuous filtration in a novel continuous filtration carousel integrated with continuous crystallization. *Org. Process Res. Dev.* **2019**, *23*, 2655–2665. [CrossRef]
18. Ottoboni, S.; Price, C.J.; Steven, C.; Meehan, E.; Barton, A.; Firth, P.; Mitchell, A.; Tahir, F. Development of a novel continuous filtration unit for pharmaceutical process development and manufacturing. *J. Pharm. Sci.* **2019**, *108*, 372–381. [CrossRef] [PubMed]
19. Ottoboni, S.; Shahid, M.; Steven, C.; Coleman, S.; Meehan, E.; Barton, A.; Firth, P.; Sutherland, R.; Price, C.J. Developing a batch isolation procedure and running it in an automated semicontinuous unit: AWL CFD25 case study. *Org. Process Res. Dev.* **2020**, *24*, 520–539. [CrossRef] [PubMed]
20. Ottoboni, S.; Coleman, S.J.; Steven, C.; Siddique, M.; Fraissinet, M.; Joannes, M.; Laux, A.; Barton, A.; Firth, P.; Price, C.J.; et al. Understanding API static drying with hot gas flow: Design and test of a drying rig prototype and drying modeling development. *Org. Process Res. Dev.* **2020**, *24*, 2505–2520. [CrossRef] [PubMed]
21. Destro, F.; Hur, I.; Wang, V.; Abdi, M.; Feng, X.; Wood, E.; Coleman, S.; Firth, P.; Barton, A.; Barolo, M.; et al. Mathematical modeling and digital design of an intensified filtration-washing-drying unit for pharmaceutical continuous manufacturing. *Chem. Eng. Sci.* **2021**, *244*, 116803. [CrossRef]
22. Barton, A. Filtration Apparatus. WO 2015/033117 A1, 12 March 2015.
23. Born, S.; Dittrich, J.; Takizawa, B.T.; Mascia, S. Continuous Rotary Plate Filter and Methods of Use Thereof. WO 2017/136779 A1, 10 August 2017.
24. Testa, C.J.; Hu, C.; Shvedova, K.; Wu, W.; Sayin, R.; Casati, F.; Halkude, B.S.; Hermant, P.; Shen, D.E.; Ramnath, A.; et al. Design and commercialization of an end-to-end continuous pharmaceutical production process: A pilot plant case study. *Org. Process Res. Dev.* **2020**, *24*, 2874–2889. [CrossRef]
25. Hu, C.; Testa, C.J.; Born, S.C.; Wu, W.; Shvedova, K.; Sayin, R.; Halkude, B.S.; Casati, F.; Ramnath, A.; Hermant, P.; et al. E-factor analysis of a pilot plant for end-to-end integrated continuous manufacturing (ICM) of pharmaceuticals. *Green Chem.* **2020**, *22*, 4350–4356. [CrossRef]

26. Testa, C.J.; Shvedova, K.; Hu, C.; Wu, W.; Born, S.C.; Takizawa, B.; Mascia, S. Heterogeneous crystallization as a process intensification technology in an integrated continuous manufacturing process for pharmaceuticals. *Org. Process Res. Dev.* **2021**, *25*, 225–238. [CrossRef]
27. Steenweg, C.; Wohlgemuth, K.; Schembecker, G. Rotating-Screw Drying Reactor—Schneckenfördertrocknungsreaktor. WO2021148108 A1, 29 July 2021.
28. Steenweg, C.; Seifert, A.I.; Schembecker, G.; Wohlgemuth, K. Characterization of a modular continuous vacuum screw filter for small-scale solid–liquid separation of suspensions. *Org. Process Res. Dev.* **2021**, *25*, 926–940. [CrossRef]
29. Steenweg, C.; Seifert, A.I.; Böttger, N.; Wohlgemuth, K. Process intensification enabling continuous manufacturing processes using modular continuous vacuum screw filter. *Org. Process Res. Dev.* **2021**, *25*, 2525–2536. [CrossRef]
30. Steenweg, C.; Wohlgemuth, K. Von der Suspension zum getrockneten Produkt mittels neuartigem Vakuumschraubenfilter. *Filtr. Sep.* **2021**, *4*, 4–11.
31. Steenweg, C.; Kufner, A.C.; Habicht, J.; Wohlgemuth, K. Towards continuous primary manufacturing processes—Particle design through combined crystallization and particle isolation. *Processes* **2021**, *9*, 2187. [CrossRef]
32. FDA. *Quality Considerations for Continuous Manufacturing: Guidance for Industry*; Department of Health and Human Services: Washington, DC, USA, 2019.
33. Wakeman, R. The influence of particle properties on filtration. *Sep. Purif. Technol.* **2007**, *58*, 234–241. [CrossRef]
34. Pudasaini, N.; Upadhyay, P.P.; Parker, C.R.; Hagen, S.U.; Bond, A.D.; Rantanen, J. Downstream processability of crystal habit-modified active pharmaceutical ingredient. *Org. Process Res. Dev.* **2017**, *21*, 571–577. [CrossRef]
35. Leane, M.; Pitt, K.; Reynolds, G.; The Manufacturing Classification System (MCS) Working Group. A proposal for a drug product manufacturing classification system (MCS) for oral solid dosage forms. *Pharm. Dev. Technol.* **2015**, *20*, 12–21. [CrossRef]
36. MacLeod, C.S.; Muller, F. On the fracture of pharmaceutical needle-shaped crystals during pressure filtration: Case studies and mechanistic understanding. *Org. Process Res. Dev.* **2012**, *16*, 425–434. [CrossRef]
37. Eren, A.; Szilagyi, B.; Quon, J.L.; Papageorgiou, C.D.; Nagy, Z.K. Experimental investigation of an integrated crystallization and wet-milling system with temperature cycling to control the size and aspect ratio of needle-shaped pharmaceutical crystals. *Cryst. Growth Des.* **2021**, *21*, 3981–3993. [CrossRef]
38. Perini, G.; Salvatori, F.; Ochsenbein, D.; Mazzotti, M.; Vetter, T. Filterability prediction of needle-like crystals based on particle size and shape distribution data. *Sep. Purif. Technol.* **2019**, *211*, 768–781. [CrossRef]
39. Lekhal, A.; Girard, K.; Brown, M.; Kiang, S.; Khinast, J.; Glasser, B. The effect of agitated drying on the morphology of l-threonine (needle-like) crystals. *Int. J. Pharm.* **2004**, *270*, 263–277. [CrossRef] [PubMed]
40. Eggers, J.; Kempkes, M.; Mazzotti, M. Measurement of size and shape distributions of particles through image analysis. *Chem. Eng. Sci.* **2008**, *63*, 5513–5521. [CrossRef]
41. Kempkes, M.; Eggers, J.; Mazzotti, M. Measurement of particle size and shape by FBRM and in situ microscopy. *Chem. Eng. Sci.* **2008**, *63*, 4656–4675. [CrossRef]
42. Kempkes, M.; Vetter, T.; Mazzotti, M. Monitoring the particle size and shape in the crystallization of paracetamol from water. *Chem. Eng. Res. Des.* **2010**, *88*, 447–454. [CrossRef]
43. Briesen, H. Simulation of crystal size and shape by means of a reduced two-dimensional population balance model. *Chem. Eng. Sci.* **2006**, *61*, 104–112. [CrossRef]
44. Wohlgemuth, K.; Schembecker, G. Modeling induced nucleation processes during batch cooling crystallization: A sequential parameter determination procedure. *Comput. Chem. Eng.* **2013**, *52*, 216–229. [CrossRef]
45. Lührmann, M.-C.; Timmermann, J.; Schembecker, G.; Wohlgemuth, K. Enhanced Product Quality Control through Separation of Crystallization Phenomena in a Four-Stage MSMPR Cascade. *Cryst. Growth Des.* **2018**, *18*, 7323–7334. [CrossRef]
46. Saal, S. Begasungskristallisation in Wässrigen Aminosäuresystemen. Master's Thesis, Technische Universität Dortmund, Dortmund, Germany, 2017.
47. Heisel, S.; Ernst, J.; Emshoff, A.; Schembecker, G.; Wohlgemuth, K. Shape-independent particle classification for discrimination of single crystals and agglomerates. *Powder Technol.* **2019**, *345*, 425–437. [CrossRef]
48. Terdenge, L.-M.; Wohlgemuth, K. Impact of agglomeration on crystalline product quality within the crystallization process chain. *Cryst. Res. Technol.* **2016**, *51*, 513–523. [CrossRef]
49. Heisel, S.; Rolfes, M.; Wohlgemuth, K. Discrimination between single crystals and agglomerates during the crystallization process. *Chem. Eng. Technol.* **2018**, *41*, 1218–1225. [CrossRef]
50. De Albuquerque, I.; Mazzotti, M.; Ochsenbein, D.R.; Morari, M. Effect of needle-like crystal shape on measured particle size distributions. *AIChE J.* **2016**, *62*, 2974–2985. [CrossRef]
51. Schorsch, S.; Vetter, T.; Mazzotti, M. Measuring multidimensional particle size distributions during crystallization. *Chem. Eng. Sci.* **2012**, *77*, 130–142. [CrossRef]
52. Steger, C. *On the Calculation of Arbitrary Moments of Polygons*; Technical Report FGBV-96-05; Technische Universität München: Munich, Germany, 1996.
53. Yu, W.; Hancock, B. Evaluation of dynamic image analysis for characterizing pharmaceutical excipient particles. *Int. J. Pharm.* **2008**, *361*, 150–157. [CrossRef] [PubMed]
54. ISO 13322-2; Particle Size Analysis—Image Analysis Methods; Dynamic Image Analysis Methods. International Organization for Standardization: Geneva, Switzerland, 2006.

55. Origin-Help: Density Dots. Available online: https://www.originlab.com/doc/Origin-Help/Density_Dots (accessed on 13 October 2021).
56. Botev, Z.I.; Grotowski, J.F.; Kroese, D. Kernel density estimation via diffusion. *Ann. Stat.* **2010**, *38*, 2916–2957. [CrossRef]
57. Kleppmann, W. *Versuchsplanung: Produkte und Prozesse Optimieren*, 8th ed.; Praxisreihe Qualitätswissen, Hanser: München, Germany, 2013.
58. Papageorgiou, C.D.; Mitchell, C.; Quon, J.L.; Langston, M.; Borg, S.; Hicks, F.; Ende, D.J.A.; Breault, M. Development of a novel screening methodology for the assessment of the risk of particle size attrition during agitated drying. *Org. Process Res. Dev.* **2020**, *24*, 242–254. [CrossRef]
59. Birch, M.; Marziano, I. Understanding and avoidance of agglomeration during drying processes: A case study. *Org. Process Res. Dev.* **2013**, *17*, 1359–1366. [CrossRef]

Article

Flow Map for Hydrodynamics and Suspension Behavior in a Continuous Archimedes Tube Crystallizer

Jana Sonnenschein [1], Pascal Friedrich [1], Moloud Aghayarzadeh [2], Otto Mierka [2], Stefan Turek [2] and Kerstin Wohlgemuth [1,*]

1 Laboratory of Plant and Process Design, TU Dortmund University, Emil-Figge-Strasse 70, 44227 Dortmund, Germany; jana-maria.sonnenschein@tu-dortmund.de (J.S.); pascal2.friedrich@tu-dortmund.de (P.F.)
2 Institute of Applied Mathematics (LS III), TU Dortmund University, Vogelpothsweg 87, 44227 Dortmund, Germany; maghayar@mathematik.tu-dortmund.de (M.A.); omierka@mathematik.tu-dortmund.de (O.M.); stefan.turek@tu-dortmund.de (S.T.)
* Correspondence: kerstin.wohlgemuth@tu-dortmund.de; Tel.: +49-2301-755-3020

Abstract: The Archimedes Tube Crystallizer (ATC) is a small-scale coiled tubular crystallizer operated with air-segmented flow. As individual liquid segments are moved through the apparatus by rotation, the ATC operates as a pump. Thus, the ATC overcomes pressure drop limitations of other continuous crystallizers, allowing for longer residence times and crystal growth phases. Understanding continuous crystallizer phenomena is the basis for a well-designed crystallization process, especially for small-scale applications in the pharmaceutical and fine chemical industry. Hydrodynamics and suspension behavior, for example, affect agglomeration, breakage, attrition, and ultimately crystallizer blockage. In practice, however, it is time-consuming to investigate these phenomena experimentally for each new material system. In this contribution, a flow map is developed in five steps through a combination of experiments, CFD simulations, and dimensionless numbers. Accordingly, operating parameters can be specified depending on ATC design and material system used, where suspension behavior is suitable for high-quality crystalline products.

Keywords: continuous processing; Archimedes tube; hydrodynamics; suspension behavior; computational fluid dynamics; flow map

Citation: Sonnenschein, J.; Friedrich, P.; Aghayarzadeh, M.; Mierka, O.; Turek, S.; Wohlgemuth, K. Flow Map for Hydrodynamics and Suspension Behavior in a Continuous Archimedes Tube Crystallizer. *Crystals* **2021**, *11*, 1466. https://doi.org/10.3390/cryst11121466

Academic Editors: Heike Lorenz, Alison Emslie Lewis, Erik Temmel and Jens-Petter Andreassen

Received: 5 October 2021
Accepted: 19 November 2021
Published: 26 November 2021

Publisher's Note: MDPI stays neutral with regard to jurisdictional claims in published maps and institutional affiliations.

Copyright: © 2021 by the authors. Licensee MDPI, Basel, Switzerland. This article is an open access article distributed under the terms and conditions of the Creative Commons Attribution (CC BY) license (https://creativecommons.org/licenses/by/4.0/).

1. Introduction

Continuous crystallization is a promising operation mode for the small-scale production ($\dot{V} \leq 100\,\mathrm{mL\,min^{-1}}$) of active pharmaceutical ingredients (APIs) and fine chemicals [1–4]. Reaching the steady state in continuous operation is advantageous regarding the consistency of product quality, e.g., particle size and its distribution. In addition, higher space-time yield than in batch operation is possible because smaller equipment sizes are required [2,4]. In order to utilize these advantages, however, the steady state must first be reached and then maintained [3]. Here, encrustation and blockage of the continuous crystallizers still remain a major challenge [5].

Existing continuous crystallizer concepts can be classified regarding achievable residence time and width of residence time distribution (RTD) [6]. For process design, a narrow RTD of liquid and solid phase is preferable [7], as the position of liquid and solid elements is predetermined. If the RTD of a continuous downstream process is broad, tracing out-of-specification material is difficult [7]. Through a narrow RTD in combination with flexible residence times, defined product particle sizes with narrow particle size distribution (PSD) can be reached. In this context, we introduced the Archimedes Tube Crystallizer (ATC) as a promising apparatus concept [6]. The apparatus consists of a coiled tube that rotates around its horizontal axis. Segmented flow is achieved by a specifically designed inlet tank that is mounted on the horizontal axis and rotates with the same velocity as the coiled tube.

With each rotation, the individual liquid segments are moved through the coiled tube. As a consequence of this design, the apparatus itself works as a pump. Therefore, the usual limitations of residence time in tubular crystallizers due to pressure loss do not apply. In addition, the segmented-flow concept leads to narrow liquid- and solid-phase RTDs which was demonstrated for a prototype with volumetric flow rates from 3–15 mL min^{-1} and residence times from 4–11 min [6].

The next step towards successful particle engineering entails the understanding of hydrodynamics and suspension behavior. These phenomena are decisive in crystallization regarding agglomeration, breakage, attrition, and blockage of the employed crystallizer. In industry, API process development is carried out within set time frames and under material constraints [1]. To reduce effort in this important time period, apparatus development must provide detailed information about mixing. Then, the only remaining time-consuming step is the adaptation to a specific material system.

Hydrodynamics in coiled structures are dominated by secondary flow patterns caused by centrifugal instabilities [8]. Centrifugal and shear forces move the inner fluid in the tube outward and the outer fluid inward. The resulting so-called Dean vortices enhance radial mixing by overlapping the parabolic laminar flow profile. In single-phase flow, the intensity of secondary flow patterns in coiled structures is characterized by Dean number Dn that is calculated based on Reynolds number Re for inner tube diameter $d_{i,tube}$ and coiled tube diameter d_{ct} as $Dn = Re \cdot \sqrt{d_{i,tube}/d_{ct}}$ [8,9]. Thereby, the coiled flow inverter (CFI) represents an improved helically coiled device design with additional bends between the helical segments. Due to the additional bends, the direction of the centrifugal force is changed at each bend so that the existing flow profile is disrupted and reformed with positive impact on mixing [10].

Air-segmented flow consists, as the name suggests, of a segmented flow of two immiscible fluids, specifically air (or another immiscible gas) and the process medium. For slug formation, usually a mixing piece (e.g., T-junction) is installed to connect process medium and air supply delivered by individual peristaltic pumps [11]. The flow regime in one individual slug is characterized by internal circulations, the so-called Taylor vortices [12]. Due to the enhanced flow regime, air-segmented flow hydrodynamics have mostly been researched as process intensification technique for micro-reactors focusing on heat and mass transfer enhancement [12,13]. Here, Talimi et al. reviewed prior numerical studies of hydrodynamics (computational fluid dynamics (CFD)) and outlined the main factors on flow pattern as bubble length, capillary number Ca, channel curvature (for curved microchannels), and superficial velocities of the two phases, among others [12]. Gaddem et al. investigated segmented flow in coiled structures, specifically in a CFI [14]. For this purpose, they developed a CFD model for the superposition of Taylor and Dean flow to assess the potential improvement in mass transfer for application as a microscale reactor [14]. To describe the effect of flow superposition, the modified Dean number Dn^* was introduced [14].

There are also various studies that describe suspension behavior of particles in the previously introduced coiled structures and in segmented flow. Tiwari et al. conducted CFD simulations in a helically coiled device with particles of 1 and 3 µm diameter and volume fraction 0.1 [8]. In the simulations, a deviation in the particle settling zone from tube bottom to inner bend was observed, which was attributed to increased wall shear stress [8]. Dbouk and Habchi observed hydrodynamics and suspension behavior in helical pipes for application as static mixer [15]. The investigated particles were monodisperse (particle size 25–400 µm, initial volume fraction 0.25–0.45), spherical, and had the same density as the Newtonian liquid phase [15]. Under these conditions, better mixing was observed for increasing particle diameter [15]. Wiedmeyer et al. found a size-dependent residence time of potash alum crystals in a helically coiled tube crystallizer, that was attributed to time-dependent secondary flow patterns identified through direct numerical flow simulations [16]. Emerged in the secondary flow, smaller particles take longer flow paths and thus remain longer in the device than larger particles [16]. For a continuous CFI crystallizer with bend angle 90°, Hohmann et al. investigated the suspension behavior for three test systems with solid-phase mass fractions between 0.01 and 0.10 $g_S\, g_{Susp}^{-1}$ and particle size fractions up to 180–250 µm [17]. For the selected material systems, they classified the flow into three regimes: stagnant sediment, moving sediment flow, and homogeneous suspension flow [17]. Based on Reynolds number Re and densimetric Froude number Fr, a flow map was set up for the investigated operating points that were almost completely allocated to moving sediment flow [17].

For air-segmented or slug flow crystallizers, various studies deal with the qualitative and quantitative evaluation of crystal suspension homogeneity. In general, two research groups found that suspension homogeneity is increased at slug aspect ratios $l_{slug}/d_{i,tube}$ near 1 which also leads to smaller product crystals [18,19]. Jiang et al. used an imaging method based on a stereomicroscope and a video camera for suspension quality evaluation [18]. Later, they refined this method as a real-time imaging method [20]. Due to the microscope placed in top view, however, the influence of gravity on the growing crystals was not taken into account. Su and Gao implemented a CFD simulation to evaluate suspension state and flow trajectory of α-glycine crystals (seed main particle size 8 µm, solid mass fraction 1–3 wt. %) [19]. According to the simulations, the highest crystal flow velocities were found at high aspect ratios of 2 [19]. However, crystals are prone to sedimentation at this operating point, so that an aspect ratio of 1 remains the best compromise [19]. In addition, variation of flow rate did not change the suspension state of the crystals [19]. With the small size of the particles investigated, this result was expected. Besenhard et al. observed the homogeneity of crystal suspension from side view [21]. Here, for D-mannitol particles up to a size fraction of 150–180 µm with solid mass fraction $0.1\, g/g_{Sol}$, worse suspension was observed for bigger particles due to gravity [21]. Homogeneity of suspension was enhanced by increasing the total volume flow rate [21]. Scheiff and Agar quantitatively investigated the suspension behavior of heterogeneous catalyst particles (<100 µm) in segmented flow [22,23]. Here, they introduced the Shields parameter Θ, known from sedimentation theory as ratio of drag and gravity, as measure for suspension quality [22]. In addition, Scheiff developed an image analysis to estimate particle distribution in horizontal direction [23]. Termühlen et al. extended this image analysis approach for continuous crystallizer applications by a vertical direction [24]. With this approach, the gravitational forces on larger particles of material system L-alanine/water with size fractions up to 315–355 µm were taken into account as well [24]. Overall, they found that smaller particle sizes and higher flow rates led to better suspension [24]. Additionally, Termühlen et al. demonstrated that suspension behavior is decisive regarding a desired narrow PSD [25]. Especially for material systems that tend to agglomerate, such as the employed material system L-alanine/water, poor suspension led to high agglomeration degrees [25].

Overall, much effort has been invested into the investigation of single- or multiphase flows in coiled structures or air-segmented flow individually. Thereby, the main focus of CFD simulations so far is on hydrodynamics for process intensification, in particular on heat and mass transfer enhancement. By contrast, the evaluation of suspension behavior is mostly observed experimentally, either qualitatively or quantitatively. Here, initial approaches exist to characterize suspension behavior in available devices by flow maps and dimensionless numbers. Altogether, however, there is still a lack of strategy to link the various approaches to speed up process development.

In this contribution, we introduce our strategy for characterization of hydrodynamics and suspension behavior in the ATC. For this purpose, we combine CFD simulations of liquid and solid phases with validation experiments. Our approach to set up a flow map for the ATC is summarized into a five-step road map. Through the strategic integration of dimensionless numbers into the development process, simulations and experiments can be run directly at the appropriate operating points. As a consequence, computational and experimental effort to set up the flow map is reduced. For the flow map, we focus on estimating the operating window for well-known sample material system L-alanine/water. Thereby, we will prove the following hypotheses:

- CFD simulations of liquid and solid phase can qualitatively predict hydrodynamics and solid phase suspension behavior.
- It is possible to predict the suspension behavior with a flow map based on dimensionless numbers.
- Different suspension states in the ATC can be estimated for various operating parameters and crystal product sizes.

In Section 2, materials including the ATC prototype and experimental methods are summarized. Afterwards, the employed numerical CFD simulation model is described in Section 3. Subsequently, the CFD simulations of hydrodynamics and suspension behavior are validated by experimental investigations (see Section 4). Based on experimental and numerical results, a five-step road map to estimate suspension behavior is outlined and conducted in Section 5, resulting in a flow map for operating parameter selection.

2. Materials and Experimental Methods

2.1. Materials

Hydrodynamics is visualized by a suspension of reflective Mica flakes in water. These commercially available pigments are commonly used for the manufacture of paint and plastics [26]. In addition, Mica/water suspensions as rheoscopic fluid are employed for flow visualization [27]. Mica flakes are small (dimension around 10 µm, 0.1 µm thin [27]), which is demonstrated in the microscope image presented in Figure 1. Due to the reflective nature of Mica flakes, their nonuniform orientation and local accumulations reveal flow patterns in the liquid phase [27]. For the validation experiments, a solid content of $w_{solid} = 0.05\, g_{Mica}\, 100\, g_{Sol}^{-1}$ is used. At this solid content, best visualization results were determined in pre-experiments.

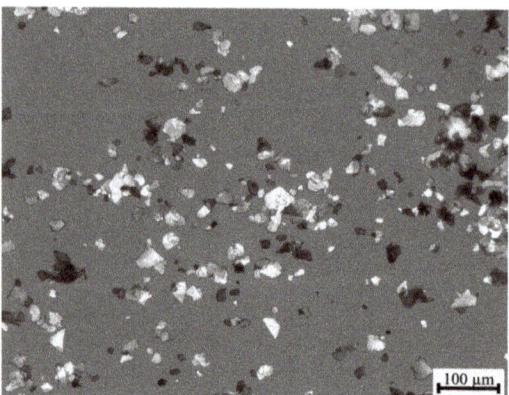

Figure 1. Microscopic image of mica powder in pale silver from *Finnabair* used in the experiments.

To observe suspension behavior, L-alanine (99.7% purity, Evonik Industries AG, Nanning, China) in ultrapure water (18.2 MΩcm) was selected as material system. The liquid phase is represented by saturated aqueous L-alanine solution at room temperature ϑ (approximately 21 °C). Thereby, the solubility c^* of L-alanine in aqueous solution depends on temperature ϑ according to Equation (1) [28].

$$c^*(\vartheta)\left[g_{Ala} \cdot g_{Sol}^{-1}\right] = 0.11238 \cdot exp(9.0849 \cdot 10^{-3} \cdot \vartheta\, [°C]) \tag{1}$$

As solid phase, L-alanine seed crystals are employed that are prepared according to Ostermann's procedure [29]. Here, practicable crystal product sizes attainable in the ATC comprise sieve fractions 100–160 µm and 250–315 µm. Figure 2 shows the particle size distributions of these sieve fractions that were determined with image analysis sensor QICPIC equipped with *Gradis* module (both Sympatec GmbH).

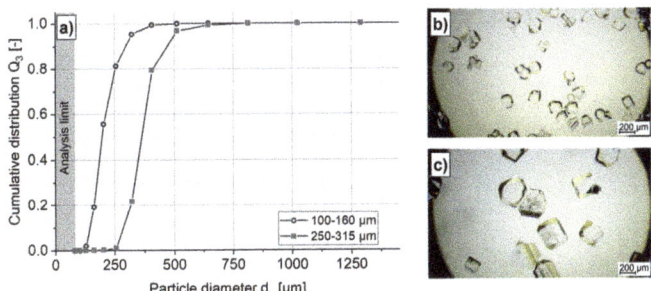

Figure 2. Seed crystal properties described by (**a**) particle size distributions Q_3 and microscopic images of (**b**) sieve fraction 100–160 µm with mode $d_{mod} = 182$ µm, and (**c**) sieve fraction 250–315 µm with mode $d_{mod} = 364$ µm.

2.2. Experimental Setup and Procedure

Figure 3a shows the geometric dimensions of the Archimedes Tube Crystallizer (ATC) and inlet tank employed. In Figure 3b, the ATC is integrated into the whole experimental setup with feed tanks, periphery, and video box. A 1 L laboratory glass bottle (Schott Duran) holds the feed solution that is stirred at 500 min^{-1} with magnetic stirrer RCT basic from IKA Labortechnik (Staufen, Germany). Mica or L-alanine suspension is provided from a stirred mixed-suspension mixed-product removal (MSMPR) tank. The borosilicate glass MSMPR tank with bottom cone and vertically adjustable overflow tube was designed by Lührmann et al. to ensure that the outlet suspension has a constant solid content [30]. Due to the bottom cone, however, liquid volume in the tank has to be kept between 200–400 mL to maintain the homogeneous suspension achieved at stirrer speed 450 min^{-1}.

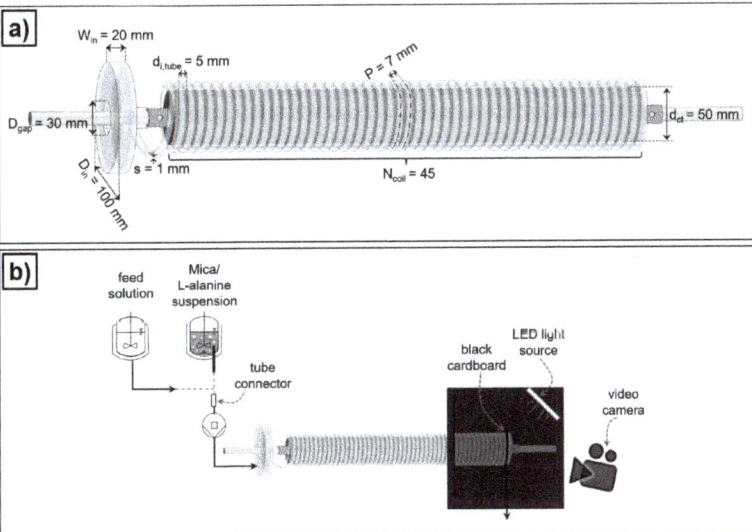

Figure 3. Schematic depiction of the (**a**) ATC dimensions (coiled tube length: 315 mm) and (**b**) experimental setup for validation experiments. For a more detailed description of experimental set-up and procedure, compare methods for solid-phase residence time distribution experiments by Sonnenschein and Wohlgemuth [6].

Feed solution and suspension are conveyed to the ATC by a peristaltic pump type Reglo-Digital MS-2/6 from Ismatec (Wertheim, Germany). Thereby, the periphery is oriented vertically for solid phase transport and composed of three parts with 4.2 mL liquid volume: tubing from tracer input to pump (silicone, $d_i = 3$ mm), pump's tubing (Tygon®, $d_i = 2.79$ mm), and tubing from pump to inlet tank (silicone, $d_i = 3$ mm).

In addition, a video box is set up to record hydrodynamics and suspension behavior in the last tube coil. Black PVC panels shield the apparatus from ambient light and a single LED light source is used for particle illumination. To prevent other tube coils from showing through the last tube coil, black cardboard is inserted between the second to last and last tube coil. Then, a video camera from Canon type EOS M6 is placed in front of the last tube. For the experiments, the ATC is operated with feed solution first until steady state is obtained. After that, the feed solution tank is disconnected to connect the feed suspension tank. Once the suspension has reached the last tube coil, the steady state is reached after another five tube coils. Then, videos of hydrodynamics and suspension behavior are recorded.

The operating window for investigations of hydrodynamics and suspension behavior (summarized in Table 1) is transferred from residence time distribution experiments at rotational speed $n_{ATC} = 4$–12 min^{-1} and filling degree $\epsilon = 0.25$–0.35 and offers residence times between $\tau = 4$–11 min [6]. Hydrodynamics is investigated by a 2^2-factorial Design of Experiments (factor levels (-1) and (1), 4 experiments in total). To investigate suspension behavior, the experimental plan is extended to include material system parameters: solid content w_{solid} is varied, whereas particle sieve fraction is kept constant. Here, w_{solid} is set to 1–5.1 $g_{Ala} \cdot 100\, g_{Sol}^{-1}$. These values represent the initial and end concentrations of a possible (seeded) cooling crystallization from 50 to 20 °C in the ATC. Initially, "worst-case" sieve fraction 250–315 µm is chosen for the experiments as larger particle sizes sediment more quickly than smaller particles due to gravity. To gain an overview of the influence of these parameters on the suspension state, a 2^{3-1} fractional factorial Design of Experiments with three center point experiments (factor levels (-1), (0), and (1), 7 experiments in total) is conducted.

Table 1. Design of Experiments to evaluate hydrodynamics and suspension behavior: rotational speed n_{ATC}, filling degree ϵ, and solid content w_{solid}.

Factor\Factor Level	(−1)	(0)	(1)
$n_{ATC}\,[\text{min}^{-1}]$	4	8	12
$\epsilon\,[-]$	0.25	0.3	0.35
$w_{solid}\,[g_{Ala} \cdot 100\, g_{Sol}^{-1}]$	1	3	5.1

3. Modeling

The numerical simulations performed in this work are related to the open-source CFD solver package of FeatFlow [31], which is a Finite Element Method (FEM)-based flow solver employing higher-order isoparametric Q_2/P_1 elements for the velocity and pressure, respectively. This flow solver has already been successfully used in the framework of numerous numerical benchmarks ranging from single phase flows [32] up to multiphase flows involving liquid and/or gaseous phases [33]. Moreover, the benchmark computations provided by Münster [34] have shown the use of the flow solver in combination with Fictitious Boundary Method (FBM) also in the framework of particulate flows by means of two-way coupled (passive) solid particles up to flows governed by the mechanical motion of the immersed (active) solid objects like micro-scallops [35].

Having the objectives of the here targeted simulation framework in mind, which is the ability to predict a suspension formation of the solid phase for the given geometrical and process parameters, the general flow solver is extended by a one-way coupled Lagrangian Particle Tracking (LPT) capable for resolving the inter-particle and wall–particle inelastic collisions. As the characteristic particle sizes subjected to the performed studies are in the order from 180 µm up to 360 µm, and only up to a very low volume fraction (<6%), the one-way coupled realization of the LPT offers itself as a reasonable compromise between computational accuracy and computational effort. A similar construction of a one-way coupled mathematical model has been recently presented by Xiao et al. [36] for the simulation of particle-laden boundary layers for the identification of particle pattern formation. According to this one-way coupled solution strategy, only particle motion is influenced by the flow of the surrounding liquid flow, but not the other way around. The particle motion respects the influence of drag and buoyancy but also the collision of the individual particles with each other but also with the physical walls bounding the liquid slug during its transportation in the Archimedes tube. Accordingly, in each time step, the particles are subjected to the force balance with respect to buoyance and drag force, as follows:

$$m_p \frac{dU_p}{dt} = \frac{1}{2}\rho_f C_D |U - U_p| (U - U_p) A_p + (\rho_p - \rho_f) g V_p \tag{2}$$

where U is the local fluid velocity, m_p is the mass of the particle, ρ_p and ρ_f are the densities of solid particle and of the fluid, respectively. V_p and A_p are the corresponding volume and area of the particle projected in the flow direction, which in case of considering the presence of strictly spherical particles, are dependent only on the particle diameter d_p. Special attention is paid only on the drag coefficient C_D which relies as well on the presence of spherical particles for which the Schiller-Naumann [37] model is applied in the form of

$$C_D = \frac{24}{Re_p}(1 + 0.15 Re_p^{0.687}) \tag{3}$$

which provides a reliable correlation for the particle Reynolds number $Re_p = \frac{\rho_f d_p |U - U_p|}{\eta_f}$, being lower than 1000, which is fulfilled in all the later considered cases.

The update of the particle center position x_p by means of its calculated velocity U_p from the force balance above is performed by a first order semi-implicit scheme, as follows:

$$x_p^{i+1} = x_p^i + U_p^{i+1} \Delta t \tag{4}$$

where the local velocity vector U is sampled at the corresponding particle center point x_p^i by the help of an octree-based algorithm identifying the respective element containing the particle center and is subsequently interpolated by taking advantage of the higher-order Q_2 finite element interpolation function.

Due to the subsequent treatment of collision mechanisms, the time steps applied in the force balance are chosen to be sufficiently small to prevent the divergence of resulting collision steps. The potentially arising collisions are carried out by means of an inelastic collision model, i.e., in case of collision of a particle pair there are no repulsive forces to be resolved, instead the particles are carried to a touching position; furthermore, the particle velocity U_p is set to the local fluid velocity U. The model described here is a rather inaccurate model in case of dense particle suspensions; however, it has the necessary accuracy in case of simulation of particle suspensions with small volume fraction of particles, matching the targeted operating conditions of this work. Additionally, the collision scheme is extended by the collision of particles with the solid walls of the simulation domain to avoid the loss of particles through the outer walls of the simulation domain. Particle collision might be strongly promoted by the dominant gravitational forces, especially in case of operating conditions characterized by small Shield's parameters. For this purpose, the surface triangulation of the fluid domain is utilized in combination with an efficient distance computation mechanism taking advantage of the related octree mechanisms.

The particular realization of the simulations is performed by means of a two-stage simulation framework. Accordingly, in the first stage, the determination of the underlying flow field is achieved for the prescribed operating conditions, which in this case is dictated by the rotational speed of the Archimedes screw. To this end, a predefined geometrical representation of the liquid slug is used, which is geometrically parametrized, on the one hand, by the walls of the Archimedes screw and by spherical surface representations at the two free-surface ends of the slugs. As a potential two-phase simulation by means of the front-tracking extension of the flow solver [38] would have required the treatment of triple phase (solid/liquid/gaseous) contact lines and considerably large computational efforts, a reduced but efficient single phase approach has been adopted by describing the gas/liquid interfaces in form of spherical surfaces. The corresponding curved surface segments Γ_{slip} have been subjected to a free slip boundary condition in terms of $u_{\Gamma,slip}$ to allow the creation of the respective recirculation patterns transporting the particles on the resulting trajectories. The boundaries of the fluid domain Γ_{wall} being aligned with the Archimedes tubing are assigned to Dirichlet boundary conditions dictated by the rotational movement of the tubing. According to the applied transformation with respect to a translational frame of reference, aside for the primary rotational speed components, also the axial velocity components are prescribed to be non-zero, as follows (see Figure 4):

$$u_{\Gamma,wall} = \begin{pmatrix} -2\pi\, y\, n_{ATC} \\ +2\pi\, x\, n_{ATC} \\ P\, n_{ATC} \end{pmatrix} \quad (5)$$

where n_{ATC} is the rotational speed and P the pitch distance of the coiled tubing.

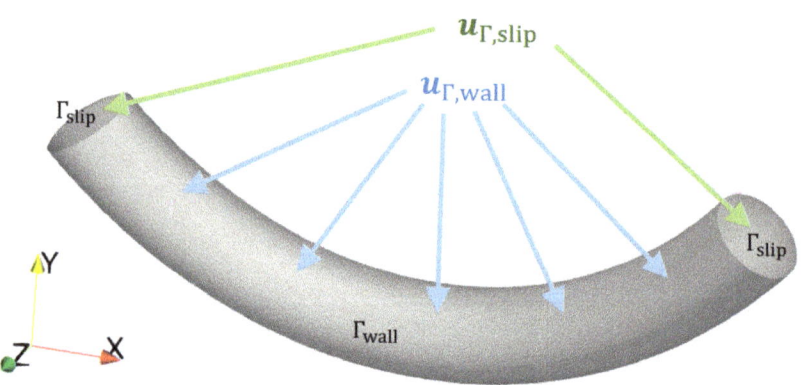

Figure 4. Geometrical representation of the liquid slug with the respective boundary conditions.

The simulations in case of low rotational speeds reach fully stationary flow fields. However, for high rotational speeds, slightly oscillating nearly periodical flow fields are attained in a spatially refined mesh convergence framework. Accordingly, a sequence of successively refined hexahedral meshes is used, so that the temporally converged solution of the coarser resolution mesh is prolongated to its finer resolution counterpart until the norm of the velocity difference between the two subsequent levels has decreased below 10^{-3}. For the simulation cases characterized by low rotational speeds (4–12 min^{-1}), a resolution level 2 solution (~10,000 elements) turned out to be sufficient, while for the cases with high rotational speeds (25–50 min^{-1}), the solution on a resolution level 3 (~80,000 elements) was necessary.

The above-described velocity solutions are applied to the subsequent particle tracing simulations (2nd stage), where the cases attributed to low rotational speeds are simulated on a stationary velocity field. The cases exhibiting instationary (but nearly periodical) behavior are simulated on the extracted periodical velocity fields. For this purpose, the corresponding flow simulation results are analyzed with respect to the periodicity of the flow. In the second step, the solutions of the individual timesteps within the estimated period are saved and provided in an infinitely looped fashion to the particle simulation tool. As the computational mesh in all cases is considered to be stationary, the velocity field between the individual outputs is linearly interpolated for the respective subtimesteps during the particle tracing simulations.

4. Validation of Simulation

In this section, hydrodynamics and suspension behavior are simulated and validated according to the experimental plans introduced in Section 2.2. In the images provided, the flow behavior is difficult to recognize in some cases. Therefore, the corresponding videos to each flow visualization image are supplied in the Supplementary Material.

4.1. Hydrodynamics

Figure 5a shows the simulated fluid velocities at the exemplary operating point at $n_{ATC} = 12\,\mathrm{min}^{-1}$ and $\epsilon = 0.25$ in the last coil of the ATC tubing. Due to the occurring pressure profile in the apparatus (compare Sonnenschein and Wohlgemuth [6]), the liquid segment in the last tubular coil is displaced by 25° in rotating direction. This displacement is irrelevant for the non-gravity consideration of hydrodynamics but is considered for the simulation of suspension behavior in Sections 4.2 and 5.2. The scale in Figure 5a displays velocity magnitude, whereas the arrows indicate flow direction and magnitude by different arrow lengths. The CFD simulation shows fluid entrainment at the tube wall in rotating direction. At the rear interface, the entrained fluid flow is decelerated, reversed, and then accelerated at the tube's center in reverse direction. This behavior is reflected by the fluid flow visualization experiment in Figure 5b. Here, green arrows and overlay were used to emphasize the flow in rotational direction, whereas a yellow arrow and overlay indicate flow in reverse direction. Both visualizations show a flow regime dominated by Taylor vortices typical for slug flow. In the CFD simulation, another detail that can be seen is that the center of the reverse flow is shifted towards the outer bend at the front side of the slug. This observation is typical for Dean vortices and becomes more pronounced at higher flow velocities. This deviation, however, is too small to be recognized in the experiment. Overall, flow behavior of simulated and experimental case are similar for this and the other investigated operating points (provided as Figures S1–S4 accompanied by videos in the Supplementary Material). Thus, the CFD simulation is a valid representation of hydrodynamics.

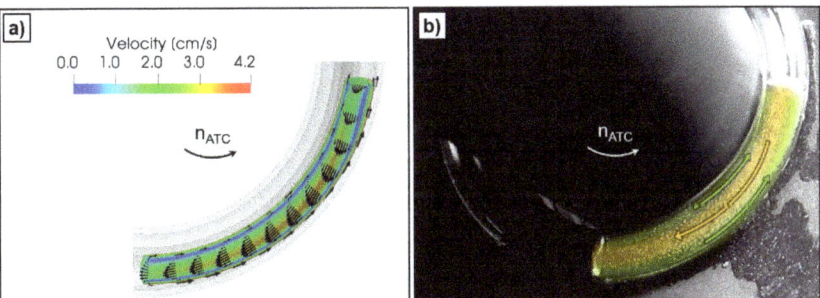

Figure 5. Hydrodynamics at exemplary operating point $n_{ATC} = 12\,\mathrm{min}^{-1}$ and $\epsilon = 0.25$ in panel (**a**) CFD simulation and (**b**) experiment with Mica powder. The corresponding video for panel (**b**) is supplied in the Supplementary Material.

4.2. Suspension Behavior

Figure 6 displays the respective suspension behavior in the last tube coil at the operating points specified in Table 1. Thereby, Figure 6a,b shows CFD simulations and experiments for particle size $d_p = 364\,\mathrm{\mu m}$, respectively. Figure 6c presents additional simulations for smaller particle size $d_p = 182\,\mathrm{\mu m}$. Over the whole investigated operating window, the CFD simulations predict accumulation and settling of the particles at the slug's rear end, independent from solid content, rotational speed, and filling degree (compare

Figure 6a). At the rear end, the particles recirculate as they follow the hydrodynamic flow field: At the tube wall, particles are entrained in rotational direction until they reach the rear interface. Here, the particles are accelerated in reverse direction. Variations in filling degree ϵ lead neither to an improvement nor to a deterioration of the suspension state. The same particle behavior is observed in the experiments, as visualized in Figure 6b. In addition to the recirculation zone, particles aggregate at the rear interface. This behavior may result from small velocities at the interface in combination with particle inertia. Another reason could be particle adsorption to the interface. This phenomenon is caused by interfacial tensions and has been investigated for liquid/liquid slug flow applications with phase-boundary catalysis [39] or water/air slug flow with polystyrene spheres [40]. In the simulations, the particles are distributed over a wider horizontal range than in the experiments. This observation can be explained by the simplifications made for the CFD simulations due to

1. the assumed uniform (and spherical) particle size distribution of the dispersed phase which might contradict to the experimental realization, and
2. the insufficient support of the adopted one-way coupled model for the accurate description of the particle dynamics in case of formation of locally dense particle patters, which then results in lacking the feedback of the dispersed phase back on the continuous fluid phase.

The validity of the simulation results is restricted only for the respective parameter spaces where the particle dynamics is corresponding to the dispersed flow scenarios. The numerical simulations characterized by the particular operation conditions, resulting in a formation local accumulation of particles, are clearly outside of the validity of the model, and therefore the dynamics of the particles experienced under such conditions is to be treated with caution. However, the dynamics before reaching these critical flow patterns is marginally still covered by the respective one-way coupled realization, as the hydrodynamic forces acting on the individual particles together with the particle–particle and particle–wall interactions are decisive for the formation of the resulting particle dynamics, which in the final consequence is then modeled without a sufficient support of the here adopted one-way coupling model.

Overall, CFD simulations can be employed to describe suspension behavior. However, the selected operating range is insufficient for the objective to gain an overview of suspension states. Thus, further targeted CFD simulations are conducted to specify the operating range. Particle size d_p (correlating to in-flow direction projected particle area A_p and particle volume V_p) has a major impact on the particle force balance (compare Equation (2)). Reducing the particle size by half ($d_p = 182$ μm), however, still leads to particle recirculation zones at the slug's rear end, where the particles accumulate (Figure 6c). Another major impact on the force balance is the local fluid velocity U_p, that depends on the hydrodynamics induced by rotational speed n_{ATC} for a fixed ATC design. In the investigated operating window, the flow profile is dominated by Taylor vortices typical for air-segmented flow. By increasing rotational speed, this profile might be superpositioned by Dean vortices leading to increased suspension. Here, process understanding is necessary to evaluate these effects and estimate reasonable ATC operation and design parameters.

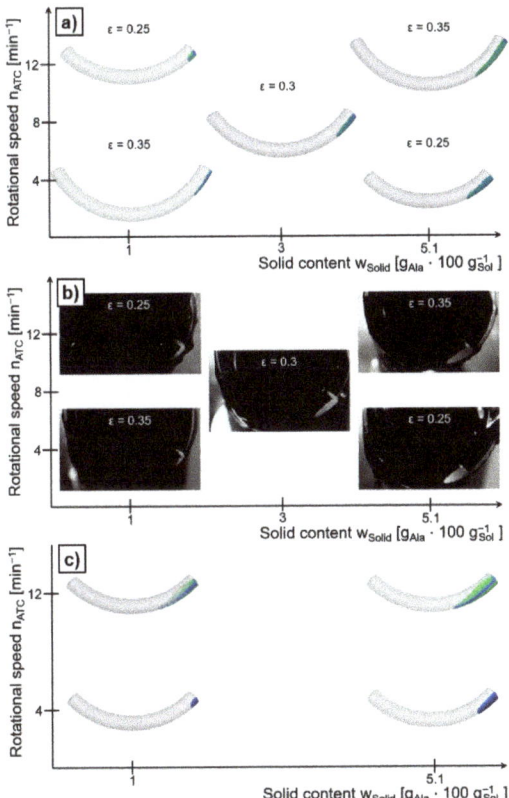

Figure 6. Suspension behavior observed (**a**) in CFD simulations with $d_p = 364$ µm at varying filling degrees ϵ, (**b**) in experiments with $d_{mod} = 364$ µm (sieve fraction 250–315 µm) at varying filling degrees ϵ, and (**c**) in CFD simulations with $d_p = 182$ µm at filling degree $\epsilon = 0.25$. Corresponding individual figures and accompanying videos are supplied in the Supplementary Material as (**a**) Figures S5–S9, (**b**) Figures S10–S14, and (**c**) Figures S15–S18.

5. Five-Step Road Map to Set Up a Flow Map for Suspension Behavior Estimation

The previously presented experiments and simulations uniformly showed the same result: The accumulation of particles at the slug's rear end. Thus, suspension behaviour in the investigated operating window can be summarized as qualitative suspension state "particle accumulation". However, to set up a flow map for the Archimedes Tube Crystallizer (ATC), information regarding further suspension states is necessary. In the apparatus development of the ATC, the challenge is to gain process understanding as early as possible, with as little effort as required. As a solution, the simulations of hydrodynamics and suspension behavior are combined with dimensionless numbers. Thereby, the dimensionless numbers provide a good first shot at possible operating regions and thus reduce computational and experimental effort. Additionally, the gained process understanding can later be used to transfer the flow map to other ATC designs and additional material systems. Our strategic approach to this flow map is summarized in five steps as visualized in Figure 7.

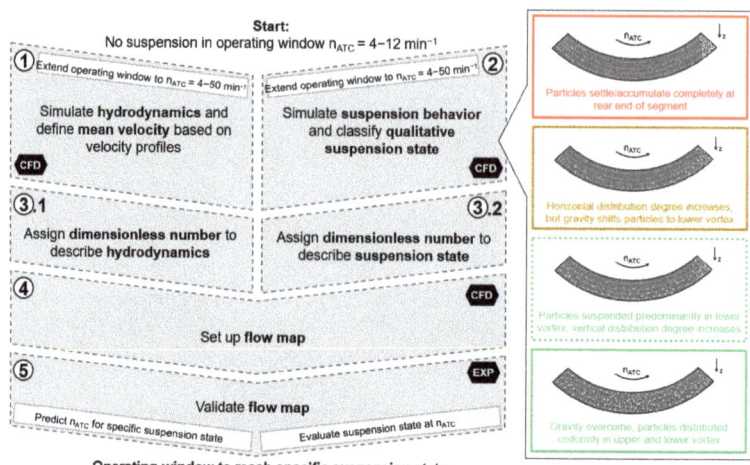

Figure 7. Five-step road map to estimate suspension behavior. The individual steps are carried out by computational fluid dynamics simulations (CFD) or experimentally (EXP).

In Step 1, hydrodynamics are simulated to understand air-segmented flow in coiled structures. Here, the impact of both flow regimes (separately and in combination) on particle suspension is discussed based on extracted velocity profiles. In Step 2, the suspension behavior is simulated and all previous and current results are classified into four qualitative suspension states analogous to Scheiff's definitions for slug flow applications [23]. Step 3 includes the assignment of dimensionless numbers to describe hydrodynamics and suspension state. Through this approach, the observed flow phenomena become calculable and can be used to set up a flow map for the ATC in Step 4. Finally, Step 5 contains the validation of the flow map.

5.1. Step 1: Simulate Hydrodynamics and Evaluate Velocity Profiles

This step's aim is the extraction of velocity profiles to understand superposition of Taylor and Dean flow and possible impacts on suspension behavior. Therefore, hydrodynamics are simulated in the extended operating window for $n_{ATC} = 4$–$50\,\text{min}^{-1}$. Apart from the previous simulations at 4 and $12\,\text{min}^{-1}$, further simulations are conducted at 25, 37.5, and $50\,\text{min}^{-1}$ to save computational effort and cover the whole operating window at the same time.

Figure 8 shows the velocity profiles at different angular positions in the slug for filling degree $\epsilon = 0.25$, exemplarily. Thereby, the total slug length at filling degree $\epsilon = 0.25$ corresponds to angles 0–90°. For the comparison between rotational speeds, velocity is normalized with the average circumferential speed $v_{circ,av}$ calculated according to Equation (6). Thereby, $v_{circ,av}$ is the product of rotational speed n_{ATC} and corrected coiled tube diameter d_{ck}, which accounts for curvature by including coiled tube diameter d_{ct}, inner tube diameter $d_{i,tube}$, wall thickness s, and pitch distance P in the calculations [6].

$$v_{circ,av} = n_{ATC} \cdot \pi \cdot d_{ck} = n_{ATC} \cdot \pi \cdot (d_{ct} + d_{i,tube} + 2s) \cdot \left(1 + \left(\frac{P}{\pi \cdot (d_{ct} + d_{i,tube} + 2s)}\right)^2\right) \quad (6)$$

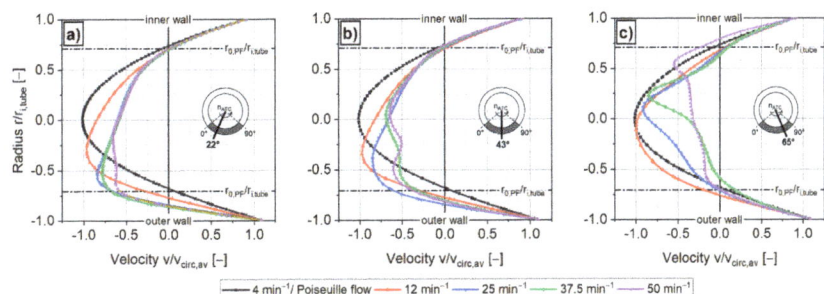

Figure 8. Velocity profiles for different rotational speeds n_{ATC} at filling degree $\epsilon = 0.25$ for angular positions (**a**) 22°, (**b**) 43°, and (**c**) 65°.

For segmented flow, the basic velocity profile is the Poiseuille flow profile that can be calculated according to Equation (7) [41]. Here, the velocity v_{PF} at a certain radius r depends on the average circumferential speed $v_{circ,av}$ and inner tube radius $r_{i,tube}$.

$$v_{PF}(r) = 2 \cdot v_{circ,av} \cdot \left(1 - \frac{r^2}{r_{i,tube}^2}\right) - v_{circ,av} \qquad (7)$$

From this equation, radius $r_{0,PF}$ is determined as radius at which velocity is equal to zero (compare Equation (8)).

$$r_{0,PF} = \frac{1}{\sqrt{2}} r_{i,tube} \qquad (8)$$

Poiseuille velocity profile and radius $r_{0,PF}$ are included in Figure 8 as gray line and horizontal dash-dotted line, respectively. For rotational velocity $n_{ATC} = 4\,\text{min}^{-1}$, the ATC velocity profile is identical to the Poiseuille velocity profile at all angular positions. Thus, segmented flow is the dominating flow regime at this operating point.

At angular position 22° (Figure 8a), it can be seen that the maximum velocity peak is shifted towards the outer tube wall with increasing velocity. The same tendency is observed at angular position 43° (Figure 8b). By contrast, the maximum velocity peak is shifted towards the inner tube wall with increasing velocity at angular position 65° (Figure 8c). These observations represent the superposition of the Poiseuille flow profile by Dean vortices and consequently an increase in mixing efficiency. At angular position 22°, the overlaying Dean vortices could be beneficial for suspension behavior. Here, r_0, the radius at which velocity is equal to zero, is shifted closer to the outer tube wall compared to $r_{0,PF}$ from rotational speed $n_{ATC} = 25\,\text{min}^{-1}$ upwards. Thus, the zone for particle entrainment in rotational direction is reduced by half. However, this angular position is not decisive for particle transport as the particles mostly accumulate at the rear end of the segment (compare Section 4.2). There, at angular position 65°, r_0 is reduced at the inner tube wall, where the impact on the particles is smaller. Nevertheless, the maximum velocity shift towards the outer tube wall is already visible in the center of the slug at angular position 43°. Thus, the positive effect of the Dean vortices on mixing might be larger than expected. In addition, the presented velocity profiles have only been extracted in radial direction. In other directions, increase of mixing efficiency by Dean vortices is expected as well.

5.2. Step 2: Simulate Suspension Behavior and Classify Qualitative Suspension State

In this step, suspension behavior is simulated and classified into qualitative suspension states for the extended operating window from $n_{ATC} = 4$ to $50\,\text{min}^{-1}$. Thereby, the investigated rotational velocities are selected analogously to Section 5.1 for filling degree $\epsilon = 0.25$. The previously employed particle diameters $d_p = 182\,\mu\text{m}$ and $d_p = 364\,\mu\text{m}$ and solid fractions $w_{solid} = 1$–$5.1\,g_{Ala} \cdot 100\,g_{Sol}^{-1}$ (see Section 4.2) are selected again for comparison as possible crystal product diameters and solid contents of a cooling crystallization process.

According to the proposed qualitative suspension states, all previous experiments and simulations (compare Figure 6) are classified as *not suspended* (red), as the particles settled and accumulated completely at the rear end of the segment. Figure 9 shows the simulation results at higher rotational speeds for the large (d_p = 364 µm, Figure 9a) and small (d_p = 182 µm, Figure 9b) particle diameters.

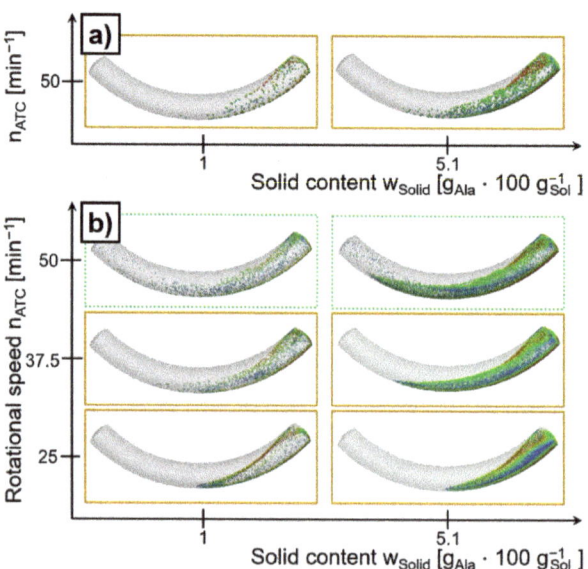

Figure 9. Suspension behavior observed in CFD simulations for (**a**) d_p = 364 µm and (**b**) d_p = 182 µm. Corresponding individual figures and accompanying videos are supplied in the Supplementary Material as (**a**) Figures S19 and S20 and (**b**) Figures S21–S26.

The simulations for particle size d_p = 364 µm show an increasing distribution of the crystals in horizontal direction (yellow) at the highest investigated rotational speed of n_{ATC} = 50 min^{-1} (compare Figure 9a). For all other simulated cases, the particles still accumulate at the rear end of the slug (red). These results are similar, even for different solid contents. For particle size d_p = 182 µm, the increasing distribution in horizontal direction (yellow) is already visible at n_{ATC} = 25 min^{-1} (compare Figure 9b). For n_{ATC} = 50 min^{-1}, the particles even start to distribute along the vertical axis (dotted green) as well. In all cases, a particle recirculation vortex can be identified at the rear half of the slug. The particles follow the fluid streamlines while sedimenting due to gravity. As soon as the particles reach the entrainment zone at the tube wall, they are accelerated towards the rear interface in rotational direction. In summary, it seems that particle gravitation has such a strong influence that only small particles can be suspended.

5.3. Step 3: Assign Suitable Dimensionless Numbers to Describe Hydrodynamics and Suspension Behavior

Using the simulation results for hydrodynamics and suspension behavior, suitable dimensionless numbers are calculated and evaluated in this section. The initial focus is on hydrodynamics. In the ATC, it is assumed that the combination of hydrodynamics in coiled structures and air-segmented flow has to be considered.

In coiled structures, the magnitude of secondary flow patterns is described by Dean number Dn as ratio between inertial and centripetal to viscous forces (Equation (9)) [8].

Dn depends on Reynolds number Re and the geometry of the coiled tubing given by inner tube diameter $d_{i,tube}$ and coiled tube diameter d_{ct}.

$$Dn = Re \cdot \sqrt{\frac{d_{i,tube}}{d_{ct}}} \qquad (9)$$

For the ATC, Re is calculated according to Equation (10) with average circumferential velocity $v_{circ,av}$, fluid density ρ_f, and viscosity η_f.

$$Re = \frac{\rho_f \cdot v_{circ,av} \cdot d_{i,tube}}{\eta_f} \qquad (10)$$

A time scale of Dean flow can be calculated based on the average velocity of the Dean vortices v_{Dn} and a representative secondary flow path l_{Dn} [14]. This time scale is beneficial for the later discussion on superposition of Dean and air-segmented flow. Equation (11) was derived semi-empirically by Bayat and Rezai for v_{Dn} [42] and also employed by Gaddem et al. [14] in the description of a coiled flow inverter with segmented flow.

$$v_{Dn} = 0.031 \cdot \frac{\eta_f}{\rho_f \cdot d_{i,tube}} \cdot Dn^{1.63} \qquad (11)$$

The representative circulation path l_{Dn} is determined according to Equation (12) found in [14]. Hereby, the Dean vortex shape is approximated by a half-circle with diameter $d_{i,tube}/\sqrt{2}$.

$$l_{Dn} = \frac{d_{i,tube}}{\sqrt{2}} \cdot \left(1 + \frac{\pi}{2}\right) \qquad (12)$$

With these quantities, Dean flow recirculation time τ_{Dn} is calculated as described by Equation (13), also from the work in [14].

$$\tau_{Dn} = \frac{l_{Dn}}{v_{Dn}} \qquad (13)$$

The flow pattern in air-segmented flow is described as Taylor flow. Taylor flow recirculation time τ_{Taylor} depends on the average circumferential velocity $v_{circ,av}$ and slug length l_{slug} (compare Equation (14)) from in [14,41]. l_{slug}, in turn, is given by filling degree ϵ and corrected coiled diameter d_{ck}.

$$\tau_{Taylor} = \frac{2 \cdot l_{Slug}}{v_{circ,av}} = \frac{2 \cdot \epsilon \cdot \pi \cdot d_{ck}}{v_{circ,av}} = \frac{2 \cdot \epsilon}{n_{ATC}} \qquad (14)$$

Gaddem et al. combined the time scales of Dean and Taylor flow to the modified Dean number Dn^* for segmented flow in coiled structures (compare Equation (15)) found in [14].

$$Dn^* = \frac{\tau_{Taylor}}{\tau_{Dn}} \cdot Dn \qquad (15)$$

Dn^* compares mixing in radial direction (Dean vortices) to mixing in angular direction (Taylor vortices). If $\tau_{Dn} > \tau_{Taylor}$, mixing in angular direction is faster than in radial direction. If $\tau_{Dn} < \tau_{Taylor}$, mixing in radial direction is faster.

The correlations just outlined are summarized in Figure 10 for the considered operating window. By plotting the ratio of Dn^*/Dn, it becomes directly apparent, according to Equation (15), which mixing process is predominant: angular mixing for $Dn^*/Dn < 1$ and radial mixing for $Dn^*/Dn > 1$. For filling degree $\epsilon = 0.25$, mixing in angular direction is faster than in radial direction below 12 min^{-1}. This observation coincides with the velocity profiles presented in Figure 8, where the typical Dean vortex shape is discernible above this rotational speed. For filling degree $\epsilon = 0.35$, mixing in radial direction is already faster than angular mixing below 12 min^{-1}. This observation is due to the longer slug length (higher filling degree) compared to the same representative circulation path of the Dean vortex. The superposition of Dean vortices on Taylor vortices might also affect suspension behavior through the increase in mixing efficiency. Thus, qualitative suspension states in the ATC flow regime could be reached at lower rotational speeds than predicted on the basis of segmented flow alone.

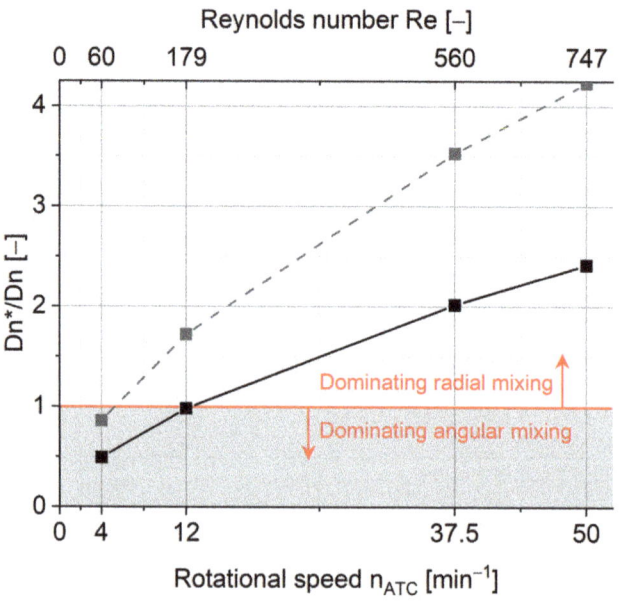

Figure 10. Rotational speed n_{ATC} and ratio of Modified Dean number Dn^* to Dean number Dn for filling degree $\epsilon = 0.25$ (solid black line) and $\epsilon = 0.35$ (dashed gray line). Reynolds number Re is calculated for water as liquid phase.

Suspension behavior in segmented flow can be evaluated based on Shield's parameter Θ, the ratio of drag, and gravity, provided in Equation (16) [23]. Thereby, Θ depends on fluid viscosity η_f, density difference between fluid and particle $\Delta\rho_{pf}$, average circumferential velocity $v_{circ,av}$, particle diameter d_p, and gravity g.

$$\Theta = \frac{9\,\eta_f\,v_{circ,av}}{(d_p/2)^2 \Delta\rho_{pf}\,g} \qquad (16)$$

Scheiff investigated the qualitative suspension behavior of catalyst particles with $d_p = 10\text{–}100\,\mu\text{m}$ and $\rho_p \gg \rho_f$ [23]. For these material system properties, he specified the following boundaries for Shield's parameter Θ to estimate the four qualitative suspension states mentioned above (compare Figure 7) [23]:

- $\Theta \ll 10$: Particles settle/accumulate completely at rear end of segment (red).
- $10 \leq \Theta \leq 30$: Horizontal distribution degree increases, but gravity shifts particles to lower vortex (yellow).
- $30 \leq \Theta \leq 180$: Particles suspended predominantly in lower vortex, vertical distribution degree increases (dotted green).
- $\Theta > 180$: Gravity overcome, particles distributed uniformly in upper and lower vortex (green).

Figure 11 shows these limits in combination with the already classified suspension states (red, yellow, dotted green, green) for the particle sizes investigated in Sections 4.2 and 5.2.

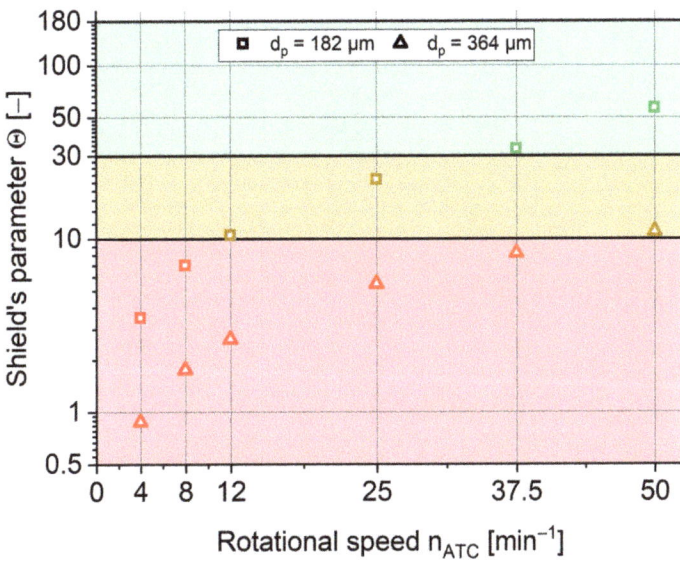

Figure 11. Rotational speed n_{ATC} and Shield's parameter Θ for selected particle sizes at filling degree $\epsilon = 0.25$. Suspension states are classified according to Figure 7.

For particle diameter $d_p = 182\,\mu\text{m}$, qualitative suspension states up to increasing vertical distribution (dotted green) can be reached within the specified operating window. For particle diameter $d_p = 364\,\mu\text{m}$, rotational speed must be higher than $50\,\text{min}^{-1}$ to even achieve a horizontal distribution of the particles (yellow). The ideal, homogeneous suspension state (green) is not achieved in either case. Overall, the observed suspension states fit the proposed boundaries. Nevertheless, the Shield's parameter Θ has limitations: Neither solid content, particle shape, slug length (filling degree), or influences of the coiled structure are taken into account.

5.4. Step 4: Set Up Flow Map

To set up a flow map, the ratio of Modified Dean number to Dean number Dn^*/Dn and Shield's parameter Θ are combined. Dn^*/Dn from hydrodynamics is shown complimentary to point out areas where enhanced suspension behavior is expected due to superpositioned secondary flow patterns, whereas Θ is used to describe the suspension state. Both dimensionless parameters depend on material system, ATC design, and operation.

Dn^*/Dn is calculated according to Equation (17) that results from inserting Equations (13) and (14) into Equation (15). Here, the highest impact on the magnitude of Dn^*/Dn is given by filling degree ϵ and corrected coiled tube diameter d_{ck}. d_{ck} is calculated based on coiled tube diameter d_{ct} according to Equation (6).

$$\frac{Dn^*}{Dn} = 0.22 \cdot \underbrace{\left(\frac{\rho_f}{\eta_f}\right)^{0.63}}_{\text{Material system}} \cdot \underbrace{d_{i,tube}^{0.445} \cdot d_{ck}^{0.815}}_{\text{Design}} \cdot \underbrace{\epsilon \cdot n_{ATC}^{0.63}}_{\text{Operation}} \qquad (17)$$

Equation (18) represents the Shield's parameter adapted to the ATC by inserting Equation (6) into Equation (16). Particle size d_p is the decisive factor for suspension quality, followed by corrected coiled tube diameter d_{ck} and rotational speed n_{ATC}.

$$\Theta = 113.1 \cdot \underbrace{\frac{\eta_f}{\Delta \rho_{pf}} \cdot d_p^{-2} \cdot g^{-1}}_{\text{Material system}} \cdot \underbrace{d_{ck}}_{\text{Design}} \cdot \underbrace{n_{ATC}}_{\text{Operation}} \quad (18)$$

The flow map visualized in Figure 12 shows the dependencies of the selected dimensionless parameters on rotational speed n_{ATC} and particle diameter d_p for coiled tube diameters (a) $d_{ct} = 50$ mm and (b) $d_{ct} = 200$ mm. As already explained in Section 5.3, ratios $Dn^*/Dn < 1$ imply faster angular than radial mixing. Thus, in this operating range, the previously defined boundaries for Shield's parameter Θ to estimate the qualitative suspension state are directly transferable. For ratios $Dn^*/Dn > 1$, faster radial than angular mixing is expected. Here, enhanced suspension behavior might be observable. In addition, the necessary rotational speed for amplified radial mixing decreases with increased coiled diameter (compare Figure 12a,b). Here, slug length increases due to the larger d_{ct}, with direct impact on the ratio of slug to Dean vortex length. Overall, whether ratios $Dn^*/Dn > 1$ are already sufficient to improve suspension has not been determined yet.

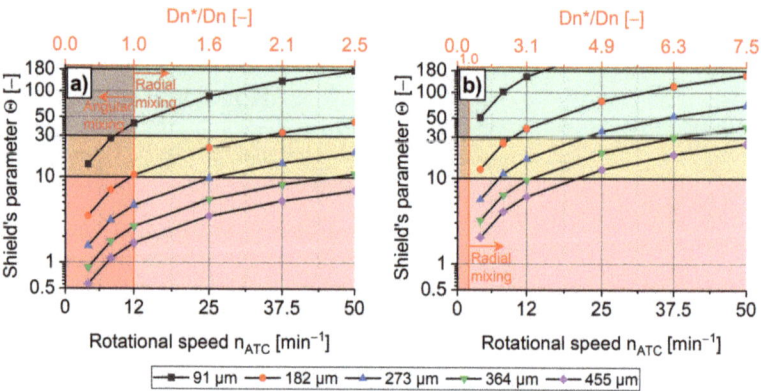

Figure 12. Calculated ratio of Modified Dean number Dn^* to Dean number Dn versus Shield´s parameter Θ in dependence of rotational speed n_{ATC} and particle diameter d_p according to Equations (17) and (18) for (a) $d_{ct} = 50$ mm and (b) $d_{ct} = 200$ mm; constant values: filling degree $\epsilon = 0.25$, $\eta_f = 0.001$ Pa s, $\rho_f = 1043$ kg m^{-3} and $\rho_p = 1420$ kg m^{-3}.

Whereas small particles with $d_p = 91$ μm can already be distributed horizontally (green) at rotational speed $n_{ATC} = 12$ min^{-1}, large particles with $d_p \geq 364$ μm are expected to settle over the whole operating window of the ATC with $d_{ct} = 50$ mm (compare Figure 12a). To reach a higher suspension state for a given material system with specified particle size, rotational speed n_{ATC} or coiled tube diameter d_{ct} needs to be increased. However, an increase in rotational speed causes a decrease of residence time in the apparatus and thus leads to a shorter crystal growth time. Therefore, increasing the coiled tube diameter is the preferable choice as visualized in Figure 12b. To suspend particles with double the size, coiled tube diameter must be enlarged by factor 4 according to Equation (18). For this ATC design, particles with $d_p = 182$ μm reach horizontal distribution (green) at rotational speeds below $n_{ATC} = 12$ min^{-1}.

Step 5: Validate Flow Map

Flow map validation is conducted at a suitable operating point at the transition between two qualitative suspension states. As no suspension is expected in the investigated operating region of n_{ATC} = 4–50 min^{-1} for sieve fraction 250–315 µm (d_{mod} = 364 µm), sieve fraction 100–160 µm (d_{mod} = 182 µm) is selected for the experiment. For this particle size, the transition from fully horizontally distributed (yellow) to a vertically distributed suspension state (dotted green) is estimated at Shield's parameter Θ = 35 or rotational speed n_{ATC} = 40 min^{-1}. Solid content w_{solid} is not considered in the developed flow map (compare Equations (17) and (18)). Thus, the selected operating point is investigated for two solid contents (compare Section 2.2) to exclude potential impacts.

Figure 13 illustrates the suspension state at the selected operating point.

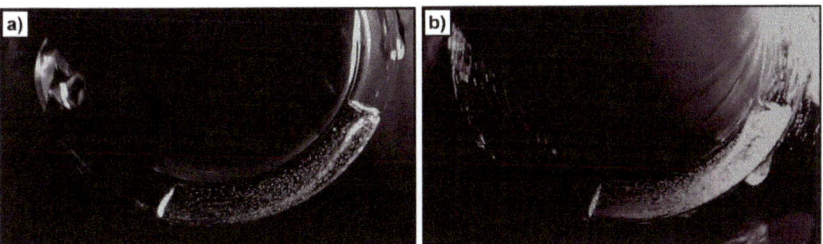

Figure 13. Experimental suspension state for d_{mod} = 182 µm at n_{ATC} = 40 min^{-1} and ϵ = 0.25 for (**a**) solid content w_{solid} = 1 g$_{Ala}$ · 100 g$_{Sol}^{-1}$ and (**b**) solid content w_{solid} = 5.1 g$_{Ala}$ · 100 g$_{Sol}^{-1}$. Corresponding individual figures and accompanying videos are supplied in the Supplementary Material as (**a**) Figure S27 and (**b**) Figure S28.

For the lower solid content (Figure 13a), particles are horizontally distributed as predicted. Furthermore, the particles are almost completely vertically distributed, thus the qualitative suspension state is close to homogeneous suspension. In addition, a particle recirculation vortex is visible at the rear half of the slug that was also calculated in the CFD simulations presented in Figure 9b. For the higher solid content (Figure 13b), the particles are no longer recognizable individually but rather seem homogeneously distributed in both directions. In this case, the higher solid content leads to particle overlapping. Thus, differences in particle distribution are difficult to record by camera. Nevertheless, higher suspension states might be reached by increasing the solid content due to particle swarm effects.

Overall, the observed suspension behavior in the validation experiments was better than expected, as almost homogeneous suspension (green) was reached at the selected operating point. This deviation reveals the suitability of the combination of the selected dimensionless numbers. The supplemental information gained from hydrodynamics with ratio Dn^*/Dn is beneficial to describe the flow enhancement through Dean vortices. For operating point n_{ATC} = 40 min^{-1}, for example, radial mixing is twice as fast as angular mixing already. As a consequence, suspension quality is improved as well.

6. Conclusions and Outlook

A study to characterize hydrodynamics and suspension behavior in the small-scale Archimedes Tube Crystallizer (ATC) was conducted. To set up a flow map for the ATC, a five-step roadmap was introduced. In this approach, Computational Fluid Dynamics (CFD) simulations and experiments were integrated with dimensionless numbers to reduce experimental and simulative effort by conducting only target-oriented investigations. The flow map developed can be applied to estimate operating parameters for a selected material system and ATC design based on the targeted crystal product size.

For a continuous cooling crystallization process in the ATC, low rotational speeds may be required to achieve the necessary residence time for sufficient crystal growth. However, the required residence time is strongly dependent on the crystal growth rates of the material system. In general, by increasing the coiled tube diameter, larger particles can be obtained with sufficient suspension as well. Nevertheless, the required qualitative suspension state to avoid agglomeration in a cooling crystallization has yet to be determined.

In the future, a dimensional analysis should be conducted to develop dimensionless numbers that include both mixing and material-specific effects, such as solid content and particle shape. In addition, the CFD simulation tool developed in this contribution could be extended with a suitable population balance model to describe interrelations between suspension behavior and agglomeration. Thereby, the simulations could be enriched to include particle size distributions and further particle effects.

Supplementary Materials: The following are available at https://www.mdpi.com/article/10.3390/cryst11121466/s1, Figures S1–S28 and corresponding videos to the figures.

Author Contributions: J.S. and K.W. conceptualized the validation experiments and flow map development, J.S. and P.F. conducted the validation. M.A. and O.M. set up and ran the CFD simulations and wrote the modeling section. K.W. and O.M. reviewed the paper that was then edited by J.S., S.T. and K.W. supervised and administrated the project. All authors have read and agreed to the published version of the manuscript.

Funding: This research received no external funding.

Institutional Review Board Statement: Not applicable.

Informed Consent Statement: Not applicable.

Conflicts of Interest: The authors declare no conflict of interest.

Abbreviations

The following abbreviations are used in this manuscript:

Ala	L-alanine
API	Active pharmaceutical ingredient
ATC	Archimedes Tube Crystallizer
CFD	Computational fluid dynamics
CFI	Coiled flow inverter
EXP	Experiments
FBM	Fictitious Boundary Method
FEM	Finite Element Method
LPT	Lagrangian Particle Tracking
PF	Poiseuille flow
PSD	Particle size distribution
RTD	Residence time distribution
Sol	Solution
Susp	Suspension

References

1. Cote, A.; Erdemir, D.; Girard, K.P.; Green, D.A.; Lovette, M.A.; Sirota, E.; Nere, N.K. Perspectives on the Current State, Challenges, and Opportunities in Pharmaceutical Crystallization Process Development. *Cryst. Growth Des.* **2020**, *20*, 7568–7581. [CrossRef]
2. Chen, J.; Sarma, B.; Evans, J.M.B.; Myerson, A.S. Pharmaceutical Crystallization. *Cryst. Growth Des.* **2011**, *11*, 887–895. [CrossRef]
3. Ma, Y.; Wu, S.; Macaringue, E.G.J.; Zhang, T.; Gong, J.; Wang, J. Recent Progress in Continuous Crystallization of Pharmaceutical Products: Precise Preparation and Control. *Org. Process Res. Dev.* **2020**. [CrossRef]
4. Orehek, J.; Teslić, D.; Likozar, B. Continuous Crystallization Processes in Pharmaceutical Manufacturing: A Review. *Org. Process Res. Dev.* **2020**. [CrossRef]
5. Acevedo, D.; Yang, X.; Liu, Y.C.; O'Connor, T.F.; Koswara, A.; Nagy, Z.K.; Madurawe, R.; Cruz, C.N. Encrustation in Continuous Pharmaceutical Crystallization Processes—A Review. *Org. Process Res. Dev.* **2019**, *23*, 1134–1142. [CrossRef]
6. Sonnenschein, J.; Wohlgemuth, K. Archimedes Tube Crystallizer: Design and Characterization for Continuous Small-Scale Crystallization: Submitted (CHERD-D-21-00690). *Chem. Eng. Res. Des.* **2021**, *2021*.
7. Kleinebudde, P.; Khinast, J.; Rantanen, J. (Eds.) *Continuous Manufacturing of Pharmaceuticals*; Advances in Pharmaceutical Technology; Wiley: Hoboken, NJ, USA, 2017.

8. Tiwari, P.; Antal, S.P.; Podowski, M.Z. Three-dimensional fluid mechanics of particulate two-phase flows in U-bend and helical conduits. *Chem. Eng. Sci.* **2006**, *18*, 043304. [CrossRef]
9. Yamamoto, K.; Yanase, S.; Yoshida, T. Torsion effect on the flow in a helical pipe. *Fluid Dyn. Res.* **1994**, *14*, 259–273. [CrossRef]
10. Kurt, S.K.; Gelhausen, M.G.; Kockmann, N. Axial Dispersion and Heat Transfer in a Milli/Microstructured Coiled Flow Inverter for Narrow Residence Time Distribution at Laminar Flow. *Chem. Eng. Technol.* **2015**, *38*, 1122–1130. [CrossRef]
11. Termühlen, M.; Strakeljahn, B.; Schembecker, G.; Wohlgemuth, K. Characterization of slug formation towards the performance of air-liquid segmented flow. *Chem. Eng. Sci.* **2019**, *207*, 1288–1298. [CrossRef]
12. Talimi, V.; Muzychka, Y.S.; Kocabiyik, S. A review on numerical studies of slug flow hydrodynamics and heat transfer in microtubes and microchannels. *Int. J. Multiph. Flow* **2012**, *39*, 88–104. [CrossRef]
13. Muzychka, Y.S.; Walsh, E.J.; Walsh, P. Heat Transfer Enhancement Using Laminar Gas-Liquid Segmented Plug Flows. *J. Heat Transf.* **2011**, *133*. [CrossRef]
14. Gaddem, M.R.; Ookawara, S.; Nigam, K.D.; Yoshikawa, S.; Matsumoto, H. Numerical modeling of segmented flow in coiled flow inverter: Hydrodynamics and mass transfer studies. *Chem. Eng. Sci.* **2021**, *234*, 116400. [CrossRef]
15. Dbouk, T.; Habchi, C. On the mixing enhancement in concentrated non-colloidal neutrally buoyant suspensions of rigid particles using helical coiled and chaotic twisted pipes: A numerical investigation. *Chem. Eng. Process. Process Intensif.* **2019**, *141*, 107540. [CrossRef]
16. Wiedmeyer, V.; Anker, F.; Bartsch, C.; Voigt, A.; John, V.; Sundmacher, K. Continuous Crystallization in a Helically Coiled Flow Tube: Analysis of Flow Field, Residence Time Behavior, and Crystal Growth. *Ind. Eng. Chem. Res.* **2017**, *56*, 3699–3712. [CrossRef]
17. Hohmann, L.; Schmalenberg, M.; Prasanna, M.; Matuschek, M.; Kockmann, N. Suspension flow behavior and particle residence time distribution in helical tube devices. *Chem. Eng. J.* **2019**, *360*, 1371–1389. [CrossRef]
18. Jiang, M.; Zhu, Z.; Jimenez, E.; Papageorgiou, C.D.; Waetzig, J.; Hardy, A.; Langston, M.; Braatz, R.D. Continuous-Flow Tubular Crystallization in Slugs Spontaneously Induced by Hydrodynamics. *Cryst. Growth Des.* **2014**, *14*, 851–860. [CrossRef]
19. Su, M.; Gao, Y. Air–Liquid Segmented Continuous Crystallization Process Optimization of the Flow Field, Growth Rate, and Size Distribution of Crystals. *Ind. Eng. Chem. Res.* **2018**, *57*, 3781–3791. [CrossRef]
20. Jiang, M.; Braatz, R.D. Low-Cost Noninvasive Real-Time Imaging for Tubular Continuous-Flow Crystallization. *Chem. Eng. Technol.* **2018**, *41*, 143–148. [CrossRef]
21. Besenhard, M.O.; Neugebauer, P.; Scheibelhofer, O.; Khinast, J.G. Crystal Engineering in Continuous Plug-Flow Crystallizers. *Cryst. Growth Des.* **2017**, *17*, 6432–6444. [CrossRef]
22. Scheiff, F.; Agar, D.W. Solid Particle Handling in Microreaction Technology: Practical Challenges and Application of Microfluid Segments for Particle-Based Processes. In *Micro-Segmented Flow: Applications in Chemistry and Biology*; Köhler, J.M., Cahill, B.P., Eds.; Springer: Berlin/Heidelberg, Germany, 2014; pp. 103–148. [CrossRef]
23. Scheiff, F. Fluiddynamik, Stofftransport und chemische Reaktion der Suspensionskatalyse bei der Flüssig/flüssig-Pfropfenströmung in Mikrokanälen. Ph.D. Thesis, TU Dortmund University, Dortmund, Germany, 2015. [CrossRef]
24. Termühlen, M.; Strakeljahn, B.; Schembecker, G.; Wohlgemuth, K. Quantification and evaluation of operating Parameters' effect on suspension behavior for slug flow crystallization. *Chem. Eng. Sci.* **2021**, *243*, 116771. [CrossRef]
25. Termühlen, M.; Etmanski, M.M.; Kryschewski, I.; Kufner, A.C.; Schembecker, G.; Wohlgemuth, K. Continuous slug flow crystallization: Impact of design and operating parameters on product quality. *Chem. Eng. Res. Des.* **2021**, *170*, 290–303. [CrossRef]
26. Borrero-Echeverry, D.; Crowley, C.J.; Riddick, T.P. Rheoscopic fluids in a post-Kalliroscope world. *Phys. Fluids* **2018**, *30*, 087103. [CrossRef]
27. Goto, S.; Kida, S.; Fujiwara, S. Flow visualization using reflective flakes. *J. Fluid Mech.* **2011**, *683*, 417–429. [CrossRef]
28. Wohlgemuth, K.; Schembecker, G. Modeling induced nucleation processes during batch cooling crystallization: A sequential parameter determination procedure. *Comput. Chem. Eng.* **2013**, *52*, 216–229. [CrossRef]
29. Ostermann, M.C.; Termühlen, M.; Schembecker, G.; Wohlgemuth, K. Growth Rate Measurements of Organic Crystals in a Cone-Shaped Fluidized-Bed Cell. *Chem. Eng. Technol.* **2018**, *41*, 1165–1172. [CrossRef]
30. Lührmann, M.C.; Termühlen, M.; Timmermann, J.; Schembecker, G.; Wohlgemuth, K. Induced nucleation by gassing and its monitoring for the design and operation of an MSMPR cascade. *Chem. Eng. Sci.* **2018**, *192*, 840–849. [CrossRef]
31. FeatFlow Homepage. Available online: www.featflow.de (accessed on 30 July 2021).
32. Bayraktar, E.; Mierka, O.; Turek, S. Benchmark computations of 3D laminar flow around a cylinder with CFX, OpenFOAM and FeatFlow. *Int. J. Comput. Sci. Eng. (IJCSE)* **2012**, *7*, 253–266. [CrossRef]
33. Schlüter, M.; Kexel, F.; von Kameke, A.; Hoffmann, M.; Herres-Pawlis, S.; Klüfers, P.; Ossberger, M.; Turek, S.; Mierka, O.; Kockmann, N.; et al. *Reactive Bubbly Flows, Fluid Mechanics and Its Applications*; Chapter Visualization and Quantitative Analysis of Consecutive Reactions in Taylor Bubbles Flows; Springer: Cham, Switzerland, 2021; ISBN 978-3-030-72361-3.
34. Münster, R.; Mierka, O.; Turek, S. Finite element-fictitious boundary methods (FEM-FBM) for 3D particulate flow. *Int. J. Numer. Meth. Fluids* **2012**, *69*, 294–313. [CrossRef]
35. Qiu, T.; Lee, T.C.; Mark, A.G.; Morozov, K.I.; Münster, R.; Mierka, O.; Turek, S.; Leshansky, A.M.; Fischer, P. Swimming by reciprocal motion at low Reynolds number. *Nat. Commun.* **2014**, *5*, 5119. [CrossRef]
36. Xiao, W.; Jin, T.; Luo, K.; Dai, Q.; Fan, J. Eulerian–Lagrangian direct numerical simulation of preferential accumulation of inertial particles in a compressible turbulent boundary layer. *J. Fluid Mech.* **2020**, *903*, A19. [CrossRef]

37. Schiller, L.; Naumann, Z. A Drag Coefficient Correlation. *Zeitschrift des Vereins Deutscher Ingenieure*; VDI-Verlag: Düsseldorf, Germany, 1935; pp. 77–318.
38. Turek, S.; Mierka, O. Chapter 7-Numerical simulation and benchmarking of drops and bubbles. In *Geometric Partial Differential Equations-Part II*; Handbook of Numerical Analysis; Bonito, A., Nochetto, R.H., Eds.; Elsevier: San Diego, CA, USA, 2021; Volume 22, pp. 419–465. [CrossRef]
39. Nur, H.; Ikeda, S.; Ohtani, B. Phase-Boundary Catalysis of Alkene Epoxidation with Aqueous Hydrogen Peroxide Using Amphiphilic Zeolite Particles Loaded with Titanium Oxide. *J. Catal.* **2001**, *204*, 402–408. [CrossRef]
40. Pieranski, P. Two-Dimensional Interfacial Colloidal Crystals. *Phys. Rev. Lett.* **1980**, *45*, 569–572. [CrossRef]
41. Thulasidas, T.C.; Abraham, M.A.; Cerro, R.L. Flow patterns in liquid slugs during bubble-train flow inside capillaries. *Chem. Eng. Sci.* **1997**, *52*, 2947–2962. [CrossRef]
42. Bayat, P.; Rezai, P. Semi-Empirical Estimation of Dean Flow Velocity in Curved Microchannels. *Sci. Rep.* **2017**, *7*, 13655. [CrossRef] [PubMed]

Review

Crystallization in Fluidized Bed Reactors: From Fundamental Knowledge to Full-Scale Applications

Marcelo Martins Seckler

Department of Chemical Engineering, Polytechnic School, University of São Paulo, São Paulo 05508-010, Brazil; marcelo.seckler@usp.br; Tel.: +55-11-9996-35142

Abstract: A review is presented on fifty years of research on crystallization in fluidized bed reactors (FBRs). FBRs are suitable for recovery of slightly soluble compounds from aqueous solutions, as it yields large, millimeter sized particles, which are suitable for reuse and permits low liquid residence times in the timescale of minutes. Full-scale applications for water softening have been applied since the 1980s, and since then, new applications have been developed or are in development for recovery of phosphorus, magnesium, fluoride, metals, sulfate, and boron. Process integration with membrane, adsorption, and biological processes have led to improved processes and environmental indicators. Recently, novel FBR concepts have been proposed, such as the aerated FBR for chemical-free precipitation of calcium carbonate, the seedless FBR to yield pure particulate products, a circulating FBR for economic recovery and extended use of seeds, as well as coupled FBRs for separation of chiral compounds and FBRs in precipitation with supercritical fluids. Advances are reported in the understanding of elementary phenomena in FBRs and on mathematical models for fluid dynamics, precipitation kinetics, and FBR systems. Their role is highlighted for process understanding, optimization and control at bench to full-scale. Future challenges are discussed.

Keywords: fluidized bed reactor; homogeneous granulation; crystallization from solutions; precipitation; wastewater treatment; water softening; phosphorus removal; struvite; chiral separations

1. Introduction

Industrial crystallization from solutions is applied as a separation operation or as a means of synthesizing particulate products. Four crystallization methods exist, depending on how the solution interacts with its environment to promote the formation of the solid phase: cooling, evaporative, antisolvent and chemical reaction crystallization. The choice of the crystallization method is primarily based on thermodynamics of multiphase systems [1], meaning the solubility of the crystallizing compound plays a dominant role. Moderately and highly soluble compounds are preferably processed by either cooling or evaporative crystallization because respectively a high yield and a low energy consumption are feasible. Unacceptably low yields would result for slightly soluble compounds, so for this class of compounds either chemical reaction (precipitation) or antisolvent crystallization is applied. The latter finds use only when high valued compounds are involved, as the cost of separating and recycling the antisolvent is high, so most slightly soluble compounds are processed by chemical reaction crystallization.

Precipitation equipment usually involves rapid mixing of the reactants (timescale of seconds or lower) followed by a long period (timescale of hours) when particles are allowed to develop until they meet requirements for downstream separation from the liquid. The rapid mixing step takes place either in a static mixing device or in a high turbulence zone within the crystallizer. The crystallizer may be a gently stirred tank or a static vessel. The logic behind this process arrangement is related to the kinetics of elementary phenomena of crystallization of slightly soluble compounds. In the mixing step fast primary nucleation (timescales << 1 s) and slow crystal growth (typically 10^{-8} to 10^{-10} ms^{-1}) take place, leading to particles in the nanometer to micrometer size ranges. In the crystallizer, such small particles ripen, i.e., they recrystallize, often as a different polymorph, and form large settleable agglomerates. Such agglomerates are difficult to dewater and retain impurities originally present either on their surface or in entrapped solution.

In fluidized bed reactors, the undesirable processing and quality issues of nanometer- and micrometer-sized particles are circumvented by the introduction of large, millimeter-sized seeds in fluidization. Precipitates are either formed directly upon the seeds or formed in solution and subsequently adhere to the seeds. Due to its large particle size, the product may be easily separated from the solution while retaining little moisture. The precipitate forms an important proportion of particulate, so economic recovery of the material is often possible. The liquid residence time is of only a few minutes, so the process is suitable for treatment of large streams.

An important issue in FBR design is the need to maximize retention of the precipitate on the seeds, thus minimizing solute loss as fine particles with the exiting solution. Much like in other precipitation options, FBR process design depends on specific features of each crystallizing system. Early industrial applications of FBRs were based on calcium carbonate crystallization for water softening [2,3], in calcium phosphate precipitation of phosphorus removal from domestic wastewaters [4,5] and in metal salts precipitation for heavy metal removal from industrial wastewaters [6,7]. Full-scale experiences have also been developed with an improved design, the circulating FBR, for softening of industrial water [8] and drinking water [9]. In the last decade, the interest in the topic has much increased, as Figure 1 illustrates. A few reviews have been done on specific applications of FBRs, such as in water softening [10,11] and phosphorus removal from waste streams by struvite crystallization [12,13]. An extensive and comprehensive review on FBRs has been published recently [14]. In this contribution, the fundamentals of the processes and process modeling work will be presented first. Then a range of applications for FBRs in the classic configuration as briefly explained above will be explored. Next, it will be shown how FBRs may be integrated with other unit operations to yield solutions to specific separation problems. Improved FBR concepts that have been proposed in recent years will be highlighted. Finally, challenges for the future will be discussed. Compared to previous publications, process fundamentals are covered in more depth, whereas integration of FBRs with other unit operations and improved reactor design concepts are new, so the present contribution offers a fresh improved cross section of FBR technology.

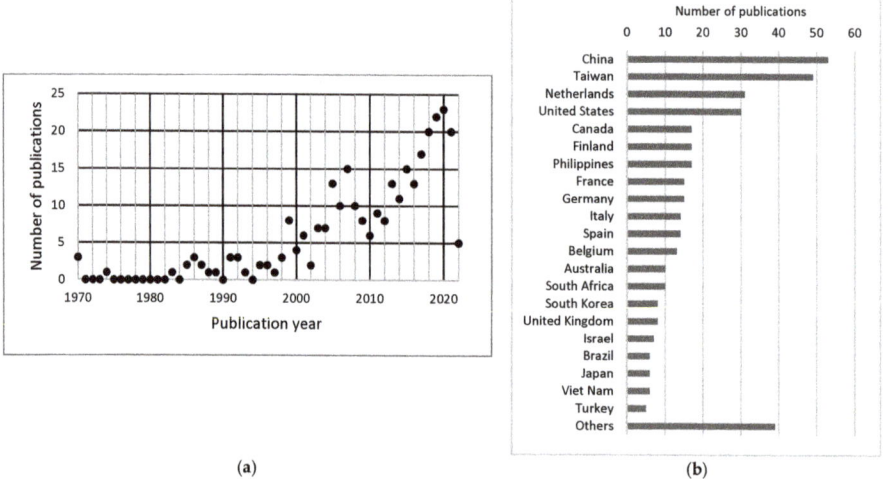

Figure 1. Number of peer reviewed publications by year (**a**) and by country (**b**). Source: Scopus, August 2022.

2. Fundamentals of FBRs and Mathematical Modeling

2.1. Main Features of FBRs and Performance Indicators

FBRs are used to separate a target compound from a liquid stream by chemical reaction crystallization. In its classic configuration (Figure 2a), reactants are mixed with the stream at the bottom of a fluidized bed. The bed is fed with seeds of a suitable size both for fluidization and for easy separation from the liquid downstream from the FBR. The seeds composition is often different from the crystallizing compound. Crystallization of the target compound takes place either directly upon the seed particles or in solution. In the latter case, most of the freshly formed precipitate adheres to the seeds. Two indicators conveniently characterize FBR performance (Figure 2b). The conversion χ is the proportion of the incoming target compound that converts to the solid phase and the recovery η gives the proportion of the incoming target compound that is retained in the seeds. Consequently, the target compound not retained in the seeds leave the system either in solution with mass fraction $1 - \chi$ or as fine particles with mass fraction $\chi - \eta$. It is desired to attain the highest possible conversion and a recovery that approaches the conversion, that is, most of the converted compound attaches to the grains. The conversion is controlled by the amount of chemicals added. Since FBRs are applied for low solubility compounds, conversions approaching 1.0 are feasible, but in some cases lower values result because a supersaturated solution leaves the crystallizer or because the formation of fines is minimized for a condition of partial conversion. Terminology in the literature is sometimes confusing, the terms removal, efficiency, crystal efficiency, granulation, among others, are used to mean either recovery or conversion. When the seeds are of the same composition as the crystallizing compound, or in case of an unseeded FBR, the term homogeneous FBR or homogeneous granulation FBR are commonly used. Grown seeds are usually called pellets, grains or simply particles.

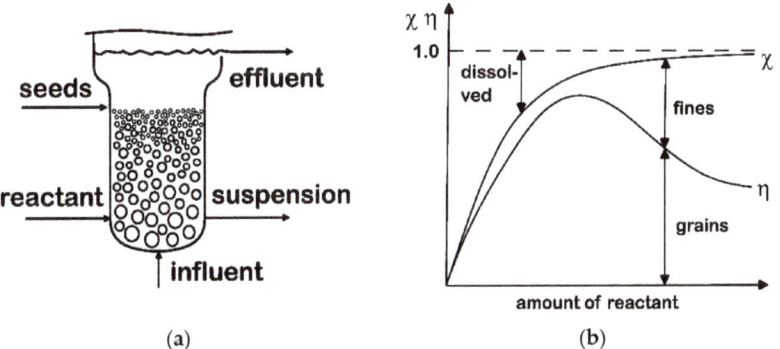

Figure 2. Fluidized bed reactor (**a**) and the performance indicators conversion χ and recovery η (**b**).

2.2. Elementary Processes of Crystallization

The first step in precipitation is the generation of supersaturation by rapid mixing of reactants with the influent at the lower part of the FBR. Two limiting behaviors may be distinguished with respect to the genesis of particle formation in FBRs depending on the supersaturation in the region where reactants are mixed and the metastable zone width of the solution. In the first one, the local supersaturation falls within the metastable zone width, so the main elementary process in the FBR is crystal growth upon the grains, possibly associated with epitaxial growth. This behavior is approached in practice for crystallization of calcium carbonate [15,16]. As an example, Figure 3 shows calcite crystals in a pellet from a full-scale reactor. In the second type of behavior, the local supersaturation exceeds the metastable zone, and primary heterogeneous nucleation takes place in solution at the bottom of the bed, so recovery occurs by aggregation of the resulting fine particles with the grains as they ascend within the bed. This behavior is usually found for compounds with extremely low solubility, such as amorphous calcium phosphate [17], nickel sulfide and copper sulfide [18], and nickel/cobalt sulfide mixtures [19]. Figure 4 illustrates this mechanism by the measurement of the fine particles' concentration profile in the fluidized bed: the fines are formed at the bottom and subsequently disappear as they aggregate upon the grains as they ascend. Struvite appears to exhibit intermediate behavior [20,21]. Fine particles may also be created by grains abrasion, particularly for seeds of large size or in regions of high turbulence, such as in the zone of reactants mixing. Abrasion may be easily noticed by a rounded shape and smooth surface of the grains, as illustrated in Figure 5a for grains collected from the bottom of a FBR. Figure 5b shows that grains at the top of the same bed, where turbulence is lower. In the absence of abrasion, particles display a rough, irregular surface. During start up, when the bed is filled with bare seeds, heterogeneous nucleation of the target compound upon the seeds surface controls recovery, which is lower than at steady operation. Such induction period usually last a few minutes [22] to a few hours [17].

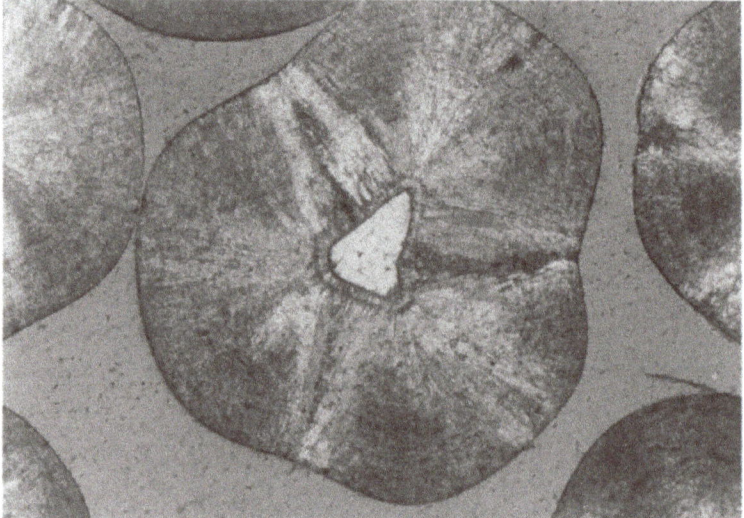

Figure 3. Photomicrograph in thin section of a pellet from a full-scale FBR. The small, clear particle at the center is the seed. The rest of the particle are calcite crystals displaying concentric layering and stellate clusters of coarser calcite. Field of view: 4 mm. Source [16].

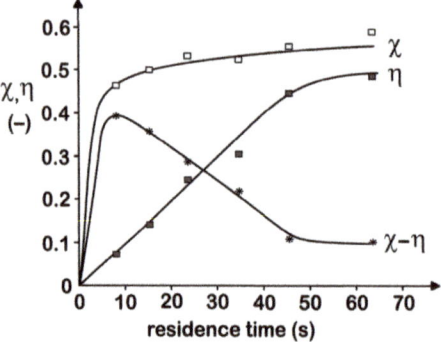

Figure 4. Conversion (χ), recovery (η), and fines ($\chi - \eta$) along bed height (expressed as residence time in the bed), suggesting that fines are formed at the bottom and subsequently adhere to the grains. Source: [23].

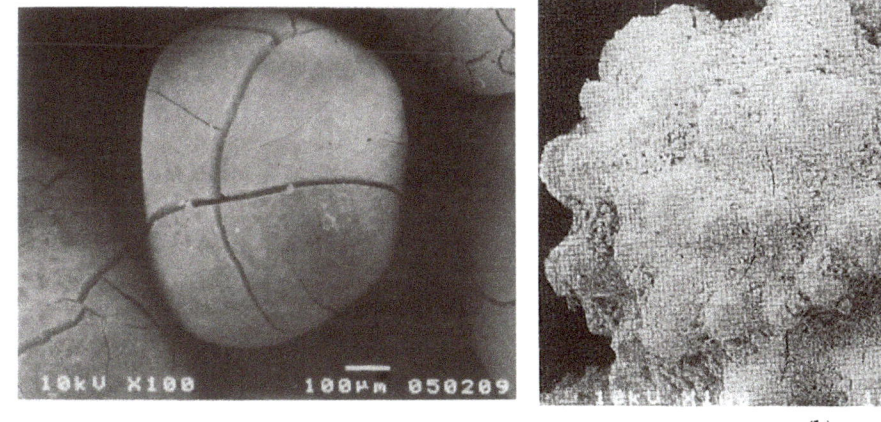

Figure 5. SEM view of (**a**) a highly abraded grain collected from the bottom of a FBR and (**b**) a non-abraded grain collected from the top of the same bed. The compound is amorphous calcium phosphate. Source: [24].

The specific surface area of the particles, expressed in m² per m³ fluidized bed is an important parameter in FBR design because a high value allows fast consumption of supersaturation that would otherwise promote undesired primary nucleation in solution as opposed to crystal growth upon the seed grains. Additionally, a high superficial velocity implies efficient use of the reactor volume for crystallization.

Mathematical models for elementary phenomena of a FBR are summarized in Table 1. Models that consider thermodynamic equilibrium and chemical speciation in solution are useful for establishing conditions for crystallization of the desired target compound [25] without coprecipitation of unwanted compounds [24] and to assess the influence of ligands on precipitation [26]. They also help to determine whether effluents leave the crystallizer as a supersaturated solution [27]. Particle growth and crystal growth kinetics in FBRs have been determined which are helpful for design, be it a kinetic expression for a two-step crystal growth as function of supersaturation [28,29] or an empirical correlation for the particle growth rate that includes dependencies for supersaturation, particle size and superficial velocity [30,31]. A kinetic model for aggregation of amorphous calcium phosphate allowed an improved understanding of the FBR operation [17]. A CFD-based population balance model has been developed to represent the nucleation and crystal growth at the bottom of a FBR [32]. The model predicts local supersaturations and particle sizes for calcium phosphate.

Table 1. Mathematical models for elementary phenomena in FBRs. Equilibrium: indicates that solubility, supersaturation and in some cases solution speciation are considered. Growth: indicates whether molecular crystal growth (indicated as "crystal") or overall particle growth, which encompasses crystal growth, aggregation and abrasion (indicated as "particle") are considered.

Crystallizing Compound	Elementary Phenomena	Authors	Year	Refs.
CaCO$_3$	Equilibrium, crystal growth	Hu et al.	2017	[15]
CaCO$_3$	Equilibrium, crystal growth	Rankin et al.	1999	[16]
Ca phosphate	Equilibrium, crystal growth, aggregation	Seckler	1994	[17]
NiS–CoS	Equilibrium, particle growth, aggregation	Lewis et al.	2006	[19]
Struvite	Equilibrium, metastable zone width, crystal growth	Bhuiygan	2008	[20]
Ca phosphate	Equilibrium	Seckler et al.	1996	[24]

Table 1. Cont.

Crystallizing Compound	Elementary Phenomena	Authors	Year	Refs.
Ca phosphate	Equilibrium	Montastruc et al.	2003	[25]
Struvite	Equilibrium	Iqbal et al.	2008	[26]
Struvite	Equilibrium	Xu et al.	2019	[27]
CaCO$_3$	Equilibrium, crystal growth	Tai et al.	1999	[28,29]
CaF$_2$	Equilibrium, particle growth	Aldaco et al.	2007	[30,31]
Ca phosphate	Equilibrium, nucleation, crystal growth, macromixing	Seckler et al.	1995	[32]

2.3. Liquid Solid Fluidization

Liquid solid fluidization is encountered in elutriation, leaching, crystallization of moderately soluble compounds, etc. Generally, a homogeneous bed of increasing height develops for superficial velocity's varying between the minimum fluidization velocity and the terminal falling velocity of the particles [33]. Table 2 summarizes mathematical models for fluidization in FBRs. Kramer et al. [34] have compared several models for predicting the voidage of monodispersed calcite particles to a wide range of experimental conditions. The best fitting was found for their explicit model of the form $\varepsilon = \varepsilon(Re, Fr)$, with ε, Re and Fr representing the bed voidage, the Reynolds and the Froude number, respectively. They have also suggested that such an explicit, simple model is the most suitable approach for monitoring and control of full-scale processes. Ye et al. [35] have used a CFD model to show that particles recirculate within the bed.

Table 2. Mathematical models for fluidization in FBRs. [1] Dimension 0D indicates that voidage is calculated in a homogeneous bed; 1D and 3D indicate particle segregation in one dimension (bed height) and three dimensions, respectively. [2] Particle segregation indicates whether particles are homogeneously distributed within the FBR, segregated by size, or by both size and density.

Crystallizing Compound	Dimensions [1]	Particle Segregation [2]	Authors	Year	Reference
CaCO$_3$	0D	None	Kramer et al.	2020	[34]
Struvite	3D	None	Ye et al.	2017	[35]
	1D	Size	Toyokura et al.	1973	[36]
NaBO$_3$·4H$_2$O	1D	Size	Frances et al.	1994	[37]
	1D	Size	Shiau et al.	2001	[38]
	3D	Size	Al-Rashed et al.	2009	[39]
	3D	Size	Al-Rashed et al.	2013	[40]
Struvite	3D	Size	Rahaman et al.	2018	[41]
Ca phosphate	1D	Size and density	Seckler	1994	[17]
CaCO$_3$	1D	Size and density	van Schagen et al.	2008	[42]
Enantiomers	3D	Size	Kerst et al.	2017	[43]

Particles of different sizes usually segregate, larger particles occupying lower portions of the fluidized bed. Segregation however is not perfect, so polydisperse materials occupy each axial position within the bed. This behavior has been addressed by Toyokura et al. [36] with an axial-dispersion model, whereas Frances et al. [37] and Shiau et al. [38] have used a mixed-tank in series approach. Al-Rashed et al. [39,40] have indicated with CFD simulations that the voidage and particle size changes both with axial and radial position within Oslo-fluidized bed crystallizers. In addition, upon scaleup at constant superficial velocity, voidage decreases due to cessation of fluid circulation flows. In a cylindrical bench scale crystallizer, both CFD and experiments have yielded a size-segregated bed with little dispersion [41].

If particle size and density vary, as is the case when the seeds and the crystallizing compound have different densities, stratification is more complex. Kennedy and Bretton [44] have considered that the driving force for segregation is the difference between the slip velocity of a given particle and the upward fluid velocity and included an axial dispersion coefficient that counteracts segregation. This basic idea has been pursued later with improvements in solution and accuracy [45,46]. Gibilaro et al. [47] established that the particles distribute in the bed in such a way that the potential energy is minimized. Seckler [17] has developed a mathematical model using this concept to adequately describe the composition and size of particles as functions of the height in a bench scale bed with sand pellets covered with amorphous calcium phosphate. Van Schagen et al. [42] have noticed that fluidization models do not accurately represent experimental data for calcium carbonate crystallization in pilot and industrial units. They have assumed that particles which size and density promote a higher pressure drop occupy a lower position in the bed. They have concluded that the calibrated Richardson Zaki model is the most accurate. Using this model, operational constraints have been established in terms of minimal and maximal pellet size at the bottom, for given superficial velocity and temperature. They have found that such operational constraints are often violated in practice, so they have recommended to use the model and to measure pressure drop at the bottom of the bed, instead of pressure drop of the total bed, as is usual.

2.4. Mathematical Models for the FBR Process

Models that describe the FBR process are summarized in Table 3. They combine fluid dynamics, mass balances, equilibrium, and crystallization kinetics, with time and special coordinates as independent variables. Models that describe the recovery and the grain size as functions of time and bed height have been proposed for crystallization of calcium carbonate [48], struvite [49–51], potassium struvite [52], calcium phosphate [53,54], gypsum [55,56], and tetrahydrate sodium perborate [37]. Precipitation is often modelled with the surface-controlled crystal growth rate with an expression of the type,

$$G = k_g(S-1)^g \quad (1)$$

where S is the supersaturation ratio, $S = c/c_{eq}$, with c and c_{eq} (kg·m^{-3}) as the active component concentrations in the supersaturated solution and in equilibrium, respectively, k_g and g are fit parameters, the latter being the growth order. Sometimes diffusion is considered the controlling mechanism for particle growth, so Equation (1) takes the form,

$$G = k_d(c - c_{eq}) \quad (2)$$

The mass transfer coefficient k_d has both been determined experimentally in a FBR [20] or estimated with semi-empirical correlations, such as the Froessling equation that relates the numbers of Sherwood, Reynolds and Schmidt [28]:

$$Sh = 0.6 Re^{\frac{1}{2}} Sc^{\frac{1}{3}} \quad (3)$$

Most often, particle abrasion and fines aggregation upon the grains are lumped into Equation (1), whereas primary nucleation is neglected. The removal of the active component from solution relates to the particle (overall) growth rate G with the mass balance for a horizontal slice of the bed,

$$\frac{dc}{dh} \sim \rho_{solid} G u^{-1} dA \quad (4)$$

where h (m) is the bed height, dA (m^2) is the interfacial area of the grains, u (m·s^{-1}) is the superficial velocity and ρ_{solid} is the solid density (kg·m^{-3}). The value of dA is derived from a fluid dynamics model, for which a fully segregated bed is the most straightforward option.

The calculations are performed for each differential slice of the bed at each time interval starting at the moment of seeds addition, implying that G, c and A are functions of h and t, $G = G(h,t)$, $c = c(h,t)$, $A = A(h,t)$. The recovery after a certain period of operation is calculated from the value of c at the top of the bed, whereas the grains size $L = L(h,t)$ is calculated from the seeds size and integrating $G(h,t)$ with time for each horizontal slice. Fines leaving the bed from the top are neglected. This type of model is useful for design when complemented by experimental data on bench or pilot-scale FBRs, as it is restricted to an operating range in which primary nucleation is negligible. It is suitable for systems in which recovery largely depends on crystal growth, such as in calcium carbonate and gypsum-based systems, but not for systems in which primary nucleation in solution and subsequent aggregation with the grains are the dominant phenomena, such as calcium phosphates.

Table 3. Mathematical models for FBRs. [1] Growth: indicates whether molecular crystal growth (indicated as "crystal") or overall particle growth, which encompasses crystal growth, aggregation and abrasion (indicated as "particle") are considered. [2] Dimensions 0D indicates that the independent variables are respectively particle size (L) and time (t); 1D indicates that they are L, t, and the bed height; 3D relates to models in L, t, and the three spatial dimensions. [3] Particle segregation indicates whether particles are homogeneously distributed within the FBR or segregated by size, or by both size and density.

Crystallizing Compound	Growth [1]	Dimensions [2]	Particle Segregation [3]	Other Phenomena	Authors	Year	References
Struvite	Particle	0D	None	aggregation	Bhuiyan et al.	2008	[20]
$CaCO_3$	Crystal	0D	None		Tai et al.	1999	[28]
$CaCO_3$	Crystal	1D	Size and density		Van et al.	2008	[48]
Struvite	Particle	1D	Size		Rahaman et al.	2014	[49]
Struvite	Particle	1D	Size		Rahaman et al.	2008	[50]
K struvite	Crystal	1D	Size		Zhang et al.	2017	[52]
Ca phosphate	Crystal	1D	Size	Fluid mixing, aggregation	Montastruc et al.	2003 2004	[53,54]
Gypsum	Crystal	0D	None		Choi et al.	2021	[55,56]
Ca phosphate	Crystal	3D	Size	Macromixing	Seckler et al.	1995	[32]
$CaCO_3$	Crystal	1D	Size and density		van Schagen et al.	2006	[57,58]
Struvite	No	3D	None	Nucleation, Macromixing	Ye et al.	2018	[59]
Struvite	Crystal	3D	Size	Macromixing	Rahaman et al.	2018	[41]
$BaCO_3$	Crystal	3D	Size	Nucleation, Micromixing, aggregation	Moguel et al.	2010	[60]
$CaCO_3$	Crystal	0D	None	CO_2 stripping	Segev et al.	2013	[61]
Enantiomers	Crystal	1D	Size	Nucleation, Breakage	Mangold et al.	2017	[62]
Enantiomers	Crystal	1D	Size	Nucleation, Breakage	Gänsch et al.	2021	[63]

Van Schagen et al. [57,58] have developed a model for process control of softening in full-scale FBRs based on crystal growth rate and complete grains segregation. A nonlinear model-predictive controller determines the values of the manipulated variables to keep the pellet size and bed height within desired range under varying operational conditions. They have found optimal values for the pellet size and for the bypass ratio (part of the influent is diverted because the FBR exceeds softening specifications) while maximizing bed height (to reduce residual supersaturation). The approach has been considered robust to inaccurate measurements.

Ye et al. [59] have developed a model for a FBR in three dimensions at unsteady state. The multiphase flow was modelled with a commercial Reynolds-averaged Navier-stokes

CFD code and particle mean sizes were determined with the moments transformation of the population balance. Crystal growth and primary nucleation kinetics with the literature parameters for struvite have been adopted. Phosphorus removal at various process conditions compared favorably with bench scale experiments, and a split inlet of reactants was suggested as useful for improved control of the supersaturation. The model is useful for performance evaluation and geometry optimization of FBRs in which aggregation is not an important phenomenon. It does consider the formation of small particles in solution. Rahaman et al. [41] have also developed a CFD model for struvite crystallization and compared it to pilot experiments. Moguel et al. [60] have developed a CFD model for barium carbonate precipitation in a FBR that included micromixing limited chemical acid–base instantaneous reaction, solution speciation, phase equilibrium, and kinetics for nucleation, growth, and aggregation. The model has been compared to pilot-scale experiments.

3. Applications

3.1. Calcium Carbonate

Calcium carbonate crystallization is applied in central softening of drinking water as it reduces scaling in household appliances and surfaces and reduces the need for cleaning agents and laundry detergents. Even though the FBR process requires energy and chemicals, it has been shown to be environmentally positive from life cycle assessment [64] and from a cost benefits [65] perspective. In addition, carbon capture in the crystallized calcite and dissolution of CO_2 into the softened water offers net carbon benefits of softening. The net total carbon footprint of drinking water softening in the Netherlands is estimated to be -0.11 Mton CO_2 eq./yr. [66].

Calcium carbonate crystallization in a FBR is the first and most developed application of FBR technology. It has already received reviews in the nineties [10,11,67], as well as a recent one [14]. Pilot studies and industrial scale applications of FBRs started in the 1970s [2,3]. In the present day there are full-scale units located in the Netherlands [34,68,69], in Saudi Arabia, in the United States of America [70], and in China [8,9]. An overview of full-scale and pilot-scale applications of FBRs is given in Tables 4 and 5, respectively. A full-scale circulating FBR has been used for softening of tower water in a thermal power plant [8], where softening enabled an increase in water circulation ratio from 4.5 to 9, to reduce anti scaling additive consumption by 30% and to use the pellets in the desulfurization unit of the power plant. The same FBR design has led to improved performance of full-scale in terms of hardness removal, time interval for pellets withdrawal, and cost [9].

The process is driven by adding a base to increase the water pH to values between 9 and 11. Calcium hydroxide is often used because of its low cost and to provide calcium if carbonate is originally in excess. If calcium is in excess, some sodium carbonate may be added [71]. Pellets are mainly constituted by calcite. This was for example the case for 16 water types at eight Danish drinking water treatment plants [72]. Besides calcite, they contained impurities, such as strontium, magnesium, iron, and sodium, each contributing with up to 1.3% of the pellet mass. The influent water composition did not change the calcium carbonate mineralogy but did correlate with the concentrations of many (but not all) impurities in the pellets. Quartz sand seeding material contributed with up to 15% of the pellet mass. The grains' low specific surface area of ≤ 0.32 m^2/g limits its potential use as soil amendment in agriculture [72]. In another study [71], garnet seeds were covered with calcium carbonate (97%) and magnesium hydroxide (3%). The polymorph aragonite was the main mineral in grains produced in a homogeneous (unseeded) FBR [73].

Rankin and Sutcliffe [16] have characterized the microstructure of calcite covering sand seeds in an industrial softening FBR. They have identified concentric banding inside the particles associated with perturbations in growth conditions induced by changes in pH inside the reactor. They have also found that individual "calcite crystals grow orthogonally from the surface of the seeds and in optical continuity across the concentric bands" and that "stellate clusters of larger crystals cross-cut these bands". Consequently, crystal growth is the dominant process in particle formation, with a growth rate of the order of 0.1 mm per day. A two-step growth model has been suggested [28], as well as semi-empirical rate laws for growth rate dependency of the supersaturation [15,74]. Most calcium and total hardness crystallize at the lowest part of the bed, within a bed height of about 1 m [8,11].

Fines lost with the effluent may result from abrasion of the pellets or from heterogeneous nucleation in solution. In a study with eight real groundwater types in a circulating FBR [71], fines were found to be predominantly calcium carbonate and magnesium carbonate particles. Fines formation could be hindered if Ca^{2+} < 180 mg/L, Mg^{2+} < 70 mg/L, and pH < 11.

Mathematical models have been developed for several aspects of a softening FBR, such as fluidization; see Section 2 in this manuscript.

3.2. Calcium Phosphates

Calcium phosphates crystallization is mostly applied to remove phosphate from wastewaters that would otherwise cause eutrophication of surface waters. Early studies on calcium phosphate precipitation in a FBR have been developed from bench to industrial scale [4,5,75]. The calcium phosphate compound formed in FBRs is an amorphous phase that may be represented by the chemical formula $Ca_3(PO_4)_2$ [23]. The process is controlled by aggregation of primary particles upon the seeds [23], so P recovery is improved by using multiple reactant inlet points to avoid high local supersaturations and to enhance aggregation [76]. A poorly crystalline hydroxyapatite is formed sometimes [77], probably by recrystallization of amorphous calcium phosphate at a low supersaturation and a long residence time [78]. A commonly used seed material is quartz sand, which provides an inert surface for phosphorus nucleation and retention in the bed. Seeding with MgO grains is more effective as it produces alkalinity near the grain surface, which promotes precipitation of carbonates and phosphates upon them [77]. Many wastewaters contain calcium and carbonate, which tend to co-precipitate with calcium phosphate, as the rise in pH is the driving force for precipitation of both compounds. The effect of carbonate on P recovery is controversial. Not only has a positive effect been found [79], but also a negative one [24]. Magnesium has been found to suppresses undesirable $CaCO_3$ coprecipitation [77]. Magnesium may also precipitate as $Mg_3(PO_4)_2 \cdot 22H_2O$, yielding a feasible process for phosphorus removal for waters with a low calcium content (Ca/P < 0.8 mol mol^{-1}) [24]. Fluorapatite, $Ca_{10}(PO_4)_6F_2$, may also precipitate in alkaline conditions upon phosphate rock and calcite seed particles [78]. Optimal conditions for phosphorus and fluoride removal are an upflow velocity of 0.12 m·min^{-1}, Ca:P:F molar ratio of 10:4:1.

A thermodynamic model for calcium phosphates precipitation for FBR conditions shows that for pH values below 7.3 dicalcium phosphate is formed, for higher pH values amorphous calcium phosphate develops [25]. Equilibrium modeling has also been applied to select conditions where coprecipitation of unwanted salts are avoided [24]. Such models offer possibilities for improving recovery, increasing product purity, and reducing effluent pH.

3.3. Struvite

Struvite ($NH_4MgPO_4 \cdot 6H_2O$) crystallization has received much attention in the context of wastewater treatment because it offers the possibility of simultaneously removing two commonly encountered nutrients, phosphorus and nitrogen. Bench scale studies have started in the turn of the millennium [80–84], and since then, pilot studies [26,85–91] as well as full-scale applications have been reported [92–94]. A FBR process is nowadays explored commercially with the tradename Ostara® [95].

Siciliano et al. [96] have recently published an extensive review that includes struvite crystallization in FBRs. The treated wastewater should not contain less than ~50 mg P/L, which roughly corresponds to the solubility of struvite. For this reason, this process has been studied for wastewaters containing high concentration of nutrients, such as swine [27,51,86,97–101] and chicken [102] wastewater, sludge lagoon wastewater [82], wastewater from vacuum toilet systems [103], wastewaters from industrial processes that use phosphoric acid, such as in the manufacture of electronics [104] and activated carbon [88], and wastewater from the dairy industry [105]. For effluents with low P content, it is possible to treat them with phosphorus-accumulating microorganisms, which subsequently release P in a higher concentration suitable for struvite crystallization, as is the case for supernatants from anaerobic digestion processes [20,49,50,80,83,84,89,90,94,106–108]. If the wastewater contains potassium, such as in source-separated urine, it is convenient to have potassium struvite ($KMgPO_4 \cdot 6H_2O$) as the target compound [52,87,109].

In the classic FBR configuration, phosphorus removal values above 90% are commonly found. Residence times vary within the range of a few minutes to a few hours due to recirculation of the effluent. Superficial velocities are generally within the range of 1 to 4 m·min^{-1}. Alkaline pH values, typically 8 to 10, are required to shift the chemical equilibrium of the reactions involving H^+, H_3PO_4, $H_2PO_4^-$, HPO_4^{2-} and PO_4^{3-} toward the latter species, which make up the crystal lattice of struvite. Magnesium is added stoichiometrically or in some excess, i.e., Mg:P molar ratios of one or slightly above it, to assure low concentration of the target compounds N and P. Seeding may be performed with struvite itself [104], with sand [81,110] or calcium phosphate [111], to yield a sub-millimetric particulate struvite product. Many process variations have been proposed to improve recovery, reduce chemicals' consumption, improve the quality of the struvite product, and reduce the residence time. They are addressed in Sections 4 and 5 in this manuscript.

Bhuiygan et al. [20] have studied the kinetics of struvite crystallization for conditions relevant for FBRs. They have determined the induction time as a function of supersaturation and the metastable zone width. Additionally, crystal growth rate kinetics have been determined in a bench scale fluidized bed. They have concluded that nucleation may be triggered at the bottom section of the bed and that the fresh crystals ascend to the top section, where a lower relative velocity should be provided so they do not escape the bed. As they grow, they gradually descend to lower portions of the bed. Nucleation and crystal growth rates have been determined in a mixed suspension crystallizer [112]. Ye et al. [113] have analyzed the morphology of struvite aggregates formed in unseeded jar tests. They have found that large, compact, mechanically strong aggregates are formed at a low phosphorus concentration and mild alkalinity. Fromberg and collaborators [21] have experimentally determined induction times and zeta potentials in struvite crystallization. They have concluded that a high recovery in FBRs might be attained by keeping a low supersaturation, because it results in low zeta potential values that favor agglomeration of primary nucleated struvite crystals upon the seeds. Consistent with these predictions, Gosh and collaborators [114] have experimentally determined that struvite recovery increases as the supersaturation decreases. Fang et al. [115] have conducted batchwise precipitation of struvite to determine its settleability, which is important for fines formation in FBRs. Ostwald ripening of struvite is suggested to explain why larger crystals develop at higher temperatures.

3.4. Other Phosphates

Besides calcium phosphate and struvite, other phosphates have been crystallized in FBRs. Barium hydrogen phosphate, $BaHPO_4$, and barium oxide, BaO, have been formed [116]. A Ba recovery of 98% was achieved at a pH of 8.5, a Ba/P molar ratio of one, and a superficial velocity of 0.46 m·min^{-1}. The addition of seeds had no effect on Ba recovery. Shih and collaborators [117] have added several alkaline earth metals but no seeds to remove phosphate from an electronics manufacturing wastewater with a high inlet P concentration of 100 ppm. It has been found that the pH determines the pseudopolymorph formed: a mixture of monetite ($CaHPO_4$) and brushite ($CaHPO_4 \cdot 2H_2O$) is recovered at pH 4.5, whereas hydroxyapatite ($Ca_5(PO_4)_3(OH)$) is formed at pH 7.5. Strontium and barium hydrogen phosphate ($SrHPO_4$ and $BaHPO_4$) crystalize at pH 4, whereas $Sr_5(PO_4)_3(OH)$ and $Ba_5(PO_4)_3(OH)$ are produced at pH 8. Magnesium phosphate crystallizes as newberite ($MgHPO_4$) and bobierrite ($Mg_3(PO_4)_2$). Cobalt precipitates as $Co_3(PO_4)_2 \cdot 8H_2O$ and a mixture of cobalt and nickel, with the latter crystallizing as $Cu_2(PO_4)OH$, by mixing two wastewaters from the semiconductor industry [118], one containing cobalt, the other one phosphate. EDTA and citrate ions, which are commonly found in these wastewaters, reduce recovery substantially.

3.5. Fluoride

Effluents containing fluoride are generated in fertilizer production (580–1680 ppm), optoelectronic and aluminum electroplating (100–6500 ppm), in municipal sewage treatment plants (5000–10,000 ppm), in aluminum trifluoride manufacture, and in groundwater (references given in [119]). The target compound chosen depends on the effluent specification in terms of fluoride concentration. Fluorite (CaF_2) is convenient for many industrial wastes as it yields effluents with ~10 ppm F^- at room temperature or ~5 ppm F^- at 80 °C. Cryolite (Na_3AlF_6) is sometimes considered as an intermediate step as it yields effluents with concentrations an order of magnitude higher. Fluorapatite has been considered for even lower effluent concentrations of about 2 ppm, such as for groundwater treatment aiming at drinking water supply [78,120]. A life cycle assessment study shows that fluorite (CaF_2) precipitation in a FBR compares favorably with precipitation in mixed tanks [121].

Calcium fluoride precipitates upon heterogeneous seeds by addition of a calcium source (calcium chloride or calcium hydroxide) at pH values in the range of 3 to 12. A neutral pH is preferred to avoid coprecipitation of compounds, such as phosphate and carbonate, which would take place at an alkaline pH. As an example, a FBR for calcium fluoride precipitation has been applied for wastewater from a rare-earth smelting wastewater [22]. Using silica sand as seeds, recoveries above 90% have been achieved with a particle size of about 1.5 mm. The superficial velocity was 2.4 m·min^{-1}, the recycle ratio of the effluent was 0.8.

The formation of CaF_2 fines is sensitive to supersaturation [122]. It has been largely avoided by keeping the saturation index (defined as the logarithm of the ratio of ion product to solubility product) at the bottom of the reactor below 3.5 [123]. Alternatively, it has been suggested to limit the supersaturation by using fluoride concentrations lower than 150 ppm at the bottom of the FBR after mixing of reactants [124]. If the influent concentration is higher, a recirculation stream may be applied, provided that fines in this stream are removed by filtration. If reintroduced into the FBR, the fines reduce recovery because they do not aggregate with the pellets, instead they promote crystallization of CaF_2 onto their surfaces [122]. Calcium fluoride precipitation has been applied even for high influent concentrations of 5.000 ppm F^- [119]. In such cases, another option, which has been applied at a pilot-scale, is to use two FBRs in a series, the first one recovers 70% of the fluorite as cryolite, the latter separates 90% of the remaining fluoride as its calcium salt [125]. Cryolite is formed upon addition of sodium hydroxide and aluminum hydroxide at pH values in the range of 2.5 to 7 [125].

Silica sand is suitable as a substrate for CaF_2 crystallization but not for economic use of the pellets in HF production, as part of the fluoride is lost as silicon tetrafluoride gas. Consequently, crystallization upon calcite grains has been proposed [124]. If conducted at an adequate pH, the seeds partly dissolve to form additional CaF_2 and release CO_2. After 230 h operation, it is possible to recover pellets containing 97% CaF_2 and only 3% calcite. This process has been yielded F^- recoveries of 70 to 80% [126].

Table 4. Full-scale investigations with FBRs.

Crystallizing Compound	Wastewater Source	Inlet Concentration (mM)	Superficial Velocity (m/min)	Bed Diameter (cm)	Recovery (%)	Seed Material	Grain Size (mm)	Note	Author Year	Refs.
$CaCO_3$	thermal power plant	0.4	0.83–1.8	240	0.5	garnet	0.1–0.3	(2)	Hu et al., 2019	[8]
$CaCO_3$	drinking water	1.9	1.0–1.7	160	0.9	garnet	0.53	(2)	Hu et al., 2018	[9]
$CaCO_3$	drinking water	n.a.	n.a.	n.a.	n.a.	n.a.	n.a.		several	[34,48,57,68,69]
$CaCO_3$	drinking water	n.a.	1.0–1.7	260	n.a.	garnet	1.2		van Schagen et al., 2008	[58]
$CaCO_3$	inland desalination	n.a.	n.a.	n.a.	n.a.	n.a.	n.a.		van Houwelingen et al., 2010	[70]
Ca phosphate	domestic, aerobic	0.16–0.32	0.67	250	77%	quartz	0.4		Seckler et al., 1990	[5]
Struvite	domestic, anaerobic	n.a.	n.a.	n.a.	n.a.	n.a.	0.5–1.0		Ueno et al., 2001	[92]
Struvite	domestic, anaerobic	1.1–2.2	1.3–2.6 min (1)	n.a.	77% P	struvite	0.5–5.0	(1)	Crutchik et al., 2017	[93]
Struvite	domestic, anaerobic	1.6	n.a.	1–8 m³/h (6)	60–70% P	quartz	0.2–0.35	(6)	Battistoni et al., 2006	[94]
Struvite	n.a.	n.a.	n.a.	n.a.	n.a.	n.a.	n.a.		Ostara et al., 2022	[95]

¹ Notes: n.a. not available, (1) hydraulic retention time, (2) circulating FBR, (6) influent flow rate.

Table 5. Pilot-scale investigations with FBRs.

Crystallizing Compound	Wastewater Source	Inlet Concentration (mM)	Superficial Velocity (m/min)	Bed Diameter (cm)	Recovery (%)	Seed Material	Grain Size (mm)	Note	Author Year	Refs.
CaCO$_3$	drinking water	n.a.	0.67–1.7	30	n.a.	garnet	0.25–1.8		van Schagen et al., 2008	[42]
CaCO$_3$	drinking water	10–400	1.8	60	70	garnet	1	(2)	Hu et al., 2021	[71]
CaCO$_3$	drinking water	3.2–6.1	1.3	9.9		quartz	1.0–1.2		Tang et al., 2019	[72]
CaCO$_3$	desalination	36	0.096	14	0.95	quartz	0.38 (seed)	(2)(5)	Segev et al., 2011	[127]
CaCO$_3$	drinking water	7.5	7.5 h (1)	7–38	0.75	Iron oxide	0.005	(1)(3)(5)	Li et al., 2004	[128]
CaCO$_3$	drinking water	n.a.	n.a.	n.a.			n.a.		several	[2,3,15]
CaCO$_3$	agriculture, industry and drinking water	10	20 min (1)	76	0.85	tartar scale, sand		(1)(4)	Neitzi et al., 2004	[129]
CaCO$_3$, Mg Silicate	coal gasification	10–400	0.25–0.50	130	50–82%	seedless	-		Hu et al., 2021	[130]
Ca phosphate	domestic, aerobic	0.24	0.9	20	80%	quartz	0.1–0.6	(2)	van Dijk et al., 1985	[4]
Ca phosphate	n.a.	0.06–0.74	3.1 min (1)	10	80%	phosphate rock	0.31 (seed)	(1)	Hirasawa I, Toya Y., 1990	[75]
CuCO$_3$ Cu(OH)$_2$	chemical, mining, and plating industry	0.47–4.7	15 min (1)	10	88–96%	quartz	0.26	(1)(5)	Lv et al., 2018	[131]
Fluorite	semiconductor industry	~1	n.a.	3 m^3/h	0.95	sand	-		van den Broeck et al., 2003	[123]
Fluorite	Fluoride industry	47–89	0.06	40–70	0.95	caco3	0.08–0.15		Jiang et al., 2017	[125]
Metal (basic) carbonates	plating and chemical industry	40–60,000 ppm metal	n.a.	10	>97%	n.a.	n.a.		Scholler et al., 1987	[7]
Struvite	domestic, anaerobic	n.a.	4	n.a.	75–85% P	struvite	0.5–3.5		Iqbal et al., 2008	[26]
Struvite	domestic, anaerobic	n.a.	n.a.	n.a.	75–90% P	struvite			Several	[49,85,89,90]
Struvite	domestic, anaerobic	6.5–10	n.a.	25–35	0.94	struvite	0.4		Shimamura et al., 2007	[91]
Struvite	domestic, anaerobic	n.a.	n.a.	1 m^3/h	95% P	n.a.	0.56		Meng et al., 2021	[106]
Struvite	source-separated human urine	10	1.0	21 L (7)	95% P	struvite	1.5–3.5	(7)	Zamora et al., 2017	[87]
Struvite	activated carbon industry	29–96	0.9–1.5	5000 L (7)	50–90% P	struvite	0.5	(7)	Ye et al., 2021	[88]
Struvite	swine	2.4–4.8	1.5–4.5	100 L (7)	>95% P	struvite	2.5	(7)	Ye et al., 2016	[86]
Struvite	swine	n.a.	1–5 h (1)	35	0.91	seedless	0.07	(1)(2)	Shim et al., 2020	[97]

^{1}Notes: n.a. not available, (1) hydraulic retention time, (2) circulating FBR, (3) FBR integrated with membrane operation, (4) aerated FBR, (5) synthetic wastewater, (6) influent flow rate, (7) reactor volume.

3.6. Metal Carbonates, Hydroxides, and Oxides

Application of FBRs for metals removal at full-scale has been realized since the 1980s to treat wastewaters from the metal plating, chlor-alkali, and other industries [6,7]. Relevant target metals were Cu, Ni, Co, Zn, among others. Metal precipitation as a carbonate is convenient because high recoveries are attainable, the metal may be readily recovered by acidification of the grains, and carbonate has a positive buffering effect on the treated water [6]. Recirculation is required for high influent concentrations. If the wastewater contains suspended solids, filtration upstream of the FBR is needed; otherwise, such solids act as seeds that promote formation of fine particles that are not retained in the FBR bed [6]. A list of the studies on metal removal with a FBR and some key processes and equipment parameters are given in Table 6.

Lewis [19,132] has found that both molecular growth and aggregation contribute to recovery from measurements of fines concentration along the bed height for the aqueous nickel hydroxy-carbonate system. She has also found that it is possible to favor growth by increasing the number of feed points and the effluent circulation rate, thereby increasing the recovery from 60 to 95%. For copper sulfide, the supersaturation was much higher, so this strategy was not effective, and recoveries were about 10% [19]. She has also found that low supersaturation results in decreased particle roughness, which reduces fines generation by abrasion of the grains. Lee et. al. [133] have studied the formation of copper carbonate in a FBR. They have found that particles formed by primary nucleation in solution aggregate to the seeds, forming a deposit that is porous and prone to abrasion. They have proposed to use low superficial velocities to reduce turbulence in the bed and to increase the specific surface of the grains. They have found that primary nucleation could be reduced by effective mixing of reactants at the bottom. A recovery of 96% was achieved. Lertratwattana et al. [134] have recovered copper from wastewater as the hydroxycarbonate of copper, malachite. The homogeneous FBR operated in two conditions to increase recovery: in the first 6 days a pH of 6.5 provided a low supersaturation suitable for initial development of the seeds, followed by a pH of eight. Lv et al. [131] have used a two-stage FBR to attain 96% recovery on a quartz seeded FBR. Wei et al. [135] have investigated $CuCO_3$ precipitation in a bench scale seeded FBR for 174 days. The removal efficiency was stable at 95%, but when the inlet supersaturation was allowed to increase, the efficiency decreased to 60–80% due to the generation of fines by primary nucleation in solution and seeds breaking. Adjusting of process conditions brought the system to the previous performance, showing the long-term feasibility of copper recovery. Copper and phosphate present in a wastewater from the semiconductor industry were removed as the basic phosphate Cu_2PO_4OH by increasing the solution pH to 6 in a homogeneous FBR [136].

Nickel hydroxy carbonate with chemical formula $Ni(OH)CO_3$ has been formed in FBRs at a pH of 9.7 via primary nucleation in solution followed by aggregation with seeds as well as by crystal growth upon the seeds. Reducing local supersaturation with multiple reactant feed points has increased recovery up to 99% [19,132]. At a higher pH value of 10.7 nullaginite, $Ni_2(OH)_2CO_3$ has also led to a high recovery in a homogeneous FBR [137].

Lead recoveries of 99% have been found attainable as the carbonate salt at pH values of 8 to 10 and a carbonate-to-lead molar ratio of 3, using either heterogeneous seeds [138,139] or lead carbonate seeds with 0.057 mm in size prepared in a separate batch setup [140]. The same recovery has been found for a much less concentrated solution (0.19 mM) at the pH and carbonate-to-lead ratio just mentioned. Sand has been used as seed at a recycle ratio of 0.67.

A limited number of studies exist for other metals. Zinc has been precipitated as a mixture of hydrozincite, $Zn_5(CO_3)_2(OH)_6$, and amorphous Zn compounds in a homogeneous FBR at a CO_3:Zn molar ratio of 1.2 and a pH of 7.2 [141]. It has been found that ferric and chloride ions result in less crystalline solids and lower recoveries. The homogeneous FBR has yielded higher recoveries than FBRs seeded with hydrozincite and with silica sand. Wastewater from the aluminum industry has been treated with a homogeneous FBR to precipitate crystalline aluminum oxide with stoichiometry $Al_{2.66}O_4$ at pH 10.4 [142];

see details in Section 5 of this manuscript. Other metals removed with a FBR are silver carbonate [143] and titanium dioxide [144].

For wastewaters containing a mixture of heavy metals, effluent concentrations are slightly higher than in the case of a single metal [6]. Zhou et al. [145] have removed the metals Cu, Ni and Zn in a single FBR as the metal hydroxides at pH 9 with recovery of 95%. Lee et al. [146] have recovered the metals Cu, Pb, and Ni as carbonates and hydroxides. The pH value in the effluent of the FBR ranged from 8.7 to 9.1. Sequential FBRs have been used, with most of the metal ions being collected in the first reactor. Depending on solubilities and concentrations, it may be possible to selectively separate one or more metals [6].

3.7. Metal Sulfides

Metal sulfides precipitation in FBRs is deemed difficult because they involve extremely high supersaturations that promote the formation of colloidal, highly charged particles [18]. In spite of these difficulties, high recoveries of copper sulfide have been attained using a semicontinuous mode and calcium coated seeds [147]. In addition, arsenic trisulfide, As_2S_3, has been removed with a high recovery after jar tests and experimental determination of the metastable zone width of the process [148]. Mixtures of cobalt and nickel sulfides were allowed to react with Na_2S in a FBR to yield recoveries of about 90% [149] without recirculation. Recoveries could be increased by using gaseous H_2S as reactant in a bubble FBR because slow gas dissolution has prevented local high supersaturation [149]. The process has been found suitable for wastewaters containing ~200 ppm metals but not for ~2.000 ppm.

3.8. Sulfate

Sulfate is commonly found in surface waters and in industrial waters, such as in ore beneficiation, paper mills, and fertilizer facilities. It forms scaling compounds that negatively interfere with water processing and with wastewater treatment with membrane operations. Sulfate also causes corrosion in sewers and negatively interferes with anaerobic wastewater treatment due to the action of sulfate reduction to sulfide by microbial action [150].

Sulfate removal in FBRs may be conducted with gypsum crystallization by reaction with added calcium chloride [151]. Recovery of gypsum in FBRs is probably controlled by crystal growth and abrasion of the grains [152]. Either alkaline or acid conditions may be chosen, which offers a good window to control coprecipitation of other impurities in the wastewater. For example, an influent solution containing sulfates of calcium, magnesium, and sodium with 1.1 mM total sulfate, has yielded grains containing both gypsum and magnesium hydroxide [153]. Ammonium or aluminum in solution favors a higher conversion [150]. Bare sand grains are not effective as a growth/aggregation substrate for gypsum, so in batch operation an initial period of low recovery may take place [154], while for continuous operation this material may be used. Another option is to use gypsum as seeds [151], as crystals of about 100 μm are easily developed in the FBR itself or in another crystallizer type. Seeding with silica or gypsum are equally effective for recovery [153]. For an influent concentration of 160 mM, optimal conditions are a Ca/SO_4 molar ratio of 1.48, a superficial velocity of 0.032 m·min^{-1} and a recirculation-to-feed ratio of 32, which yields a sulfate conversion of 82% and recovery of 67% [150]. Gypsum crystallization in FBRs is sometimes applied aiming at calcium removal [56].

3.9. Boron

Boron removal may be accomplished by adding hydrogen peroxide and a divalent metal to convert boric acid into a slightly soluble perborate salt. This chemistry has been applied to a homogeneous FBR to precipitate calcium perborate compounds [155]. Additionally, barium from another wastewater has been used as a reactant, having either silica sand or a waste-derived mesoporous aluminosilicate. The latter seed type was found superior due to its ability to adsorb barium ions on its surface, which promote heterogeneous nucleation of the target boron compound on its surface [156].

Table 6. Metals removal in FBR reactors.

Target Element	Inlet Concentration (mM)	Superficial Velocity (m/min)	Bed Diameter (cm)	Recovery (%)	Crystallizing Compound	Seed Material	Grain Size (mm)	Note	References
Al	12	0.25–0.6	2	97%	$Al_{2.66}O_4$	Seedless	0.25		[142]
Al	6.6	0.22	2	99%	$Al(OH)_3$ Bayerite	Seedless	0.5		[157]
As	500–1000 ppm	0.02	2	97%	As_2S_3 Arsenic trisulfide	Silica sand	-	Aerated FBR	[148]
B	185	0.42	3.5	60%	$CaB(OH)_3OOB(OH)_3$ and $Ca(B(OH)_3OOH)_2$ Amorphous calcium perborate	Seedless	1		[155]
B	200	0.05	2	93%	$BaB2(OO)_2(OH)_4$ barium peroxoborate	Mesoporous aluminosilicate		Barium adsorbs on seed surface	[156]
Ba	-	0.48	-	98%	$BaHPO_4$ and BaO at pH < 10 $Ba_3(PO_4)_2$ at pH 11	Seedless	0.8–1.0	Seeding had no effect	[116]
Ca	10	0.2	2	90%	$CaCO_3$	Seedless	1–2		[73,158]
Cu	10 ppm	0.41	3	96%	$CuCO_3$	Quartz	-		[133]
Cu	6–8	0.032	2	92%	$Cu_2CO_3(OH)_2$ malachite	Seedless	<0.3	Dual pH	[134]
Cu	20–50 ppm	0.22	3	95%	Probably $Cu_2CO_3(OH)_2$	Quartz	0.3	174 days	[135]
Cu P	6.5	0.032	2	96%	Cu_2PO_4OH	Seedless	1		[136]
Ni	100 ppm	0.24	2.5	99%	$Ni(OH)CO_3$	Quartz	-		[132,159,160]
Ni	200	0.16	2	98%	$Ni_2(OH)_2CO_3$ Nullaginite	Seedless	0.15–0.5	granules visible in 4–5 days	[137]
Pb	0.19	0.37		99%	$PbCO_3$	sand			[138].
Pb	2.4	0.16	4	99%	Cerussite $PbCO_3$	garnet			[139]

Table 6. Cont.

Target Element	Inlet Concentration (mM)	Superficial Velocity (m/min)	Bed Diameter (cm)	Recovery (%)	Crystallizing Compound	Seed Material	Grain Size (mm)	Note	References
Pb	48	5.6	5.2	99%	$PbCO_3$	$PbCO_3$		Seeds prepared in a batch reactor	[140]
Zn	7.6	0.9–5.4	2	86–92%	$Zn(CO_3)_2(OH)_6$ Hydrozincite	Seedless	0.04–0.15	Cl^- and Fe^{3+} decrease granule size	[141]
Cu Ni Zn	10–20 ppm	0.31	1	95%	Mixture of metal carbonates with metal hydroxides	Quartz			[145]
Cu Pb Ni	130 to 250 ppm			93 to 98%	Mixture of metal carbonates and hydroxides			FBRs in series	[146]
Co, Cu	30	0.7	2	96%	$Co_3(PO_4)_2 \cdot 8H_2O$ and $Cu_2(PO_4)OH$	Seedless	0.7		[118]
Ni Co					sulfides				[19,149]
F	0.5–250	0.07–0.6	2	98%	CaF_2	Seedless	<0.2		[119]
Oxalate	150–450			95%	$CaC_2O_4 \cdot H_2O$	Seedless	0.15		[161]
Oxalate	3	0.1	2	82%	$CaC_2O_4 \cdot H_2O$	Seedless	0.3–1.0		[162]
P	16	0.06	2	90%	$CaHPO_4 \cdot 2H_2O$ brushite	Seedless	0.5		[163]
P	100	-	-	86%	Mg_2PO_4OH and $Mg_3(PO_4)_2$	Seedless	<1	Recovery with seeds was 93%	[111]
S	80–160		5.2	66%	Gypsum $CaSO_4 \cdot 2H_2O$	Sand, gypsum			[150,151]

4. Integration with Other Unit Operations

4.1. FBR and Membrane Operations

Membrane operations, such as reverse osmosis and electrodialysis, are applied for water reuse within industry and as part of wastewater treatment. The water recovery is often limited by scaling of slightly soluble contaminants upon the membrane surfaces. Consequently, removing scaling components upstream of the membrane operation or between membrane stages leads to improved water recoveries. Precipitation techniques are so far the only options developed to a full-scale, the FBR has the advantages of a reduced footprint, chemical demand, and sludge generation, in spite of a higher capital cost [164]. The combination of FBRs with membranes has been applied for the treatment of wastewater [165], brackish groundwater [55,56,127,164,166], and drinking water [167,168]. Usually, the target compound is calcium carbonate [70,128,167,169,170], but other compounds may also be separated, such as calcium sulfate dihydrate [56], a mixture of calcium carbonate and nickel hydroxide [171], a mixture of calcium carbonate and magnesium compounds [165], as well as calcium phosphate [172].

Membranes have also been used to improve the performance of the FBR. In analogy with the membrane assisted suspension crystallizer originally proposed by Sluys et al. [173], Li and collaborators [128] have immersed an ultrafiltration membrane in the fluidized bed, so only the solution leaves the crystallizer.

Nanofiltration of seawater has been used to produce a low-cost magnesium solution, as this element is present at high concentrations in the oceans (~1400 mg/L) and is efficiently rejected by nanofiltration membranes. The solution has been fed to a FBR to crystallize struvite, $NH_4MgPO_4 \cdot 6H_2O$, reducing the cost of the process for treating the supernatant of a domestic-sludge dewatering facility [107]. As seawater contains other impurities for which nanofiltration rejection is variable, FBR pH and magnesium dosage had to be adjusted to avoid both undesirable calcium phosphate precipitation, and sodium and chloride excessive wastewater salinity. For a nanofiltration rejection of 90%, more than 90% of P removal was achieved, and the struvite purity was ~95%. Struvite precipitation was conducted in the presence of an antifouling agent, which was added to the NF brine to prevent the clogging of the NF membrane.

4.2. FBR and Adsorption

FBR and adsorption are combined by using an adsorber as seeds in the FBR. The method allows faster adsorption than in a fixed bed and potentially uses a waste as an adsorbate. The disadvantage is that the adsorbed compound is not recovered in concentrated form. Lee et al. [174] have seeded a FBR with the sludge from a water clarifier. Over 90% of the copper in the wastewater was removed by simultaneous adsorption and precipitation of copper hydroxide upon the sludge particles. During removal from organic compounds by Fenton oxidation and adsorption, simultaneous precipitation of Fe(III) hydroxides upon the seeds substantially reduced downstream sludge formation [175–178].

4.3. FBR and Absorption for Carbon Capture from Air

A proprietary process has been proposed [179] in which CO_2 is absorbed from air with a KOH solution to form a K_2CO_3 solution; see Figure 6. This solution is treated with $Ca(OH)_2$ in a FBR to separate CO_3 as $CaCO_3$ and to recover the KOH solution. The FBR yields the calcium carbonate as large particles that upon calcination are converted into the desired concentrated CO_2 suitable for further processing or storage and a solid CaO stream. This oxide then converted into $Ca(OH)_2$ that returns to the FBR. The FBR differs from those used in water softening because the carbonate concentration is much larger, ~0.5 against ~0.005 M, and the pH is higher. It has been found that for pH values in the range of 12 to 14 vaterite is formed, whereas for lower values calcite is the dominant polymorph, but carbonate recovery is not significantly affected by the pH [180].

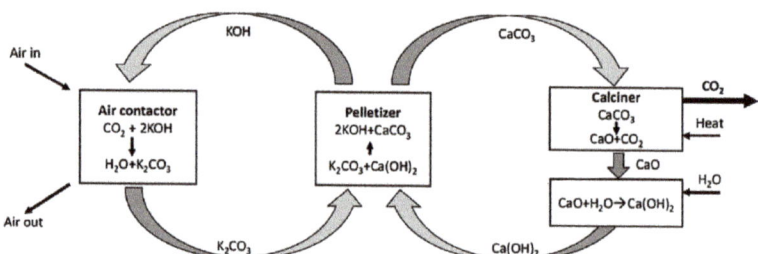

Figure 6. Process for carbon capture from air involving a FBR. Source [179].

4.4. FBR and Biological Process for Treatment of Acid Mine Drainage

Wastewater derived from mining operations are typically acidic and contain sulfate and metals that must be removed before discharge to surface waters. Sulfate reducing bacteria may produce hydrogen sulfide that subsequently reacts with metals to form metal sulfide precipitates. The reader is referred to nice reviews on this topic [181,182], here just a brief description is given. Biological sulfate reduction and chemical reaction crystallization may be simultaneously conducted in a fluidized bed bioreactor. This configuration has advantages in relation to other bioreactors because it offers a large surface area for microorganism growth, high mass transfer rate, among others. Much like in a conventional FBR, the metal sulfide precipitate may be found on the surface of the (biofilm) seeds, in the effluent as fine particles or in solution. The concentration in solution is usually very low because of the low metal sulfides solubilities. When an inverted fluidized bed design is used, the precipitates may be collected separately from the biomass. The inverted bed is designed with biofilm carriers of low density, so the influent flows downward, with the biomass being collected at the top and the metal precipitate at the bottom. It is also possible to precipitate the metals in a separate reactor, for instance, upstream from the biological FBR by recirculation of the effluent containing sulfide or H_2S. Metal sulfides precipitation is advantageous in comparison to conventional metal hydroxide precipitation because the required hydraulic retention times are lower, the effluent concentration is an order of magnitude lower, and the sludge is more compact and easier to dewater.

5. Improved FBR Concepts

5.1. Aerated FBR

Segev et al. [127] have proposed to crystallize calcium carbonate in a FBR by simply inserting an air stream into the fluidized bed. This is an improvement in relation to the well-developed softening operation which promotes crystallization with the aid of a chemical, usually calcium hydroxide. Calcium carbonate precipitation relies on the pH increase promoted by the release of part of the carbonate originally in solution as CO_2 to the air stream. This route is feasible only for solutions with a high carbonate content. The authors propose to apply the process in desalination of brackish waters for production of drinking water and to aid reverse osmosis operation by removing scale-forming calcium carbonate. A mathematical model for the process has been developed that is useful for design [61]. CO_2 stripping with air has also been used for pilot-scale decarbonation of geothermal waters [129], to keep a high pH during struvite crystallization in a FBR [110], for phosphate removal from dairy and swine wastewater as struvite or mixtures of struvite and calcium carbonate [100], and as a means of improving fluid dynamics during recovery in copper removal from metal industry wastewater [131].

5.2. Circulating FBR

Hu et al. [9] have introduced an annular zone within the cylindrical fluidized bed to promote circulation of the grains; see Figure 7. Consequently, the particle size does not vary with bed height. The authors conclude that this configuration requires a reduced bed height and allows for an improved recovery and an extended discharge time of grains.

Experimental data for two industrial scale units have been presented [8,9]. Another modification of the bed, including three concentric sections, has been applied for simultaneous removal of calcium, magnesium, and silica [130] by dosing precipitation agents (NaOH and MgO) and flocculants (aluminum sulfate and polyacrylamide). Circulation of the grains within the bed is also promoted by adding one of the reactants through a nozzle in the upward direction [97]. The upper part of the bed may have a conical shape to provide a settling zone, so only the smallest particles leave the system in the overflow. Recirculation of the FBR effluent has also been applied to retain part of the fine particles in the bed [97].

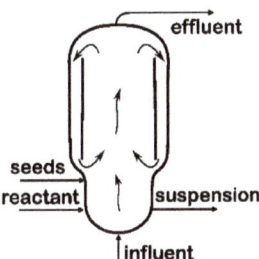

Figure 7. Circulating fluidized bed introduced by Hu et al. [9].

5.3. Homogeneous FBR

The basic idea of a classic FBR is to allow compounds that would otherwise develop as micrometer-sized particles to attach to large seed particles of a suitable support material. However, for some systems it has been shown that, even in the absence of seeds, it is possible to develop submillimeter- or millimeter-sized particles of the target compound in the FBR. During the startup period, which may last a few hours to a few days, a low superficial velocity and a high recirculation ratio of the suspension leaving the reactor may be required. Glass beads of a few centimeters in diameter are often added at the bottom at the point of the reactants addition to prevent clogging and formation of bubbles [157]. Once a fluid bed develops, operation is like a seeded FBR, with the advantage that the particles are constituted by the pure target compound. This type of crystallizer is known as homogeneous granulation FBR. Applications of homogeneous FBRs described in Table 6 encompass crystallization of fifteen compounds and their mixtures. Recoveries are generally above 90% and grains sizes range from 0.1 to 1 mm, thus similar to classic FBRs. The superficial velocities are highly variable; in general, they lie within the range of 0.05 to 0.5 m·min^{-1}. So far, the processes have been developed to the bench scale. Aluminum removal as bayerite (aluminum hydroxide) was conducted at a pH of 9.2 and a molar ratio of H_2O_2/Al of 2 [157]. A low superficial velocity 0.22 m·min^{-1} was used at the start to allow for growth of the fluidized particles, which were retained in an upper section of the bed with a diameter twice the man crystallizer body. Later, a superficial velocity of 0.5 m·min^{-1} and a recirculation-to-feed ratio of three were used. In another application, aluminum oxide with stoichiometry $Al_{2.66}O_4$ developed as 0.25 mm sized granules, but a startup period of 11 days was required [142]. Calcium fluoride has also been removed with large recovery even for high influent concentrations of 10,000 ppm F^- [119]. A calcium carbonate recovery of 90% and compact granules of 1 to 2 mm have been obtained, with the pH being the most important variable not only for recovery but also for particle size, shape, and crystallinity [73,158]. Zinc wastewater from the mechanical industry wastewater was treated to yield a recovery of about 90%, but chloride and ferric ions reduced the recovery and the granule size and wee incorporated in the particles [141]. Thin-film transistors liquid crystal display industry produces a phosphate-rich effluent, which yielded magnesium phosphate as a mixture of salts [111]. Barium removal with phosphates yielded different solid compounds depending on the pH. Homogeneous and seeded processes resulted in the same high recovery of 98% [116].

5.4. Separation of Enantiomers

Cooling crystallization is commonly applied for the separation of enantiomers from racemic mixtures. Crystallization of the desired enantiomer is promoted by seeding with the same material in a mixed suspension. Batch operation is usually applied for flexibility, but continuous configuration is potentially more economical and meets more stringent product quality demands. A continuous selective crystallization process has been proposed [183,184] that consists of two crystallizers, one for each enantiomer. The liquid recirculates among the two crystallizers, so its composition does not change much while allowing for a substantial yield. The FBR has been proposed as an alternative crystallizer [62] with some advantages. A slightly conical geometry enhances particle segregation by size, enabling extraction of the product of desired particle size in a conveniently located outlet port. Additionally, small particles that formed in solution, possibly of the undesired enantiomer, are readily removed with the upflowing liquid and redissolved in the feed tank. Further, particles larger than the product are removed from the bottom, crushed, and reinserted in the FBR as seeds, exempting the need for a dedicated unit for seeds' preparation. The process has also proven to offer higher yield (in mass of product per hour per volume crystallizer) than a batch process for the enantiomers of asparagine monohydrate [63,185]. CFD-DEM modeling has been applied to improve reactor geometry [43,186]. Besides enantiomers, also other stereoisomers (positional isomers OABA and PABA) could be continuously separated in FBRs [187].

5.5. FBR with Supercritical Fluids

Supercritical CO_2 has found increased use as an antisolvent for the production of micro- and nanoparticles in the pharmaceutical and food applications [188]. In the so-called RESS process (rapid expansion of supercritical solutions), the target compound is initially dissolved in supercritical CO_2 and precipitates as the fluid expands upon flowing through a nozzle. In the SAS process (supercritical antisolvent), the target compound is initially in a solution with a solvent and precipitates upon mixing with supercritical CO_2. The nanoparticles are not used as such, instead microencapsulated or nanoencapsulated particles are required to provide desirable bioavailability or surface properties [188]. Production of such capsules in FBRs combined with the RESS and SAS processes received attention in recent years. Fluidization with supercritical CO_2 has been found to obey well known correlations for fluidization [189]. The FBR-RESS process proposed in 1985 [190] consists of a circulating fluidized bed with an internal nozzle at the center of the riser, which provides rapid expansion of a supercritical fluid solution. The fluid in expansion provides fluidization of seeds and promotes precipitation of the target compound upon the seeds. The method has been applied for five compounds that mimic actual active pharmaceuticals on microcrystalline cellulose seeds, which is a well-known excipient, suggesting that the method may be applied to many active compounds [191]. In the FBR-SAS process, the supercritical CO_2 is added at the bottom of the fluidized bed through a sintered plate distributor to provide fluidization of excipient particles. A solution of the active compound is added at the bottom, where it mixes with CO_2 and precipitates upon the excipient particles [192]. The method has been applied to the deposition of the active compound narigin on the excipient microcrystalline cellulose [192], curcumin on lactose [193], curcumin and poly (vinyl pyrrolidone) on microcrystalline cellulose (MCC, 175 µm), corn starch (15 µm) and lactose (<5 µm) [194], and sirolimus on microcrystalline cellulose, lactose, and sucrose [195]. It has been suggested that the approach could be transferred to other industries where release is important, such as agrochemical, cosmetic, and food [192].

6. Conclusions

In fluidized bed reactors, the undesirable processing features of micrometer-sized particles are circumvented by the use of large, millimeter-sized seeds. The resulting large particles may be easily separated from the solution and used as a valuable product. Liquid residence times of only a few minutes are commonly found, so the process is suitable for treatment of large streams. Full-scale applications for softening of drinking water are in use since the 1980s. Since then, full-scale units were implemented for phosphate removal from wastewaters as struvite and calcium phosphate, as well as for metal removal from wastewater. Full-scale experiences exist with the improved design of the circulating FBR for softening of industrial water and drinking water. Additionally, new applications have been developed or are in development for recovery of fluoride, sulfate, boron, among others.

Process integration with other unit operations have been proposed to address a number sustainability demands, as shown in the following examples. In desalination and in water treatment for reuse with zero liquid discharge, metals precipitation in the FBR are combined with membrane processes to reduce scaling upon membrane surfaces, thereby reducing energy demand. FBRs seeded with an adsorbent offer perspective of improved recovery and use of waste materials as seeds. Aiming at carbon capture from air, calcium carbonate precipitation in FBRs is combined with absorption and calcination operations to form an efficient closed system. In the treatment of acid mine drainage, metals sulfide precipitation in FBRs are integrated with biogenically produced H_2S and HCO_3 with low energy and chemicals demand.

Recently, novel FBR concepts have been proposed: the aerated FBR allows chemical-free precipitation of calcium carbonate for wastewaters containing high carbonate content; a seedless FBR, known as the homogeneous granulation FBR, yields a pure particulate product; the circulating FBR offers economic recovery of the target compound and extended utilization of the seeds; coupled FBRs have been developed for separation of chiral compounds; FBRs with supercritical fluids precipitate core-shell particles suitable as pharmaceutical formulations.

Fundamental studies have elucidated the elementary steps of crystallization in FBRs. Crystal growth is the dominant process for the formation of calcium carbonate, whereas agglomeration is important for calcium phosphate, metal sulfides and struvite particles. Grains in a FBR have varying density and size, so they tend to segregate to minimize the potential energy, but they also mix due to turbulence. These phenomena have been incorporated in population, mass, and energy balances to describe certain experimental systems. Consequently, mathematical models have been extensively used for process improvement, optimization, and control of bench, pilot, and full-scale units.

7. Future Perspectives

Crystallization of a few compounds in FBRs is well established, but many opportunities exist to expand the list of crystallization systems to be applied at full-scale. To this end, many existing nice studies linking fundamentals and mathematical models with pilot-scale experiments might serve as inspiration to support the development of robust processes with sufficient maturity for scaleup.

There is room for the development of new applications of FBRs by integration with other unit operations such as membrane, adsorption, and absorption. In such cases, knowledge of FBR behavior for existing compounds will have to be extended to new conditions. When precipitation occurs simultaneously to adsorption and microorganisms' growth, little attention has so far been paid to the precipitation aspects of the process.

Recent promising developments such as the aerated FBR and the homogeneous granulation FBR have so far been developed to the bench scale, more research on a larger scale and on the fundamentals of these processes are needed. The circulating FBR has already been applied in full-scale for calcium carbonate precipitation, extension for other systems would be welcome.

The fluidization behavior of calcium carbonate seeded systems has been thoroughly studied. For other systems, more research is needed on the morphology, shape, and mechanical properties of the particles. The residence time of solids is in the order of days or even months, so long duration experiments are needed.

FBR mathematical models usually consider simple precipitation kinetics fitted to experiments with lumped parameters. A better knowledge is needed on the true kinetics of crystal growth, agglomeration, and abrasion. The roles of primary nucleation in solution, crystal growth upon the grains, and grains' abrasion on FBR performance is not always well understood. Additionally, many FBR processes include recirculation of the effluent, so fine particles are introduced in the reactor, contributing to recovery in poorly understood ways. "Single pass" performance indicators for FBRs, which would be useful for both process understanding and design, are scarce. Studies considering the true supersaturation along the bed height and about the metastable zone width would help elucidate the true kinetics in FBRs.

Funding: The author gratefully acknowledges the financial support of the RCGI—Research Centre for Greenhouse Gas Innovation, hosted by the University of São Paulo (USP) and sponsored by FAPESP—São Paulo Research Foundation (2014/50279-4 and 2020/15230-5 Project 66 GHG-3) and Shell Brazil, and the strategic importance of the support given by ANP (Brazil's National Oil, Natural Gas and Biofuels Agency) through the R&D levy regulation. The author also acknowledges the funding of FAPESP, Process numbers n° 20/15230-5, 17/19087-0, 14/14289-5.

Conflicts of Interest: The authors declare no conflict of interest.

References

1. Lewis, A.E.; Seckler, M.M.; Kramer, H.; Van Rosmalen, G.M. *Industrial Crystallization: Fundamentals and Applications*; Cambridge University Press: Cambridge, UK, 2015; ISBN 9781107280427.
2. Gledhill, E.G.; McCanlis, A.W.H. Softening of chalk well water and river water in a pellet reactor followed by upward flow sand filtration. *Water Treat. Exam* **1970**, *19*, 51–69.
3. Hilson, M.A.; Law, F. Softening of bunter sandstone waters and river waters of varying qualities in pellet reactors. *Water Treat. Exam* **1970**, *19*, 32–50.
4. Van Dijk, J.C.; Braakensiek, H. Phosphate Removal by Crystallization in a Fluidized Bed. *Water Sci. Technol.* **1985**, *17*, 133–142. [CrossRef]
5. Seckler, M.M.; Bruinsma, O.S.L.; Van Rosmalen, G.M.; Van Dijk, J.C.; Delgorge, F.; Eggers, J. Phosphate removal by means of a full scale pellet reactor. In Proceedings of the 11th Symposium Industrial Crystallization, Garmisch-Partenkirschen, Germany, 18–20 September 1990.
6. Schöller, M.; van Dijk, J.C.; van Haute, A.; Wilms, D.; Pawłowski, L.; Wasag, H. Recovery of Heavy Metals by Crystallization in the Pellet Reactor, a Promising Development. *Stud. Environ. Sci.* **1988**, *34*, 77–90. [CrossRef]
7. Scholler, M.; van Dijk, J.C.; Wilms, D. Recovery of heavy metals by crystallization. *Met. Finish.* **1987**, *85*, 31–34.
8. Hu, R.; Huang, T.; Wang, T.; Wang, H.; Long, X. Application of Chemical Crystallization Circulating Pellet Fluidized Beds for Softening and Saving Circulating Water in Thermal Power Plants. *Int. J. Environ. Res. Public Health* **2019**, *16*, 4576. [CrossRef]
9. Hu, R.; Huang, T.; Zhi, A.; Tang, Z. Full-scale experimental study of groundwater softening in a circulating pellet fluidized reactor. *Int. J. Environ. Res. Public Health* **2018**, *15*, 1592. [CrossRef]
10. Van Ammers, M.; Van Dijk, J.C.; Graveland, A.; Nuhn, P.A.N.M. State of the art of pellet softening in the Netherlands. *Water Supply* **1986**, *4*, 223–235.
11. Harms, W.D.; Robinson, R.B. Softening by fluidized bed crystallizers. *J. Environ. Eng.* **1992**, *118*, 513–529. [CrossRef]
12. Liu, Y.; Kumar, S.; Kwag, J.-H.; Ra, C. Magnesium ammonium phosphate formation, recovery and its application as valuable resources: A review. *J. Chem. Technol. Biotechnol.* **2013**, *88*, 181–189. [CrossRef]
13. Le Corre, K.S.; Valsami-Jones, E.; Hobbs, P.; Parsons, S.A. Phosphorus Recovery from Wastewater by Struvite Crystallization: A Review. *Crit. Rev. Environ. Sci. Technol.* **2009**, *39*, 433–477. [CrossRef]
14. Gui, L.; Yang, H.; Huang, H.; Hu, C.; Feng, Y.; Wang, X. Liquid solid fluidized bed crystallization granulation technology: Development, applications, properties, and prospects. *J. Water Process Eng.* **2022**, *45*, 102513. [CrossRef]
15. Hu, R.; Huang, T.; Wen, G.; Yang, S. Modelling particle growth of calcium carbonate in a pilot-scale pellet fluidized bed reactor. *Water Supply* **2017**, *17*, 643–651. [CrossRef]
16. Rankin, A.H.; Sutcliffe, P.J.C. Morphology, chemistry and growth mechanisms of calcite concretions from an industrial water-softening process: Implications for the origin of natural ooids in sediments. *Proc. Geol. Assoc.* **1999**, *110*, 33–40. [CrossRef]
17. Seckler, M.M. Calcium Phosphate Precipitation in a Fluidized Bed. Ph.D. Thesis, Delft University of Technology, Delft, The Netherlands, 1994.
18. Mokone, T.P.; van Hille, R.P.; Lewis, A.E. Metal sulphides from wastewater: Assessing the impact of supersaturation control strategies. *Water Res.* **2012**, *46*, 2088–2100. [CrossRef] [PubMed]
19. Lewis, A.E. Fines Formation (and Prevention) in Seeded Precipitation Processes. *KONA Powder Part. J.* **2006**, *24*, 119–125. [CrossRef]
20. Bhuiyan, M.I.H.; Mavinic, D.S.; Beckie, R.D. Nucleation and growth kinetics of struvite in a fluidized bed reactor. *J. Cryst. Growth* **2008**, *310*, 1187–1194. [CrossRef]

21. Fromberg, M.; Pawlik, M.; Mavinic, D.S. Induction time and zeta potential study of nucleating and growing struvite crystals for phosphorus recovery improvements within fluidized bed reactors. *Powder Technol.* **2020**, *360*, 715–730. [CrossRef]
22. Zeng, G.; Ling, B.; Li, Z.; Luo, S.; Sui, X.; Guan, Q. Fluorine removal and calcium fluoride recovery from rare-earth smelting wastewater using fluidized bed crystallization process. *J. Hazard. Mater.* **2019**, *373*, 313–320. [CrossRef]
23. Seckler, M.M.; Bruinsma, O.S.L.; Van Rosmalen, G.M. Phosphate removal in a fluidized bed—I. Identification of physical processes. *Water Res.* **1996**, *30*, 1585–1588. [CrossRef]
24. Seckler, M.M.; Bruinsma, O.S.L.; Van Rosmalen, G.M. Calcium phosphate precipitation in a fluidized bed in relation to process conditions: A black box approach. *Water Res.* **1996**, *30*, 1677–1685. [CrossRef]
25. Montastruc, L.; Azzaro-Pantel, C.; Biscans, B.; Cabassud, M.; Domenech, S. A thermochemical approach for calcium phosphate precipitation modeling in a pellet reactor. *Chem. Eng. J.* **2003**, *94*, 41–50. [CrossRef]
26. Iqbal, M.; Bhuiyan, M.I.H.; Mavinic, D.S. Assessing struvite precipitation in a pilot-scale fluidized bed crystallizer. *Environ. Technol.* **2008**, *29*, 1157–1167. [CrossRef] [PubMed]
27. Xu, K.; Ge, L.; Wang, C. Effect of upflow velocity on the performance of a fluidized bed reactor to remove phosphate from simulated swine wastewater. *Int. Biodeterior. Biodegrad.* **2019**, *140*, 78–83. [CrossRef]
28. Tai, C.Y.; Chien, W.-C.; Chen, C.-Y. Crystal growth kinetics of calcite in a dense fluidized-bed crystallizer. *AIChE J.* **1999**, *45*, 1605–1614. [CrossRef]
29. Tai, C.Y. Crystal growth kinetics of two-step growth process in liquid fluidized-bed crystallizers. *J. Cryst. Growth* **1999**, *206*, 109–118. [CrossRef]
30. Aldaco, R.; Garea, A.; Irabien, A. Modeling of particle growth: Application to water treatment in a fluidized bed reactor. *Chem. Eng. J.* **2007**, *134*, 66–71. [CrossRef]
31. Aldaco, R.; Garea, A.; Irabien, A. Particle growth kinetics of calcium fluoride in a fluidized bed reactor. *Chem. Eng. Sci.* **2007**, *62*, 2958–2966. [CrossRef]
32. Seckler, M.M.; Bruinsma, O.S.L.; Van Rosmalen, G.M. Influence of Hydrodynamics on Precipitation: A Computational Study. *Chem. Eng. Commun.* **1995**, *135*, 113–131. [CrossRef]
33. Richardson, J.F.; Backhurst, J.R.; Harker, J.H. Coulson & Richardson's Chemical engineering. In *Particle Technology and Separation Processes*; Buterworth Heinemann: Oxford, UK, 2002; Volume 2.
34. Kramer, O.J.I.; de Moel, P.J.; Padding, J.T.; Baars, E.T.; El Hasadi, Y.M.F.; Boek, E.S.; van der Hoek, J.P. Accurate voidage prediction in fluidisation systems for full-scale drinking water pellet softening reactors using data driven models. *J. Water Process Eng.* **2020**, *37*, 101481. [CrossRef]
35. Ye, X.; Chu, D.; Lou, Y.; Ye, Z.-L.; Wang, M.K.; Chen, S. Numerical simulation of flow hydrodynamics of struvite pellets in a liquid–solid fluidized bed. *J. Environ. Sci.* **2017**, *57*, 391–401. [CrossRef] [PubMed]
36. Toyokura, K.; Tanaka, H.; Tanahashi, J. Size Distribution of Crystals from Classified Bed Type Crystallizer. *J. Chem. Eng. Jpn.* **1973**, *6*, 325–331. [CrossRef]
37. Frances, C.; Biscans, B.; Laguerie, C. Modelling of a continuous fluidized-bed crystallizer effects of mixing and segregation on crystal size distribution during the crystallization of tetrahydrate sodium perborate. *Chem. Eng. Sci.* **1994**, *49*, 3269–3276. [CrossRef]
38. Shiau, L.-D.; Lu, T.-S. Interactive Effects of Particle Mixing and Segregation on the Performance Characteristics of a Fluidized Bed Crystallizer. *Ind. Eng. Chem. Res.* **2001**, *40*, 707–713. [CrossRef]
39. Al-Rashed, M.; Wójcik, J.; Plewik, R.; Synowiec, P.; Kuś, A. Multiphase CFD modeling-scale-up of Fluidized-Bed Crystallizer. *Comput. Aided Chem. Eng.* **2009**, *26*, 695–700. [CrossRef]
40. Al-Rashed, M.; Wójcik, J.; Plewik, R.; Synowiec, P.; Kuś, A. Multiphase CFD modeling: Fluid dynamics aspects in scale-up of a fluidized-bed crystallizer. *Chem. Eng. Process. Process Intensif.* **2013**, *63*, 7–15. [CrossRef]
41. Rahaman, M.S.; Choudhury, M.R.; Ramamurthy, A.S.; Mavinic, D.S.; Ellis, N.; Taghipour, F. CFD modeling of liquid-solid fluidized beds of polydisperse struvite crystals. *Int. J. Multiph. Flow* **2018**, *99*, 48–61. [CrossRef]
42. van Schagen, K.M.; Rietveld, L.C.; Babuška, R.; Kramer, O.J.I. Model-based operational constraints for fluidised bed crystallisation. *Water Res.* **2008**, *42*, 327–337. [CrossRef]
43. Kerst, K.; Roloff, C.; Medeiros de Souza, L.G.; Bartz, A.; Seidel-Morgenstern, A.; Thévenin, D.; Janiga, G. CFD-DEM simulations of a fluidized bed crystallizer. *Chem. Eng. Sci.* **2017**, *165*, 1–13. [CrossRef]
44. Kennedy, S.C.; Bretton, R.H. Axial dispersion of spheres fluidized with liquids. *AIChE J.* **1966**, *12*, 24–30. [CrossRef]
45. Patwardhan, V.S.; Tien, C. Distribution of solid particles in liquid fluidized beds. *Can. J. Chem. Eng.* **1984**, *62*, 46–54. [CrossRef]
46. Dutta, B.K.; Bhattacharya, S.; Chaudhury, S.; Barman, B. Mixing and segregation in a liquid fluidized bed of particles with different size and density. *Can. J. Chem. Eng.* **1988**, *66*, 676–680. [CrossRef]
47. Gibilaro, L.G.; Di Felice, R.; Waldram, S.P.; Foscolo, P.U. A predictive model for the equilibrium composition and inversion of binary-solid liquid fluidized beds. *Chem. Eng. Sci.* **1986**, *41*, 379–387. [CrossRef]
48. Van Schagen, K.M.; Rietveld, L.C.; Babusška, R. Dynamic modelling for optimisation of pellet softening. *J. Water Supply Res. Technol.* **2008**, *57*, 45–56. [CrossRef]
49. Rahaman, M.S.; Mavinic, D.S.; Meikleham, A.; Ellis, N. Modeling phosphorus removal and recovery from anaerobic digester supernatant through struvite crystallization in a fluidized bed reactor. *Water Res.* **2014**, *51*, 1–10. [CrossRef]

50. Rahaman, M.S.; Mavinic, D.S.; Ellis, N. Phosphorus recovery from anaerobic digester supernatant by struvite crystallization: Model-based evaluation of a fluidized bed reactor. *Water Sci. Technol.* **2008**, *58*, 1321–1327. [CrossRef]
51. Wang, J.; Ye, X.; Zhang, Z.; Ye, Z.-L.; Chen, S. Selection of cost-effective magnesium sources for fluidized struvite crystallization. *J. Environ. Sci.* **2018**, *70*, 144–153. [CrossRef]
52. Zhang, C.; Xu, K.; Li, J.; Wang, C.; Zheng, M. Recovery of Phosphorus and Potassium from Source-Separated Urine Using a Fluidized Bed Reactor: Optimization Operation and Mechanism Modeling. *Ind. Eng. Chem. Res.* **2017**, *56*, 3033–3039. [CrossRef]
53. Montastruc, L.; Azzaro-Pantel, C.; Biscans, B.; Cabassud, M.; Domenech, S.; Dibouleau, L. A General Framework for Pellet Reactor Modelling: Application to P-Recovery. *Chem. Eng. Res. Des.* **2003**, *81*, 1271–1278. [CrossRef]
54. Montastruc, L.; Azzaro-Pantel, C.; Pibouleau, L.; Domenech, S. A systemic approach for pellet reactor modeling: Application to water treatment. *AIChE J.* **2004**, *50*, 2514–2525. [CrossRef]
55. Choi, J.Y.; Kaufmann, F.; Rahardianto, A.; Cohen, Y. Corrigendum to 'Desupersaturation of RO concentrate and gypsum removal via seeded precipitation in a fluidized bed crystallizer' [Water Research 190 (2021)]. *Water Res.* **2021**, *193*, 116943. [CrossRef] [PubMed]
56. Choi, J.Y.; Kaufmann, F.; Rahardianto, A.; Cohen, Y. Desupersaturation of RO concentrate and gypsum removal via seeded precipitation in a fluidized bed crystallizer. *Water Res.* **2021**, *190*, 116766. [CrossRef] [PubMed]
57. van Schagen, K.M.; Babuška, R.; Rietveld, L.C.; Baars, E.T. Optimal flow distribution over multiple parallel pellet reactors: A model-based approach. *Water Sci. Technol.* **2006**, *53*, 493–501. [CrossRef] [PubMed]
58. van Schagen, K.; Rietveld, L.; Babuška, R.; Baars, E. Control of the fluidised bed in the pellet softening process. *Chem. Eng. Sci.* **2008**, *63*, 1390–1400. [CrossRef]
59. Ye, X.; Gao, Y.; Cheng, J.; Chu, D.; Ye, Z.-L.; Chen, S. Numerical simulation of struvite crystallization in fluidized bed reactor. *Chem. Eng. Sci.* **2018**, *176*, 242–253. [CrossRef]
60. Moguel, L.F.; Muhr, H.; Dietz, A.; Plasari, E. CFD simulation of barium carbonate precipitation in a fluidized bed reactor. *Chem. Eng. Res. Des.* **2010**, *88*, 1206–1216. [CrossRef]
61. Segev, R.; Hasson, D.; Semiat, R. Modeling $CaCO_3$ precipitation in fluidized bed CO_2 stripping desalination process. *Desalination* **2013**, *311*, 192–197. [CrossRef]
62. Mangold, M.; Khlopov, D.; Temmel, E.; Lorenz, H.; Seidel-Morgenstern, A. Modelling geometrical and fluid-dynamic aspects of a continuous fluidized bed crystallizer for separation of enantiomers. *Chem. Eng. Sci.* **2017**, *160*, 281–290. [CrossRef]
63. Gänsch, J.; Huskova, N.; Kerst, K.; Temmel, E.; Lorenz, H.; Mangold, M.; Janiga, G.; Seidel-Morgenstern, A. Continuous enantioselective crystallization of chiral compounds in coupled fluidized beds. *Chem. Eng. J.* **2021**, *422*, 129627. [CrossRef]
64. Godskesen, B.; Hauschild, M.; Rygaard, M.; Zambrano, K.; Albrechtsen, H.-J. Life cycle assessment of central softening of very hard drinking water. *J. Environ. Manag.* **2012**, *105*, 83–89. [CrossRef]
65. Van der Bruggen, B.; Goossens, H.; Everard, P.A.A.; Stemgée, K.; Rogge, W. Cost-benefit analysis of central softening for production of drinking water. *J. Environ. Manag.* **2009**, *91*, 541–549. [CrossRef] [PubMed]
66. Beeftink, M.; Hofs, B.; Kramer, O.; Odegard, I.; van der Wal, A. Carbon footprint of drinking water softening as determined by life cycle assessment. *J. Clean. Prod.* **2021**, *278*, 123925. [CrossRef]
67. van Dijk, J.C.; Wilms, D.A. Water treatment without waste material. Fundamentals and state of the art of pellet softening. *Aqua-J. Water Supply Res. Technol.* **1991**, *40*, 263–280.
68. Kramer, O.J.I.; van Schaik, C.; Hangelbroek, J.J.; de Moel, P.J.; Colin, M.G.; Amsing, M.; Boek, E.S.; Breugem, W.P.; Padding, J.T.; van der Hoek, J.P. A novel sensor measuring local voidage profile inside a fluidised bed reactor. *J. Water Process Eng.* **2021**, *42*, 102091. [CrossRef]
69. Schippers, D.; Kooi, M.; Sjoerdsma, P.; De Bruijn, F. Colour removal by ion exchange and reuse of regenerant by means of nanofiltration. *Water Sci. Technol. Water Supply* **2004**, *4*, 57–64. [CrossRef]
70. Van Houwelingen, G.; Bond, R.; Seacord, T.; Fessler, E. Experiences with pellet reactor softening as pretreatment for inland desalination in the USA. *Desalin. Water Treat.* **2010**, *13*, 259–266. [CrossRef]
71. Hu, R.; Huang, T.; Wen, G.; Tang, Z.; Liu, Z.; Li, K. Pilot study on the softening rules and regulation of water at various hardness levels within a chemical crystallization circulating pellet fluidized bed system. *J. Water Process Eng.* **2021**, *41*, 102000. [CrossRef]
72. Tang, C.; Hedegaard, M.J.; Lopato, L.; Albrechtsen, H.-J. Softening of drinking water by the pellet reactor—Effects of influent water composition on calcium carbonate pellet characteristics. *Sci. Total Environ.* **2019**, *652*, 538–548. [CrossRef]
73. Sioson, A.S.; Choi, A.E.S.; de Luna, M.D.G.; Huang, Y.-H.; Lu, M.-C. Calcium carbonate granulation in a fluidized-bed reactor: Kinetic, parametric and granule characterization analyses. *Chem. Eng. J.* **2020**, *382*, 122879. [CrossRef]
74. Sioson, A.S.; De Luna, M.D.G.; Lu, M.-C. A kinetic study of calcium carbonate granulation through fluidized-bed homogeneous process for removal of calcium-hardness from raw and tap waters. *Adv. Sci. Technol. Innov.* **2020**, 199–201. [CrossRef]
75. Hirasawa, I.; Toya, Y. *Fluidized-Bed Process for Phosphate Removal by Calcium Phosphate Crystallization*; ACS Symposium Series; ACS Publications: Washington, DC, USA, 1990; Volume Chapter 26–438, pp. 355–363. [CrossRef]
76. Seckler, M.M.; Van Leeuwen, M.L.J.; Bruinsma, O.S.L.; Van Rosmalen, G.M. Phosphate removal in a fluidized bed—II Process optimization. *Water Res.* **1996**, *30*, 1589–1596. [CrossRef]
77. Harris, W.G.; Wilkie, A.C.; Cao, X.; Sirengo, R. Bench-scale recovery of phosphorus from flushed dairy manure wastewater. *Bioresour. Technol.* **2008**, *99*, 3036–3043. [CrossRef] [PubMed]

78. Deng, L.; Wang, Y.; Zhang, X.; Zhou, J.; Huang, T. Defluoridation by fluorapatite crystallization in a fluidized bed reactor under alkaline groundwater condition. *J. Clean. Prod.* **2020**, *272*, 122805. [CrossRef]
79. Rugaika, A.M.; Van Deun, R.; Njau, K.N.; Van der Bruggen, B. Phosphorus Recovery as Calcium Phosphate by a Pellet Reactor Pre-Treating Domestic Wastewater Before Entering a Constructed Wetland. *Int. J. Environ. Sci. Technol.* **2019**, *16*, 3851–3860. [CrossRef]
80. Battistoni, P.; De Angelis, A.; Prisciandaro, M.; Boccadoro, R.; Bolzonella, D. P removal from anaerobic supernatants by struvite crystallization: Long term validation and process modelling. *Water Res.* **2002**, *36*, 1927–1938. [CrossRef]
81. Battistoni, P.; Fava, G.; Pavan, P.; Musacco, A.; Cecchi, F. Phosphate removal in anaerobic liquors by struvite crystallization without addition of chemicals: Preliminary results. *Water Res.* **1997**, *31*, 2925–2929. [CrossRef]
82. Ohlinger, K.N.; Young, T.M.; Schroeder, E.D. Postdigestion struvite precipitation using a fluidized bed reactor. *J. Environ. Eng.* **2000**, *126*, 361–368. [CrossRef]
83. Battistoni, P. Struvite crystallization: A feasible and reliable way to fix phosphorus in anaerobic supernatants. *Water Res.* **2000**, *34*, 3033–3041. [CrossRef]
84. Battistoni, P.; De Angelis, A.; Pavan, P.; Prisciandaro, M.; Cecchi, F. Phosphorus removal from a real anaerobic supernatant by struvite crystallization. *Water Res.* **2001**, *35*, 2167–2178. [CrossRef]
85. Britton, A.; Koch, F.A.; Mavinic, D.S.; Adnan, A.; Oldham, W.K.; Udala, B. Pilot-scale struvite recovery from anaerobic digester supernatant at an enhanced biological phosphorus removal wastewater treatment plant. *J. Environ. Eng. Sci.* **2005**, *4*, 265–277. [CrossRef]
86. Ye, X.; Ye, Z.-L.; Lou, Y.; Pan, S.; Wang, X.; Wang, M.K.; Chen, S. A comprehensive understanding of saturation index and upflow velocity in a pilot-scale fluidized bed reactor for struvite recovery from swine wastewater. *Powder Technol.* **2016**, *295*, 16–26. [CrossRef]
87. Zamora, P.; Georgieva, T.; Salcedo, I.; Elzinga, N.; Kuntke, P.; Buisman, C.J.N. Long-term operation of a pilot-scale reactor for phosphorus recovery as struvite from source-separated urine. *J. Chem. Technol. Biotechnol.* **2017**, *92*, 1035–1045. [CrossRef]
88. Ye, X.; Chen, M.; Wang, W.; Shen, J.; Wu, J.; Huang, W.; Xiao, L.; Lin, X.; Ye, Z.-L.; Chen, S. Dissolving the high-cost with acidity: A happy encounter between fluidized struvite crystallization and wastewater from activated carbon manufacture. *Water Res.* **2021**, *188*, 116521. [CrossRef] [PubMed]
89. Fattah, K.P.; Mavinic, D.S.; Koch, F.A.; Jacob, C. Determining the feasibility of phosphorus recovery as struvite from filter press centrate in a secondary wastewater treatment plant. *J. Environ. Sci. Health Part A* **2008**, *43*, 756–764. [CrossRef]
90. Bhuiyan, M.I.H.; Mavinic, D.S.; Koch, F.A. Phosphorus recovery from wastewater through struvite formation in fluidized bed reactors: A sustainable approach. *Water Sci. Technol.* **2008**, *57*, 175–181. [CrossRef] [PubMed]
91. Shimamura, K.; Ishikawa, H.; Tanaka, T.; Hirasawa, I. Use of a seeder reactor to manage crystal growth in the fluidized bed reactor for phosphorus recovery. *Water Environ. Res.* **2007**, *79*, 406–413. [CrossRef]
92. Ueno, Y.; Fujii, M. Three years experience of operating and selling recovered struvite from full-scale plant. *Environ. Technol.* **2001**, *22*, 1373–1381. [CrossRef] [PubMed]
93. Crutchik, D.; Morales, N.; Vázquez-Padín, J.R.; Garrido, J.M. Enhancement of struvite pellets crystallization in a full-scale plant using an industrial grade magnesium product. *Water Sci. Technol.* **2017**, *75*, 609–618. [CrossRef]
94. Battistoni, P.; Paci, B.; Fatone, F.; Pavan, P. Phosphorus Removal from Anaerobic Supernatants: Start-Up and Steady-State Conditions of a Fluidized Bed Reactor Full-Scale Plant. *Ind. Eng. Chem. Res.* **2006**, *45*, 663–669. [CrossRef]
95. Ostara. Available online: https://ostara.com/phosphates-production/ (accessed on 28 July 2022).
96. Siciliano, A.; Limonti, C.; Curcio, G.M.; Molinari, R. Advances in struvite precipitation technologies for nutrients removal and recovery from aqueous waste and wastewater. *Sustainability* **2020**, *12*, 7538. [CrossRef]
97. Shim, S.; Won, S.; Reza, A.; Kim, S.; Ahmed, N.; Ra, C. Design and optimization of fluidized bed reactor operating conditions for struvite recovery process from swine wastewater. *Processes* **2020**, *8*, 422. [CrossRef]
98. Chu, D.; Ye, Z.-L.; Chen, S.; Xiong, X. Comparative study of heavy metal residues in struvite products recovered from swine wastewater using fluidised bed and stirred reactors. *Water Sci. Technol.* **2018**, *78*, 1642–1651. [CrossRef] [PubMed]
99. Lee, C.-W.; Kwon, H.B.; Kim, Y.-J.; Jeon, H.-P. Nutrients recovery by struvite formation from wastewater in a fluidized bed reactor. *Mater. Sci. Forum* **2005**, *486–487*, 387–390. [CrossRef]
100. Rabinovich, A.; Rouff, A.A.; Lew, B.; Ramlogan, M.V. Aerated Fluidized Bed Treatment for Phosphate Recovery from Dairy and Swine Wastewater. *ACS Sustain. Chem. Eng.* **2018**, *6*, 652–659. [CrossRef]
101. Ryu, H.D.; Lim, D.Y.; Kim, S.J.; Baek, U.-I.; Chung, E.G.; Kim, K.; Lee, J.K. Struvite precipitation for sustainable recovery of nitrogen and phosphorus from anaerobic digestion effluents of swine manure. *Sustainability* **2020**, *12*, 8574. [CrossRef]
102. Al-Mallahi, J.; Sürmeli, R.Ö.; Çalli, B. Recovery of phosphorus from liquid digestate using waste magnesite dust. *J. Clean. Prod.* **2020**, *272*, 122616. [CrossRef]
103. Sun, H.; Hernandez, V.L.C.; Mohammed, A.N.; Liu, Y. Impact of Total Suspended Solids on Struvite Precipitation from Source-Diverted Blackwater. *J. Environ. Eng.* **2021**, *147*, 04020157. [CrossRef]
104. Shih, Y.-J.; Abarca, R.R.M.; de Luna, M.D.G.; Huang, Y.-H.; Lu, M.-C. Recovery of phosphorus from synthetic wastewaters by struvite crystallization in a fluidized-bed reactor: Effects of pH, phosphate concentration and coexisting ions. *Chemosphere* **2017**, *173*, 466–473. [CrossRef]

105. Rabinovich, A.; Heckman, J.R.; Lew, B.; Rouff, A.A. Magnesium supplementation for improved struvite recovery from dairy lagoon wastewater. *J. Environ. Chem. Eng.* **2021**, *9*, 105628. [CrossRef]
106. Meng, Y.; Li, Z.; Wang, L.; Li, Y. P-recovery as struvite from the enhanced anaerobic phosphorus release supernatant in a pilot-scale fluidized bed reactor and characterization of the product. *Chin. J. Environ. Eng.* **2021**, *15*, 1377–1385. [CrossRef]
107. Lahav, O.; Telzhensky, M.; Zewuhn, A.; Gendel, Y.; Gerth, J.; Calmano, W.; Birnhack, L. Struvite recovery from municipal-wastewater sludge centrifuge supernatant using seawater NF concentrate as a cheap Mg(II) source. *Sep. Purif. Technol.* **2013**, *108*, 103–110. [CrossRef]
108. Battistoni, P.; Pavan, P.; Cecchi, F.; Mata-Alvares, J. Effect of composition of anaerobic supernatants from an anaerobic, anoxic and oxic (A$_2$O) process on struvite and hydroxyapatite formation. *Ann. Chim.* **1998**, *88*, 761–772.
109. Wilsenach, J.A.; Schuurbiers, C.A.H.; van Loosdrecht, M.C.M. Phosphate and potassium recovery from source separated urine through struvite precipitation. *Water Res.* **2007**, *41*, 458–466. [CrossRef]
110. Jordaan, E.M.; Rezania, B.; Çiçek, N. Investigation of chemical-free nutrient removal and recovery from CO$_2$-rich wastewater. *Water Sci. Technol.* **2013**, *67*, 2195–2201. [CrossRef] [PubMed]
111. Su, C.-C.; Dulfo, L.D.; Dalida, M.L.P.; Lu, M.-C. Magnesium phosphate crystallization in a fluidized-bed reactor: Effects of pH, Mg:P molar ratio and seed. *Sep. Purif. Technol.* **2014**, *125*, 90–96. [CrossRef]
112. Hanhoun, M.; Montastruc, L.; Azzaro-Pantel, C.; Biscans, B.; Frèche, M.; Pibouleau, L. Simultaneous determination of nucleation and crystal growth kinetics of struvite using a thermodynamic modeling approach. *Chem. Eng. J.* **2013**, *215–216*, 903–912. [CrossRef]
113. Ye, Z.; Shen, Y.; Ye, X.; Zhang, Z.; Chen, S.; Shi, J. Phosphorus recovery from wastewater by struvite crystallization: Property of aggregates. *J. Environ. Sci.* **2014**, *26*, 991–1000. [CrossRef]
114. Ghosh, S.; Lobanov, S.; Lo, V.K. Impact of supersaturation ratio on phosphorus recovery from synthetic anaerobic digester supernatant through a struvite crystallization fluidized bed reactor. *Environ. Technol.* **2019**, *40*, 2000–2010. [CrossRef]
115. Fang, C.; Zhang, T.; Jiang, R.; Ohtake, H. Phosphate enhance recovery from wastewater by mechanism analysis and optimization of struvite settleability in fluidized bed reactor. *Sci. Rep.* **2016**, *6*, 32215. [CrossRef]
116. Su, C.-C.; Reano, R.L.; Dalida, M.L.P.; Lu, M.-C. Barium recovery by crystallization in a fluidized-bed reactor: Effects of pH, Ba/P molar ratio and seed. *Chemosphere* **2014**, *105*, 100–105. [CrossRef]
117. Shih, Y.-J.; Chang, H.-C.; Huang, Y.-H. Reclamation of phosphorus from aqueous solutions as alkaline earth metal phosphate in a fluidized-bed homogeneous crystallization (FBHC) process. *J. Taiwan Inst. Chem. Eng.* **2016**, *62*, 177–186. [CrossRef]
118. Bayon, L.L.E.; Ballesteros, F.C.; Choi, A.E.S.; Garcia-Segura, S.; Lu, M.C. Remediation of cobalt from semiconductor wastewater in the frame of fluidized-bed homogeneous granulation process. *J. Environ. Chem. Eng.* **2021**, *9*, 105936. [CrossRef]
119. Lacson, C.F.Z.; Lu, M.-C.; Huang, Y.-H. Chemical precipitation at extreme fluoride concentration and potential recovery of CaF$_2$ particles by fluidized-bed homogeneous crystallization process. *Chem. Eng. J.* **2021**, *415*, 128917. [CrossRef]
120. Deng, L.; Zhang, X.; Huang, T.; Zhou, J. Investigation of fluorapatite crystallization in a fluidized bed reactor for the removal of fluoride from groundwater. *J. Chem. Technol. Biotechnol.* **2019**, *94*, 569–581. [CrossRef]
121. Dominguez-Ramos, A.; Aldaco, R.; Irabien, A. Life Cycle Assessment as a Tool for Cleaner Production: Application to Aluminium Trifluoride. *Int. J. Chem. React. Eng.* **2007**, *5*. [CrossRef]
122. Aldaco, R.; Irabien, A.; Luis, P. Fluidized bed reactor for fluoride removal. *Chem. Eng. J.* **2005**, *107*, 113–117. [CrossRef]
123. Van den Broeck, K.; Van Hoornick, N.; Van Hoeymissen, J.; de Boer, R.; Giesen, A.; Wilms, D. Sustainable treatment of HF wastewaters from semiconductor industry with a fluidized bed reactor. *IEEE Trans. Semicond. Manuf.* **2003**, *16*, 423–428. [CrossRef]
124. Aldaco, R.; Garea, A.; Irabien, A. Fluoride Recovery in a Fluidized Bed: Crystallization of Calcium Fluoride on Silica Sand. *Ind. Eng. Chem. Res.* **2006**, *45*, 796–802. [CrossRef]
125. Jiang, K.; Zhou, K.G. Recovery and removal of fluoride from fluorine industrial wastewater by crystallization process: A pilot study. *Clean Technol. Environ. Policy* **2017**, *19*, 2335–2340. [CrossRef]
126. Aldaco, R.; Garea, A.; Fernández, I.; Irabien, A. Resources reduction in the fluorine industry: Fluoride removal and recovery in a fluidized bed crystallizer. *Clean Technol. Environ. Policy* **2008**, *10*, 203–210. [CrossRef]
127. Segev, R.; Hasson, D.; Semiat, R. Improved high recovery brackish water desalination process based on fluidized bed air stripping. *Desalination* **2011**, *281*, 75–79. [CrossRef]
128. Li, C.-W.; Jian, J.-C.; Liao, J.-C. Integrating membrane filtration and a fluidized-bed pellet reactor for hardness removal. *J. Am. Water Work. Assoc.* **2004**, *96*, 151–158. [CrossRef]
129. Nefzi, M.; Amor, M.B.; Maâlej, M. A clean technology for decarbonation of geothermal waters from Chott El Fejjej using a three-phase fluidized bed reactor-Modelling aspects. *Desalination* **2004**, *165*, 337–350. [CrossRef]
130. Hu, Y.; Liu, Z.; Huang, T.; Wen, G. Pilot test of simultaneous removal of silica, hardness, and turbidity from gray water using circulating pellet fluidized bed. *J. Water Process Eng.* **2021**, *42*, 102149. [CrossRef]
131. Lv, X.; Li, J.; Chen, H.; Tang, H. Copper wastewater treatment with high concentration in a two-stage crystallization-based combined process. *Environ. Technol.* **2018**, *39*, 2346–2352. [CrossRef] [PubMed]
132. Costodes, V.C.T.; Lewis, A.E. Reactive crystallization of nickel hydroxy-carbonate in fluidized-bed reactor: Fines production and column design. *Chem. Eng. Sci.* **2006**, *61*, 1377–1385. [CrossRef]
133. Lee, C.-I.; Yang, W.-F.; Hsieh, C.-I. Removal of Cu(II) from aqueous solution in a fluidized-bed reactor. *Chemosphere* **2004**, *57*, 1173–1180. [CrossRef]

134. Lertratwattana, K.; Kemacheevakul, P.; Garcia-Segura, S.; Lu, M.-C. Recovery of copper salts by fluidized-bed homogeneous granulation process: High selectivity on malachite crystallization. *Hydrometallurgy* **2019**, *186*, 66–72. [CrossRef]
135. Wei, Z.; Xiong, Y.; Chen, J.; Bai, J.; Wu, J.; Zuo, J.; Wang, K. Recovery of Cu(II) from aqueous solution by induced crystallization in a long-term operation. *J. Environ. Sci.* **2018**, *69*, 183–191. [CrossRef]
136. Bayon, L.L.E.; Ballesteros, F.C.; Garcia-Segura, S.; Lu, M.-C. Water reuse nexus with resource recovery: On the fluidized-bed homogeneous crystallization of copper and phosphate from semiconductor wastewater. *J. Clean. Prod.* **2019**, *236*, 117705. [CrossRef]
137. Ballesteros, F.C.; Salcedo, A.F.S.; Vilando, A.C.; Huang, Y.-H.; Lu, M.-C. Removal of nickel by homogeneous granulation in a fluidized-bed reactor. *Chemosphere* **2016**, *164*, 59–67. [CrossRef] [PubMed]
138. Chen, J.P.; Yu, H. Lead removal from synthetic wastewater by crystallization in a fluidized-bed reactor. *J. Environ. Sci. Health Part A Toxic/Hazard. Subst. Environ. Eng.* **2000**, *35*, 817–835. [CrossRef]
139. Hoang, M.T.; Pham, T.D.; Nguyen, V.T.; Nguyen, M.K.; Pham, T.T.; Van der Bruggen, B. Removal and recovery of lead from wastewater using an integrated system of adsorption and crystallization. *J. Clean. Prod.* **2019**, *213*, 1204–1216. [CrossRef]
140. De Luna, M.D.G.; Bellotindos, L.M.; Asiao, R.N.; Lu, M.-C. Removal and recovery of lead in a fluidized-bed reactor by crystallization process. *Hydrometallurgy* **2015**, *155*, 6–12. [CrossRef]
141. Udomkitthaweewat, N.; Anotai, J.; Choi, A.E.S.; Lu, M.C. Removal of zinc based on a screw manufacturing plant wastewater by fluidized-bed homogeneous granulation process. *J. Clean. Prod.* **2019**, *230*, 1276–1286. [CrossRef]
142. Vilando, A.C.; Caparanga, A.R.; Lu, M.-C. Enhanced recovery of aluminum from wastewater using a fluidized bed homogeneously dispersed granular reactor. *Chemosphere* **2019**, *223*, 330–341. [CrossRef]
143. Wilms, D.; Vercaemst, K.; Van Dijk, J.C. Recovery of silver by crystallization of silver carbonate in a fluidized-bed reactor. *Water Res.* **1992**, *26*, 235–239. [CrossRef]
144. Tsevis, A.; Sotiropoulou, M.; Koutsoukos, P.G. Preparation of titania powders in a fluidized-bed reactor. *Prog. Colloid Polym. Sci.* **2000**, *115*, 151–155. [CrossRef]
145. Zhou, P.; Huang, J.-C.; Li, A.W.; Wei, S. Heavy metal removal from wastewater in fluidized bed reactor. *Water Res.* **1999**, *33*, 1918–1924. [CrossRef]
146. Lee, C.-I.; Yang, W.-F. Heavy Metal Removal from Aqueous Solution in Sequential Fluidized-Bed Reactors. *Environ. Technol.* **2005**, *26*, 1345–1354. [CrossRef] [PubMed]
147. Chung, J.; Jeong, E.; Choi, J.W.; Yun, S.T.; Maeng, S.K.; Hong, S.W. Factors affecting crystallization of copper sulfide in fed-batch fluidized bed reactor. *Hydrometallurgy* **2015**, *152*, 107–112. [CrossRef]
148. Huang, C.; Pan, J.R.; Lee, M.; Yen, S. Treatment of high-level arsenic-containing wastewater by fluidized bed crystallization process. *J. Chem. Technol. Biotechnol.* **2007**, *82*, 289–294. [CrossRef]
149. Lewis, A.E.; Swartbooi, A. Factors Affecting Metal Removal in Mixed Sulfide Precipitation. *Chem. Eng. Technol.* **2006**, *29*, 277–280. [CrossRef]
150. De Luna, M.D.G.; Rance, D.P.M.; Bellotindos, L.M.; Lu, M.-C. A statistical experimental design to remove sulfate by crystallization in a fluidized-bed reactor. *Sustain. Environ. Res.* **2017**, *27*, 117–124. [CrossRef]
151. De Luna, M.D.G.; Rance, D.P.M.; Bellotindos, L.M.; Lu, M.-C. Removal of sulfate by fluidized bed crystallization process. *J. Environ. Chem. Eng.* **2017**, *5*, 2431–2439. [CrossRef]
152. Seewoo, S.; Van Hille, R.; Lewis, A.E. Aspects of gypsum precipitation in scaling waters. *Hydrometallurgy* **2004**, *75*, 135–146. [CrossRef]
153. Maharaj, C.; Chivavava, J.; Lewis, A.E. Treatment of a highly-concentrated sulphate-rich synthetic wastewater using calcium hydroxide in a fluidised bed crystallizer. *J. Environ. Manag.* **2018**, *207*, 378–386. [CrossRef] [PubMed]
154. Halevy, S.; Korin, E.; Gilron, J. Kinetics of Gypsum Precipitation for Designing Interstage Crystallizers for Concentrate in High Recovery Reverse Osmosis. *Ind. Eng. Chem. Res.* **2013**, *52*, 14647–14657. [CrossRef]
155. Vu, X.; Lin, J.-Y.; Shih, Y.-J.; Huang, Y.-H. Reclaiming Boron as Calcium Perborate Pellets from Synthetic Wastewater by Integrating Chemical Oxo-Precipitation within a Fluidized-Bed Crystallizer. *ACS Sustain. Chem. Eng.* **2018**, *6*, 4784–4792. [CrossRef]
156. Tsai, C.-K.; Lee, N.-T.; Huang, G.-H.; Suzuki, Y.; Doong, R. Simultaneous Recovery of Display Panel Waste Glass and Wastewater Boron by Chemical Oxo-precipitation with Fluidized-Bed Heterogeneous Crystallization. *ACS Omega* **2019**, *4*, 14057–14066. [CrossRef]
157. Le, V.G.; Vo, T.D.H.; Nguyen, B.S.; Vu, C.T.; Shih, Y.J.; Huang, Y.H. Recovery of iron(II) and aluminum(III) from acid mine drainage by sequential selective precipitation and fluidized bed homogeneous crystallization (FBHC). *J. Taiwan Inst. Chem. Eng.* **2020**, *115*, 135–143. [CrossRef]
158. De Luna, M.D.G.; Sioson, A.S.; Choi, A.E.S.; Abarca, R.R.M.; Huang, Y.-H.; Lu, M.-C. Operating pH influences homogeneous calcium carbonate granulation in the frame of CO_2 capture. *J. Clean. Prod.* **2020**, *272*, 1222325. [CrossRef]
159. Guillard, D.; Lewis, A.E. Optimization of nickel hydroxycarbonate precipitation using a laboratory pellet reactor. *Ind. Eng. Chem. Res.* **2002**, *41*, 3110–3114. [CrossRef]
160. Guillard, D.; Lewis, A.E. Nickel carbonate precipitation in a fluidized-bed reactor. *Ind. Eng. Chem. Res.* **2001**, *40*, 5564–5569. [CrossRef]
161. Mamuad, R.Y.; Caparanga, A.R.; Choi, A.E.S.; Lu, M.-C. Remediation of oxalate in a homogeneous granulation process in the frame of crystallization. *Chem. Eng. Commun.* **2022**, *209*, 378–389. [CrossRef]

162. Guevara, H.P.R.; Ballesteros, F.C.; Vilando, A.C.; de Luna, M.D.G.; Lu, M.-C. Recovery of oxalate from bauxite wastewater using fluidized-bed homogeneous granulation process. *J. Clean. Prod.* **2017**, *154*, 130–138. [CrossRef]
163. Caddarao, P.S.; Garcia-Segura, S.; Ballesteros, F.C., Jr.; Huang, Y.-H.; Lu, M.-C.; Ballesteros, F.C.; Huang, Y.-H.; Lu, M.-C. Phosphorous recovery by means of fluidized bed homogeneous crystallization of calcium phosphate. Influence of operational variables and electrolytes on brushite homogeneous crystallization. *J. Taiwan Inst. Chem. Eng.* **2018**, *83*, 124–132. [CrossRef]
164. Li, X.; Hasson, D.; Semiat, R.; Shemer, H. Intermediate concentrate demineralization techniques for enhanced brackish water reverse osmosis water recovery—A review. *Desalination* **2019**, *466*, 24–35. [CrossRef]
165. Tran, A.T.K.; Zhang, Y.; Jullok, N.; Meesschaert, B.; Pinoy, L.; Van der Bruggen, B. RO concentrate treatment by a hybrid system consisting of a pellet reactor and electrodialysis. *Chem. Eng. Sci.* **2012**, *79*, 228–238. [CrossRef]
166. Sobhani, R.; Abahusayn, M.; Gabelich, C.J.J.; Rosso, D. Energy Footprint analysis of brackish groundwater desalination with zero liquid discharge in inland areas of the Arabian Peninsula. *Desalination* **2012**, *291*, 106–116. [CrossRef]
167. Li, C.-W.; Liao, J.-C.; Lin, Y.-C. Integrating a membrane and a fluidized pellet reactor for removing hardness: Effects of NOM and phosphate. *Desalination* **2005**, *175*, 279–288. [CrossRef]
168. Kohli, H.; Emmermann, D.; Kadaj, R.; Said, H. Hybrid desalting technology maximizes recovery. *Desalination* **1985**, *56*, 61–68. [CrossRef]
169. Chen, Y.; Davis, J.R.J.R.; Nguyen, C.H.C.H.; Baygents, J.C.J.C.; Farrell, J. Electrochemical Ion-Exchange Regeneration and Fluidized Bed Crystallization for Zero-Liquid-Discharge Water Softening. *Environ. Sci. Technol.* **2016**, *50*, 5900–5907. [CrossRef] [PubMed]
170. Liu, K.; Huang, Q.; Long, X.; Cao, Z.; Zhou, B. RO concentrate softening through induced crystallization in a fluidized bed reactor. *MATEC Web Conf.* **2016**, *49*, 04003. [CrossRef]
171. Tran, A.T.K.; Mondal, P.; Lin, J.; Meesschaert, B.; Pinoy, L.; Van der Bruggen, B. Simultaneous regeneration of inorganic acid and base from a metal washing step wastewater by bipolar membrane electrodialysis after pretreatment by crystallization in a fluidized pellet reactor. *J. Membr. Sci.* **2015**, *473*, 118–127. [CrossRef]
172. Tran, A.T.K.; Zhang, Y.; De Corte, D.; Hannes, J.-B.; Ye, W.; Mondal, P.; Jullok, N.; Meesschaert, B.; Pinoy, L.; Van Der Bruggen, B. P-recovery as calcium phosphate from wastewater using an integrated selectrodialysis/crystallization process. *J. Clean. Prod.* **2014**, *77*, 140–151. [CrossRef]
173. Sluys, J.T.M.; Verdoes, D.; Hanemaaijer, J.H. Water treatment in a Membrane-Assisted Crystallizer (MAC). *Desalination* **1996**, *104*, 135–139. [CrossRef]
174. Lee, C.; Yang, W.; Chiou, C. Utilization of water clarifier sludge for copper removal in a liquid fluidized-bed reactor. *J. Hazard. Mater.* **2006**, *129*, 58–63. [CrossRef]
175. Lyu, C.; Zhou, D.; Wang, J. Removal of multi-dye wastewater by the novel integrated adsorption and Fenton oxidation process in a fluidized bed reactor. *Environ. Sci. Pollut. Res.* **2016**, *23*, 20893–20903. [CrossRef]
176. Anotai, J.; Sakulkittimasak, P.; Boonrattanakij, N.; Lu, M.-C. Kinetics of nitrobenzene oxidation and iron crystallization in fluidized-bed Fenton process. *J. Hazard. Mater.* **2009**, *165*, 874–880. [CrossRef]
177. Cai, Q.Q.; Wu, M.Y.; Hu, L.M.; Lee, B.C.Y.; Ong, S.L.; Wang, P.; Hu, J.Y. Organics removal and in-situ granule activated carbon regeneration in FBR-Fenton/GAC process for reverse osmosis concentrate treatment. *Water Res.* **2020**, *183*, 116119. [CrossRef] [PubMed]
178. Bello, M.M.; Abdul Raman, A.A.; Asghar, A. Fenton oxidation treatment of recalcitrant dye in fluidized bed reactor: Role of SiO_2 as carrier and its interaction with fenton's reagent. *Environ. Prog. Sustain. Energy* **2019**, *38*, 13188. [CrossRef]
179. Burhenne, L.; Giacomin, C.; Follett, T.; Ritchie, J.; McCahill, J.S.J.; Mérida, W. Characterization of reactive $CaCO_3$ crystallization in a fluidized bed reactor as a central process of direct air capture. *J. Environ. Chem. Eng.* **2017**, *5*, 5968–5977. [CrossRef]
180. Giacomin, C.E.; Holm, T.; Mérida, W. $CaCO_3$ growth in conditions used for direct air capture. *Powder Technol.* **2020**, *370*, 39–47. [CrossRef]
181. Papirio, S.; Villa-Gomez, D.K.; Esposito, G.; Pirozzi, F.; Lens, P.N.L. Acid Mine Drainage Treatment in Fluidized-Bed Bioreactors by Sulfate-Reducing Bacteria: A Critical Review. *Crit. Rev. Environ. Sci. Technol.* **2013**, *43*, 2545–2580. [CrossRef]
182. Kaksonen, A.H.; Puhakka, J.A. Sulfate Reduction Based Bioprocesses for the Treatment of Acid Mine Drainage and the Recovery of Metals. *Eng. Life Sci.* **2007**, *7*, 541–564. [CrossRef]
183. Qamar, S.; Galan, K.; Peter Elsner, M.; Hussain, I.; Seidel-Morgenstern, A. Theoretical investigation of simultaneous continuous preferential crystallization in a coupled mode. *Chem. Eng. Sci.* **2013**, *98*, 25–39. [CrossRef]
184. Vetter, T.; Burcham, C.L.; Doherty, M.F. Separation of conglomerate forming enantiomers using a novel continuous preferential crystallization process. *AIChE J.* **2015**, *61*, 2810–2823. [CrossRef]
185. Temmel, E.; Gänsch, J.; Seidel-Morgenstern, A.; Lorenz, H. Systematic investigations on continuous fluidized bed crystallization for chiral separation. *Crystals* **2020**, *10*, 394. [CrossRef]
186. Kerst, K.; de Souza, L.M.; Bartz, A.; Seidel-Morgenstern, A.; Janiga, G. CFD-DEM simulation of a fluidized bed crystallization reactor. *Comput. Aided Chem. Eng.* **2015**, *37*, 263–268. [CrossRef]
187. Binev, D.; Seidel-Morgenstern, A.; Lorenz, H. Continuous Separation of Isomers in Fluidized Bed Crystallizers. *Cryst. Growth Des.* **2016**, *16*, 1409–1419. [CrossRef]
188. Soh, S.H.; Lee, L.Y. Microencapsulation and nanoencapsulation using supercritical fluid (SCF) techniques. *Pharmaceutics* **2019**, *11*, 21. [CrossRef]

189. Niu, F.; Subramaniam, B. Particle Fluidization with Supercritical Carbon Dioxide: Experiments and Theory. *Ind. Eng. Chem. Res.* **2007**, *46*, 3153–3156. [CrossRef]
190. Tsutsumi, A.; Nakamoto, S.; Mineo, T.; Yoshida, K. A novel fluidized-bed coating of fine particles by rapid expansion of supercritical fluid solutions. *Powder Technol.* **1995**, *85*, 275–278. [CrossRef]
191. Leeke, G.A.; Lu, T.; Bridson, R.H.; Seville, J.P.K. Application of nano-particle coatings to carrier particles using an integrated fluidized bed supercritical fluid precipitation process. *J. Supercrit. Fluids* **2014**, *91*, 7–14. [CrossRef]
192. Li, Q.; Huang, D.; Lu, T.; Seville, J.P.K.; Xing, L.; Leeke, G.A. Supercritical fluid coating of API on excipient enhances drug release. *Chem. Eng. J.* **2017**, *313*, 317–327. [CrossRef]
193. Matos, R.L.; Lu, T.; McConville, C.; Leeke, G.; Ingram, A. Analysis of curcumin precipitation and coating on lactose by the integrated supercritical antisolvent-fluidized bed process. *J. Supercrit. Fluids* **2018**, *141*, 143–156. [CrossRef]
194. Matos, R.L.; Lu, T.; Leeke, G.; Prosapio, V.; McConville, C.; Ingram, A. Single-step coprecipitation and coating to prepare curcumin formulations by supercritical fluid technology. *J. Supercrit. Fluids* **2020**, *159*, 104758. [CrossRef]
195. Chen, T.; Liu, L.; Zhang, L.; Lu, T.; Matos, R.L.; Jiang, C.; Lin, Y.; Yuan, T.; Ma, Z.; He, H.; et al. Optimization of the supercritical fluidized bed process for sirolimus coating and drug release. *Int. J. Pharm.* **2020**, *589*, 119809. [CrossRef]

MDPI
St. Alban-Anlage 66
4052 Basel
Switzerland
www.mdpi.com

Crystals Editorial Office
E-mail: crystals@mdpi.com
www.mdpi.com/journal/crystals

Disclaimer/Publisher's Note: The statements, opinions and data contained in all publications are solely those of the individual author(s) and contributor(s) and not of MDPI and/or the editor(s). MDPI and/or the editor(s) disclaim responsibility for any injury to people or property resulting from any ideas, methods, instructions or products referred to in the content.